International Scientific and Technical Youth Conference "Perspective Materials of Constructional and Medical Purpose" (PMCP 2018)

IOP Conference Series: Materials Science and Engineering
Volume 511

Tomsk, Russia
26 – 30 November 2018

ISBN: 978-1-5108-8632-2
ISSN: 1757-8981

Printed from e-media with permission by:

Curran Associates, Inc.
57 Morehouse Lane
Red Hook, NY 12571

proceedings.com

Some format issues inherent in the e-media version may also appear in this print version.

Printed by Curran Associates, Inc. (2019)

Additional copies of this publication are available from:

Curran Associates, Inc.
57 Morehouse Lane
Red Hook, NY 12571 USA
Phone: 845-758-0400
Fax: 845-758-2633
Email: curran@proceedings.com
Web: www.proceedings.com

Table of contents

Volume 511

International Scientific and Technical Youth Conference "Perspective Materials of Constructional and Medical Purpose"

26–30 November 2018, Tomsk, Russian Federation

Accepted papers received: 8 March 2019
Published online: 24 April 2019

Preface

Papers

Preface

The International scientific and technical youth conference "Perspective Materials of Constructional and Medical Purpose" was held at Tomsk Polytechnic University, 26-30 November, 2018. The conference was organized by major Siberian institutions: National Research Tomsk Polytechnic University, Institute of Strength Physics and Materials Science SB RAS.

The Conference brought together young researchers to discuss the current issues of the development and manufacturing of construction and medical materials, improving their properties, as well as the creation of new materials based on new ideas about the nature of the behavior of a solid body under external impact.

The mains topics of the conference were as follows:

- Problems of materials reliability
- Functional materials
- Surface and coating modification
- Perspective materials of new generation for biomedical applications

More than 180 papers of young and experienced researchers from Russia and abroad were submitted which covered the topics of the conference. The best 44 papers were recommended for publication in IOP Conference Series: Materials Science and Engineering.

The Organizing Committee would like to thank leading scientists in the field of modern materials from Russia, India, Hungary and Vietnam, which made review reports on papers of the young participants of the conference.

We also appreciate the financial support from the Russian Foundation for Basic Research.

The Organizing Committee of PMC&MP 2018

Dr. Alexey Yakovlev, TPU

Dr. Ekaterina Kulinich, TPU

IOP Conf. Series: Materials Science and Engineering **511** (2019) 011002 doi:10.1088/1757-899X/511/1/011002

Peer review statement

All papers published in this volume of *IOP Conference Series: Materials Science and Engineering* have been peer reviewed through processes administered by the proceedings Editors. Reviews were conducted by expert referees to the professional and scientific standards expected of a proceedings journal published by IOP Publishing.

Alumina-based composites reinforced with single-walled carbon nanotubes

A A Leonov and E V Abdulmenova

National Research Tomsk Polytechnic University, 30 Lenin Ave., Tomsk, 634050, Russia

Email: laa91@tpu.ru

Abstract. Monolithic Al_2O_3 ceramics and Al_2O_3 – 3 vol% single-walled carbon nanotubes (SWCNTs) composites were prepared by spark plasma sintering. The influence of SWCNTs and sintering temperature on sintering behavior and mechanical properties were investigated. Nanotubes were relatively homogeneously distributed in the composite powder, although some agglomerates/bundles existed. It was found that SWCNTs addition retards slightly the sinterability of alumina by nanotubes hindering of particle rearrangement. The average microhardness of the composites was lower than that of Al_2O_3 ceramics, but there were also high microhardness values of composites (20.41 GPa). The reduction in microhardness is explained in this paper. The average values of the fracture toughness (4.94 $MPa{\cdot}m^{1/2}$) from the composite sintered at 1500 °C were approximately 6 % and 12 % higher than those from the Al_2O_3 ceramics sintered at 1500 °C and 1600 °C, respectively. In addition, the analysis of the phase composition and the parameters of the crystal structure of the samples were made.

1. Introduction

Alumina (Al_2O_3) ceramics possess wide technological applications in modern industry because of their excellent properties such as high hardness, optimal high temperature property retention, acceptable dielectric properties, chemical inertness, corrosion resistance, adsorption properties and biocompatibility. Brittleness, low fracture toughness and flexural strength of alumina ceramics largely limits their application scope [1]. Many strategies have been proposed to tackle this problem and adding reinforcing materials as a second phase in the form of particles, platelets or whiskers/fibers is a useful method [2, 3]. The improvement in toughness with fiber reinforcement, especially in a continuous form, is much greater than that with particulate reinforcement [4]. Reinforcement of the Al_2O_3 matrix can be performed different fibrous fillers such as: alumina nanofibers/whiskers [5-8], mullite whiskers [9, 10], SiC whiskers [11, 12], carbon nanofibers [13, 14], carbon nanotubes [15-18], boron nitride nanotubes [19, 20]. Most of investigations have been focused especially on the mechanical characteristics of carbon nanotubes (CNTs) since CNTs with a tubular structure exhibit almost 5 times of elastic modulus and nearly 100 times tensile strength compared to high strength steels. Therefore, it is expected that CNTs can upgrade the mechanical properties of Al_2O_3 and make it suitable for numerous advanced applications. For example, Zhan et al. [21] reported a fracture toughness of 9.7 $MPa{\cdot}m^{1/2}$ in 10 vol% single-walled carbon nanotubes manufactured by spark plasma sintering (SPS) technique versus only 3.7 $MPa{\cdot}m^{1/2}$ for the matrix, in alumina composites. Bi et al. fabricated fully dense composites of multi-walled carbon nanotubes (MWCNTs) with alumina matrix by using hydrothermal crystallization methods and followed by hot-pressing. The bending strength and fracture toughness of the sample were increased by 24.6 % and 80.3 % respectively compared with monolithic alumina [22]. Bocanegra-Bernal et al. [23] reported that for 0.1 wt% MWCNT/Al_2O_3 composites, the fracture toughness had an increase of 63 % compared with the monolithic alumina. Wang et al. [20] claimed that the alumina grain growth was remarkably inhibited by boron nitride nanotubes (BNNTs), and the fracture toughness and bending strength were increased by 31 % and 67 % for the composites containing 1.0 and 2.0 wt% of BNNTs, respectively. Given that the incorporation of SWCNTs may provide more potential for strengthening and toughening, the effect of SWCNTs content on the mechanical properties of Al_2O_3-based ceramics was investigated in this paper. Spark plasma sintering was employed to obtain Al_2O_3-SWCNTs composites, and the effect of sintering temperatures was also investigated.

2. Materials and methods

In the present work, alumina powder with the addition of nanopowder of the same composition in the amount of 6 % was used as the matrix base. Commercial SWCNTs «Tuball» were purchased from OCSiAl, Russia (content of SWCNTs ~ 75 wt%, diameter 1–3 nm and length 1–5 μm). The SWCNTs was dispersed in ethanol by means of ultrasonic treatment during 5 min and a drop of the suspension was transferred onto a copper grid and by means of transmission electron microscope JEM-2100F (JEOL) its morphology was studied. The measurements of particle size distribution of initial Al_2O_3 powder by means of laser diffraction method were performed with SALD-7101 (Shimadzu). The specific surface area of the Al_2O_3 powder, SWCNTs and Al_2O_3-SWCNTs composite powder was measured by BET technique using installation Sorbi-M (META). Before mixing, the SWCNTs were first dispersed in ethanol for 30 min by ultrasonic bath (110 W, 35 kHz) to obtain homogeneous ethanol-SWCNTs suspension. Next, an alumina was added to this suspension at the ratio of the SWCNTs 3 vol% and then ultrasonic dispersion was continued for an additional 30 min. After the expiration of the ultrasonic dispersion, the processed composite suspension was put on a magnetic stirrer and agitation was performed for 30 min. After mixing, the slurry was dried at 120 °C for 3 h. The morphology and detailed structure of composite powder were observed by scanning electron microscope JSM-7500FA (JEOL). The resulting powder mixture was placed into a graphite die with an inner diameter of 14 mm and then spark plasma sintered at 1500 °C and 1600 °C respectively for 10 min under a uniaxial load of 40 MPa in vacuum. A heating rate of 100 °C/min was employed. For comparison, monolithic Al_2O_3 ceramics were sintered under the same conditions. The bulk density of the sintered samples was measured using the Archimedes method with distilled water as the immersing medium. The relative density was calculated by dividing the bulk density by the theoretical density. Vickers indentation tests were carried out to evaluate the microhardness (H_V) and fracture toughness (K_{IC}) of sintered Al_2O_3 and composites at room temperature. Tests were performed on sample surfaces using a Vickers indenter with a load 4.9 H (PMT-3M, LOMO). The fracture toughness was calculated by using the equation given by Anstis [24]. Analysis of the phase composition and the parameters of the crystal structure of the samples were determined by X-ray diffraction (XRD) equipment diffractometer of XRD-7000S (Shimadzu) type using Cu-Kα radiations (λ = 1.54056 Å). The tube voltage and current were 40 kV and 30 mA respectively. Scan range and sampling pitch were 10 ° – 120 ° and 0.03 °, respectively. XRD data were used to calculate size of the coherently diffracting domain (CDD).

3. Results and discussion

Figure 1a shows the TEM-image of the SWCNTs. It is seen that nanotubes are bundles of various widths with sphere-shaped catalyst particles on them or agglomerates of the catalyst particles, which are also sphere-shaped. A large number of bundles have a cross-sectional dimension of 10-40 nm. However, there are even larger bundles up to 200 nm. The bundles of SWCNTs have a flattened ribbonlike shape. Individual nanotubes occur in the images very rarely. Owing to its nanometric diameter, a SWCNT has a very large specific surface area 546.45 m^2/g [25]. The SEM image of composite powder containing 3 vol% of SWCNTs is shown in Figure 1b. Disentangled SWCNTs are well dispersed among Al_2O_3 powder, however, some agglomerates/bundles could be observed. No evidence of nanotubes damage after mixing can be observed. As seen in Fig. 1b, the unique flexible nature of SWCNTs makes them bend and pass through space between particles or wrap around them. Specific surface area of alumina powder and alumina/SWCNT composite powder are 1.45 m^2/g and 5.16 m^2/g, respectively. Figure 1c shows the particle size distribution of initial Al_2O_3 powder. The sizes of alumina particles are distributed in the ranges of 97–409 nm, 453–842 nm, 933 nm–1.920 μm and 2.129–9.992 μm, which further confirms the observed result from SEM (Figure 1b) and the size with different ranges may beneficial for dense packing of particles. The average particle size of alumina powder is 1.71 ± 0.45 μm.

Figure 1. (a) TEM image showing morphology of SWCNTs, (b) SEM image showing morphology of composite powder and (c) particle size distribution of Al_2O_3 powder.

After sintering by SPS, the produced materials were designated as A-1500, A-1600, AS-1500 and AS-1600 for samples of Al_2O_3 ceramics sintered at 1500 °C and 1600 °C and Al_2O_3/SWCNT composites sintered at the same temperatures. The relative density, Vickers microhardness and fracture toughness are reported in Table 1. From Table 1 it can be seen that an increase in the sintering temperature of Al_2O_3 ceramics to 1600 °C does not lead to an increase of the relative density; it is almost the same for both samples and was > 97 %. The relative density of composites (~ 96 %) was slightly lower than that of Al_2O_3 ceramics, showing that SWCNTs inhibit the densification upon sintering, as observed in other studies [26-29]. The decrease in the sintered density upon the SWCNTs addition is due to the slightly reduced densification that leads to the increase of pores.

Table 1. Properties of the samples: ρ: relative density; H_V: Vickers microhardness and K_{IC}: fracture toughness (average, minimum/maximum values are indicated).

Sample	ρ, %	H_V (GPa)	K_{IC} (MPa·m$^{1/2}$)
A-1500	97.43	18.40 (16.94/19.70)	4.65 (4.00/5.16)
A-1600	97.20	17.64 (16.12/18.78)	4.34 (3.71/4.95)
AS-1500	95.91	16.58 (12.00/20.41)	4.94 (4.30/5.45)
AS-1600	96.10	16.60 (13.38/20.01)	3.95 (2.01/5.34)

The average microhardness of composites (16.60 GPa) is lower than that of Al_2O_3 ceramics (~ 18 GPa) and has a higher scatter of values. The decrease in the microhardness by the addition of 3 vol% SWCNTs was attributed mainly to the decrease in the composite density. In addition, nanotubes are a soft phase. Although the hardness of SWCNTs was unknown, given that SWCNTs come from the graphitic family of carbon with dominantly sp^2 bonds mixed with sp^3 bonds, their hardness should be lower than that of Al_2O_3. Another factor is that the reinforcing phase usually creates tensile stresses in the crystals of the matrix and at their boundaries. Further, not very high sintering temperature also resulted in poor SWCNT/Al_2O_3 interface that led to insufficient load sharing between filler and matrix in nanocomposites. However, it is worth noting that composites have high microhardness values (20.41 GPa and 20.01 GPa) that exceed the values of H_V for alumina ceramics (Table 1). For a significant increase in the microhardness of composites with CNT required higher sintering temperature to achieve better consolidation and proper interface performance.

The fracture toughness of the AS-1500 composite is higher than that of samples A-1500 and A-1600 by 6 % and 12 %, respectively. The role of SWCNTs was indeed substantial, because the high strength, impressive flexibility and very high aspect ratio of the CNTs offered great potential for them to hold and resist the crack opening through elastic deformation. In this manner, SWCNTs worked efficiently for dissipating energy via transferring the stresses from one grain to another during the entire loading event, and for direct absorbing energy. Zharikov et al. [30] showed that alumina composite with 3 vol% CNTs have fracture toughness (6.6 $MPa\cdot m^{1/2}$) two times higher as compared to the initial ceramic material (3.2 $MPa\cdot m^{1/2}$). Kim et al. [31] have reached the maximum improvement in fracture toughness (about 40%, 4.68 $MPa\cdot m^{1/2}$ vs 3.32 $MPa\cdot m^{1/2}$) in 3 vol% CNTs/alumina composites. Zhang et al. [32] obtained the maximum increase in fracture toughness (24 %, 4.1 $MPa\cdot m^{1/2}$ vs 3.3 $MPa\cdot m^{1/2}$) by addition of 1 vol% CNTs. However, they reported only 6 % (3.5 $MPa\cdot m^{1/2}$) enhancement in fracture toughness for 3 vol% CNTs/Al_2O_3 composites.

Figure 2 shows X-ray diffractograms and phase identification of powders SWCNTs (a), Al_2O_3 (b) and sintered samples A-1500 and A-1600 (c, d), AS-1500 and AS-1600 (e, f). It can be seen that the phase composition of the Al_2O_3 powder corresponds to the corundum (α-Al_2O_3) phase of the trigonal syngony (JCPDS No. 01-078-2426) [33], the CDD size of 48 nm (Figure 2b). Phase composition of the SWCNTs powder (Figure 2a) consists corresponds to graphite phase (JCPDS No. 01-075-1621) [33, 34], the CDD size of 8 nm. The existence of (110) lattice planes indicates a developed in-planar graphitic structure and (101) planes can be considered to be to the foreign of three-dimensional stacking, mainly caused by the lack of defects in the SWCNT [34]. In addition, SWCNTs powder contains catalyst residues corresponding to the bcc structure, in the apparent, of Fe (JCPDS No. 01-085-1410) with a lattice parameter of $2.88686\cdot 10^{-10}$ and Fe_3C (JCPDS No. 35-0772).

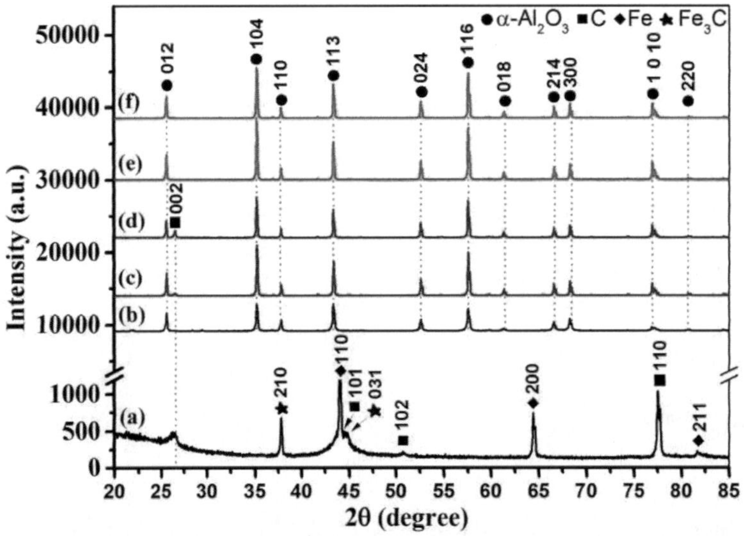

Figure 2. X-ray diffractograms and phase identification of powders SWCNTs (a), Al_2O_3 (b) and sintered samples A-1500 and A-1600 (c, d), AS-1500 and AS-1600 (e, f).

After sintering of Al_2O_3 powder (A-1500 and A-1600), a two-phase state is formed with the creation of a second carbon-like structure, the integral intensity of which is 7 % and 19 %, respectively (Figure 2c, d). Apparently, this may be due to the contamination of the powder during the granulation of Al_2O_3 or with graphite paper residue after SPS sintering. The CDD size of the main phase α-Al_2O_3 is 81 nm (A-1500) and 80 nm (A-1600), the second carbon-containing phase is 33 nm (A-1500) and 41 nm (A-1600). Only peaks corresponding to crystalline α-Al_2O_3 were in the diffractograms of the

AS-1500 and AS-1600, whereas CNT peaks were absent (Figure 2e, f). Owing to small quantity of SWCNTs in the composites (3 vol%) and strong matrix peaks, identification of sp^2 carbon phase was not possible [35-37]. The CDD size of α-Al_2O_3 with increasing sintering temperature was constant and amounted to 82 nm.

4. Conclusions

The effect of SWCNTs on sintering behavior and mechanical properties behavior of Al_2O_3/SWCNTs composites obtained by spark plasma sintering was investigated. The following conclusions can be drawn from this study:

- Sintered densities ~ 96 % of theoretical density were obtained in the composites.
- The average microhardness of the composites was lower than that of Al_2O_3 ceramics, but there were also high microhardness values of composites (20.41 GPa).
- The fracture toughness (4.94 MPa·$m^{1/2}$) of the AS-1500 composite was higher than that of Al_2O_3 ceramics samples A-1500 and A-1600 by 6 % and 12 %, respectively.
- The phase composition of the composites corresponds to the crystalline α-Al_2O_3, SWCNT peaks were absent. The CDD size of α-Al_2O_3 with increasing sintering temperature was constant and amounted to 82 nm.

References

[1] Osayande L and Okoli O I 2008 Fracture toughness enhancement for alumina system: a review *Int. J. App. Ceram. Technol.* **5** 313–323
[2] Krenkel W 2008 *Ceramic Matrix Composites: Fiber Reinforced Ceramics and their Applications* (Weinheim: Wiley-VCH Verlag GmbH & Co.KGaA) p 443
[3] Chawla K K 2012 *Composite Materials: Science and Engineering* (New York: Springer-Verlag)
[4] Estili M and Sakka Y 2014 Recent advances in understanding there inforcing ability and mechanism of carbon nanotubes in ceramic matrix composites *Sci. Technol. Adv. Mater.* **15** 064902
[5] Hussainova I, et al. 2016 Wear performance of hierarchically structured alumina reinforced by hybrid graphene encapsulated alumina nanofibers *Wear* **368–369** 287–295
[6] Dudkin B N, et al. 2010 Corundum/lanthanum hexaaluminate/alumina nanofiber ceramic composite *Inorg. Mater.* **46** 445–448
[7] Tamura Y, et al. 2017 Spark plasma sintering of fine-grained alumina ceramics reinforced with alumina whiskers *Ceram. Int.* **43** 658–663
[8] Tamura Y, et al. 2018 Is an alumina-whisker-reinforced alumina composite the most efficient choice for an oxidation-resistant high-temperature ceramic? *J. Eur. Ceram. Soc.* **38** 1812–18
[9] Robertson T, et al. 2016 High temperature performance of mullite whisker-reinforced ZTA *J. Compos. Mater.* **50(26)** 3719–29
[10] Lu Z L, et al. 2016 Effect of in situ synthesised mullite whiskers on the high-temperature strength of Al_2O_3-based ceramic moulds for casting hollow turbine blades *Ceram. Int.* **42** 18851–58
[11] Wu W, et al. 2017 The effect of residual stress on whisker reinforcements in SiC_w-Al_2O_3 composites during cooling *J. Alloy. Compd.* **725** 639–643
[12] Lee D Y and Yoon D H 2014 Properties of alumina matrix composites reinforced with SiC whisker and carbon nanotubes *Ceram. Int.* **40** 14375–83
[13] He P, et al. 2016 Preparation and mechanical performance of ductile C_{sf}/Al_2O_3-BN composites, Part 1: effects of fiber length and sintering temperature *Ceram. Int.* **42** 9821–29
[14] Jia J, et al. 2018 Preparation and mechanical properties of short carbon fibers reinforced α-Al_2O_3-based composites *Ceram. Int.* **44** 19345–51
[15] Castillo-Rodrígueza M, et al. 2018 Creep study on alumina and alumina/SWCNT nanocomposites *J. Eur. Ceram. Soc.* **38** 5497–02
[16] Leonov A A, et al. 2017 Spark plasma sintering of ceramic matrix composite based on alumina,

reinforced by carbon nanotubes *IOP Conf. Ser.: Mater. Sci. Eng.* **286** 012034
[17] Asiq Rahman O S, et al. 2018 Synergistic effect of hybrid carbon nanotube and graphene nanoplatelets reinforcement on processing, microstructure, interfacial stress and mechanical properties of Al_2O_3 nanocomposites *Ceram. Int.* **44** 2109–22
[18] Golovin Y I, et al. 2018 Effect of carbon nanotubes on strength characteristics of nanostructured ceramic composites for biomedicine *Nanotechnol Russia* **13** 168–172
[19] Wang W L, et al. 2011 Fabrication of alumina ceramic reinforced with boron nitride nanotubes with improved mechanical properties *J. Am. Ceram. Soc.* **94** 3636–40
[20] Wang W L, et al. 2011 Microstructure and mechanical properties of alumina ceramics reinforced by boron nitride nanotubes *J. Eur. Ceram. Soc.* **31** 2277–84
[21] Zhan G D, et al. 2003 Single-wall carbon nanotubes as attractive toughening agents in alumina-based nanocomposites *Nat. Mater.* **2** 38–42
[22] Bi S, Hou G, Su X, et al. 2011. Mechanical properties and oxidation resistance of α-alumina/multi-walled carbon nanotube composite ceramics *Mater. Sci. Eng. A* **528** 1596–1601
[23] Bocanegra-Bernal M H, et al. 2017 Effect of low-content of carbon nanotubes on the fracture toughness and hardness of carbon nanotube reinforced alumina prepared by sinter, HIP and sinter + HIP routes *Mater. Res. Express* **4** 085004
[24] Anstis G R, et al. 1981 A critical evaluation of indentation techniques for measuring fracture toughness: I, direct crack measurements *J. Am. Ceram. Soc.* **64** 533–538
[25] Peigney A, et al. 2001 Specific surface area of carbon nanotubes and bundles of carbon nanotubes *Carbon* **39** 507–514
[26] Ahmad I, et al. 2018 Synergic influence of MWCNTs and SiC nanoparticles on the microstructure and properties of Al_2O_3 ceramic hybrid nanocomposites *Curr. Appl. Phys.* **16** 1649–58
[27] Hanzel O, et al. 2017 Mechanical and tribological properties of alumina-MWCNTs composites sintered by rapid hot-pressing *J. Eur. Ceram. Soc.* **37** 4821–31
[28] Aguilar-Elguézabal A and Bocanegra-Bernal M H 2014 Fracture behaviour of α-Al_2O_3 ceramics reinforced with a mixture of single-wall and multi-wall carbon nanotubes *Compos. Part B* **60** 463–470
[29] Ghobadi H, et al. 2014 Improving CNT distribution and mechanical properties of MWCNT reinforced alumina matrix *Mater. Sci. Eng. A* **617** 110–114
[30] Zharikov E V, et al. 2016 Reinforcement of Al_2O_3–MgO composite materials by multiwalled carbon nanotubes *Inorg. Mater. Appl. Res.* **7** 163–169
[31] Kim S W, et al. 2009 Improvement of flexure strength and fracture toughness in alumina matrix composites reinforced with carbon nanotubes *Mater. Sci. Eng. A* **517** 293–299
[32] Zhang S C, et al. 2010 Pressureless sintering of carbon nanotube–Al_2O_3 composites *J. Eur. Ceram. Soc.* **30** 1373–80
[33] Ahmad I, et al. 2010 Carbon nanotube toughened aluminium oxide nanocomposite *J. Eur. Ceram. Soc.* **30** 865–873
[34] Estili M, et al. 2008 The homogeneous dispersion of surfactantless, slightly disordered, crystalline, multiwalled carbon nanotubes in a-alumina ceramics for structural reinforcement *Acta Mater.* **56** 4070–79
[35] Shahzad H K, et al. 2018 Synthesis and characterization of alumina-CNT membrane for cadmium removal from aqueous solution *Ceram. Int.* **44** 17189–98
[36] Shin J H, et al. 2018 Comparative study on carbon nanotube- and reduced graphene oxide-reinforced alumina ceramic composites *Ceram. Int.* **44** 8350–57
[37] Sikder P, et al. 2015 Indentation size effect and wear characteristics of spark plasma sintered, hard MWCNT/Al_2O_3 nanocomposites *Adv. Appl. Ceram.* **114** 448–455

IOP Conf. Series: Materials Science and Engineering **511** (2019) 012002 doi:10.1088/1757-899X/511/1/012002

Application of the Cabaret scheme in task of shock-wave loading

V A Kolotilov

Siberian Branch of Russian Academy of Sciences Khristianovich Institute of Theoretical and Applied Mechanics, 630090, Novosibirsk
Novosibirsk State University, 630090, Novosibirsk

kolotilov1992@gmail.com

Abstract. A method for calculating problems of uniaxial shock-wave loading of material is proposed in this paper. The method is based on Cabaret finite difference scheme. The mathematical model of the impact loading is described in Lagrangian formulation in the one-dimensional case. The task is approximated by the scheme. A comparison is made with the analytical solution. Verification of the approach is carried out on experimental data.

1. Introduction

The problem of high-velocity interaction of solids is accompanied by various physical phenomena, such as heating, melting and evaporation, physical hardening of materials, destruction of materials, etc. The problems of cratering and penetration of materials are associated with shock-wave loading of bodies. There are many publication on this problem, including generalizing, for example, monograph [1]. However, the practical importance of the task makes this problem relevant and incomplete today. There are many approaches to the description of a deformable body, but all these approaches are divided into three types [2]: continual, microstructural and atomistic. It was considered only a first approach in present paper. The numerical implementation of the first approach is carried out by finite difference methods. Up-to-date way of realization of the problem is the family of Cabaret scheme. The initial version of the difference scheme was proposed independently in [3] and [4]. Today, the scheme includes a large number of different implementations [5]. Two-layer time variable scheme is widely used in hydrodynamic problem [6].

2. The system of equations for task of shock-wave loading of bodies. Lagrange formulation

The Maxwell model for describing a solid medium [7] in a one-dimensional formulation was considered. The system of conservation laws look like this:

$$\rho_t = -\rho\frac{\partial v}{\partial x};\; v_t = \frac{1}{\rho}\frac{\partial \sigma}{\partial x};\; (E+\frac{u^2}{2})_t = \frac{1}{\rho}\frac{\partial(u\sigma)}{\partial x} - \frac{1}{\rho}\frac{\partial W}{\partial x};\; \frac{dx}{dt} = v. \tag{1}$$

Density $\rho(x, t)$, velocity $v(x, t)$ and internal energy $E(x, t)$ are desired values in the system. The parameter σ is a stressed state of the material. It is described by state of equation (2). The parameter $\boldsymbol{W}$ is the heat-flux density. It's defined by the Fourier law (6). The equation of state in the tensor form is as follows:

$$\sigma_{ij} = -P\delta_{ii} + S_{ij} \tag{2}$$

Here P is spherical and S_{ij} is deviator components of the stress tensor. Spherical tensor has following form equation of state:

$$P = P_x + P_H; E = E_x + E_H \tag{3}$$

Here P_x is the pressure of elastic compression or cold pressure component. This parameter depends only on the specific volume of the substance and is equal to the total pressure at absolute zero temperature. P_H is the hot component of pressure, depending on the intensity of the movement of molecules and thermal excitation by electrons. For the task of shock-wave loading were chosen the equation of state of the condensed material in the Theta form (4) and relationship between hot components pressure and internal energy through the specific heat (5)

$$P_x = A(\left(\frac{\rho}{\rho_0}\right)^n - 1) \quad . \tag{4}$$

$$P_\Gamma = \gamma\rho E_T = \gamma\rho C_v\left(T - T_0\right) \tag{5}$$

Here A and n are parameters of material. They are determined by experiments. ρ_0 is reference value of density, γ is the Gruneisen parameter, C_v is specific heat and T_0 is the temperature value at the initial moment of time.

Fourier's law is as follows:

$$W = -\chi T_x \tag{6}$$

Here χ is thermal conduction coefficient. It is a function of the thermodynamic state of the medium: $\chi = \chi(\rho, T)$. Here T is temperature of medium. The hyperbolicity of the system of equations is violated if the heat-flux density (6) is not equal to zero on the right in the energy conservation law (1). There are two methods to solve this problem:

- numerical solution of the problem using implicit difference schemes. However, the disadvantage of this approach is low accuracy compared with the Cabaret scheme [4];
- Application of the splitting method for this problem [9].

The temperature change in the process of shock-wave loading is negligible in this work. As a result, the mathematical model of the medium is as follows (7):

$$\rho_t = -\rho\frac{\partial v}{\partial x};\ v_t = \frac{1}{\rho}\frac{\partial \sigma}{\partial x};\ \sigma = -P_x = -A(\left(\frac{\rho}{\rho_0}\right)^n - 1);\ \frac{dx}{dt} = v. \tag{7}$$

It is necessary to determine the characteristics and invariants of a system of differential equations in order to use the Cabaret scheme considered in [6].The characteristics λ_1 and λ_2 and invariants w_1 and w_2 of the system are as follows:

$$\begin{aligned} \lambda_1 &= -c \quad w_1 = v - \frac{2}{n-1}c \\ \lambda_2 &= c \quad w_2 = v + \frac{2}{n-1}c \end{aligned} \tag{8}$$

Here c is critical velocity, which is determined by the formula (9) for the equation of state:

$$c = \sqrt{\frac{\partial P_x}{\partial \rho}} = \sqrt{An\frac{\rho^{n-1}}{\rho_0^n}} \tag{9}$$

3. Modified Cabaret scheme

Two calculation areas were defined. Construct a rectangular differential grid for the drummer and for the target:

$$\left\{x_m, t^n\right\} : x_i = ih; t^{n+1} = t^n + \tau^n \tag{10}$$

Here h is constant grid spacing and τ^n is time step. It is determined from the condition of stability:

$$\tau^n = \frac{zh}{\max\limits_{i,j} \left| (\lambda_j)^n_{i+1/2} \right|} \tag{11}$$

Here z = 0.5 is Courant number. The characteristic values $(\lambda_j)^n_{i+1/2} = \lambda_j(\vec{U}^n_{i+1/2})$ are defined in the center of the cell on the n-th time layer. The minimum time step in both cases was determined for time synchronization in both computing areas. All variables are divided into streaming and conservative in the Cabaret scheme. The first variables are determined at the cell boundaries $\vec{u}^n_i = u(x_i, t^n)$, and the second ones are responsible for the propagation of disturbances in the medium and are located in the centers of the cells $\vec{U}^n_{i+1/2} = U(x_{i+1/2}, t^n)$. The numerical solution of problem (7) was determined in four stages. At the first stage, conservative variables were found on the semi-integral time layer using difference equations (12):

$$\begin{cases} \dfrac{\mathrm{P}^{n+1/2}_{i+1/2} - \mathrm{P}^n_{i+1/2}}{\tau^n / 2} + \mathrm{P}^n_{i+1/2} \dfrac{v^n_{i+1} - v^n_i}{h} = 0 \\ \dfrac{V^{n+1/2}_{i+1/2} - V^n_{i+1/2}}{\tau^n / 2} - \dfrac{1}{\mathrm{P}^n_{i+1/2}} \dfrac{\sigma^n_{i+1} - \sigma^n_i}{h} = 0 \end{cases} \tag{12}$$

In the second stage, the values of the flow variables $v_i^{n+1/2}$ and $\sigma_i^{n+1/2}$ were calculated on the half-integral time layer. At the third stage, the values of conservative variables at the new time layer are calculated according to the following difference equations (13):

$$\begin{cases} \dfrac{\mathrm{P}^{n+1}_{i+1/2} - \mathrm{P}^n_{i+1/2}}{\tau^n} + \mathrm{P}^{n+1/2}_{i+1/2} \dfrac{v^{n+1/2}_{i+1} - v^{n+1/2}_i}{h} = 0 \\ \dfrac{V^{n+1}_{i+1/2} - V^n_{i+1/2}}{\tau^n} - \dfrac{1}{\mathrm{P}^{n+1/2}_{i+1/2}} \dfrac{\sigma^{n+1/2}_{i+1} - \sigma^{n+1/2}_i}{h} = 0 \end{cases} \tag{13}$$

Values of the flow variables are calculated at the new time layer in the fourth stage. In conclusion, the new location of the impactor and target was determined by the following equations:

$$\begin{cases} x^{n+1}_{i+1} = x^n_i + dt \cdot v^{n+1}_i \\ X^{n+1}_{i+1/2} = X^n_{i+1/2} + dt \cdot V^{n+1}_{i+1/2} \end{cases} \tag{14}$$

A collision of impactor with a target is defined as an ideal mechanical contact [10]. The moment of collision is determined by the equality of the coordinates of the nodes of the extreme cells of the impactor and the target.

4. Results and verification

The problem of shock-wave loading for two metallic homogeneous materials was considered to test the obtained calculation algorithm, which corresponds to uniaxial loading in the initial stage of the process.

The first calculation was carried out for the case of a collision with a velocity of 660 m / s of two identical materials on the example of copper. The thickness of impactor is 0.5 mm, and the target – 2 mm. The number of grid nodes of the impactor is 20 and the target - 200. The convergence on a fine grid was confirmed: 40 nodes on the impactor and 400 nodes on the target. Black circles set the centers of the impactor cells. Red circles set the centers of the target cells. The comparison was carried out with the analytical solution and is shown in figure 1 using the velocity profile as an example. The graph shows an accurate coincidence of the calculation with the analytical solution. The scheme smoothes the values in front of the shock front both in the impactor and in the target. The rest of the calculation accurately repeats the analytical solution.

The next step was the verification of the method with the software package. We used the software package LS-Dyna. The comparison was carried out on the dependence of the velocity of the contact surface of the target on time. The LS-Dyna software package does not include the equation of state in theta form; therefore, for comparison, some of the simplest hydrodynamic and elastic-plastic models of materials were considered. Impact took place for two identical materials. Impactor thickness 0.1 m, target thickness - 1 m. Impact velocity - 500 m / s. Comparison of calculations is shown in figure 2. At the initial moment, there is a qualitative convergence of the calculation by the Cabaret scheme with the calculations in LS-Dyna. However, further discrepancies are observed. It is due to the interference of shock waves. It's possible to solve only the 3D formulation of the problem in this software package. Therefore, the results after 5 ms can't be compared. More details on the models can be found in [11]. The data of the all numerical experiments are summarized in table 1.

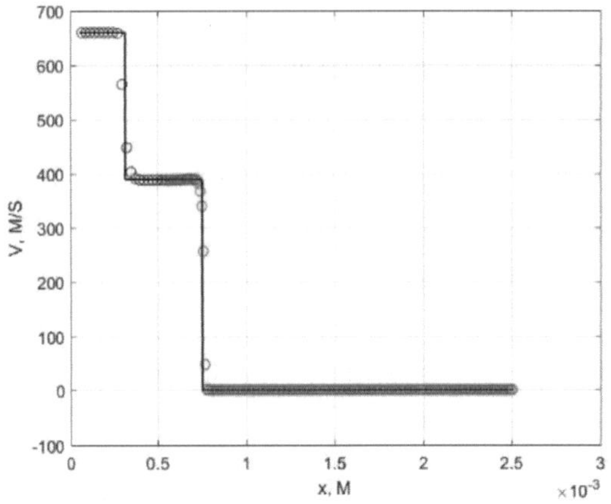

Figure 1. Velocity profile with uniaxial loading of a copper target with a copper plate. Comparison with analytical solution.

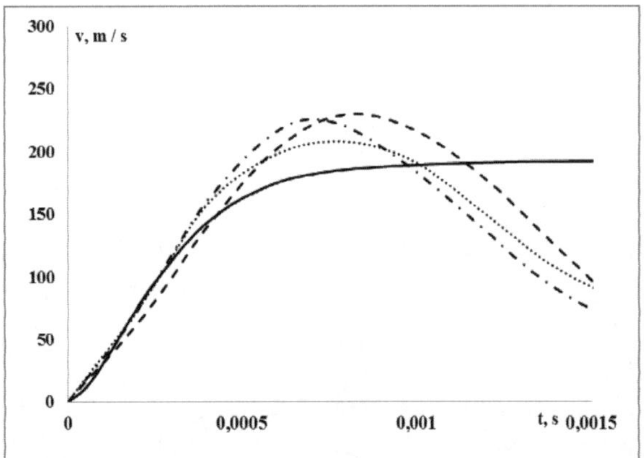

Figure 2. Velocity profile of the contact surface. Comparison of results with the LS-Dyna package. The solid line is a Cabaret solution. The point line is a Elastic fluid model in LS-Dyna. The stippled line is Plastic kinematic model in LS-Dyna. The dash-and-dot line is the Elastic-Plasitc Hydro model in LS-Dyna

The final stage of verification was a comparative analysis of the Cabaret scheme with experimental data [12]. The following experiment was performed numerically. Aluminum plate 0.2 mm thick swoops at a velocity of 660 m / s on a target of homogeneous copper 0.7 mm thick. Comparison will be carried out along the velocity profile of the contact surface. A comparison was made on the velocity profile of the contact surface depending on time in figure 3. There is a qualitative coincidence of the result with experimental data in the first moments of time. However, later it is necessary to take into account the parameters of elasticity and plasticity of metals.

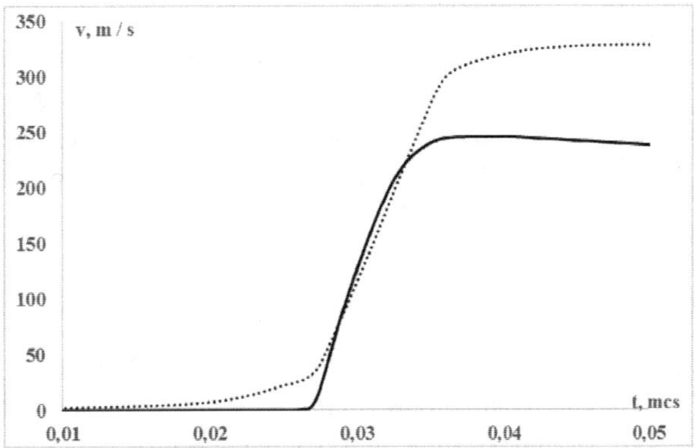

Figure 3. Velocity profile of the contact surface. Comparison of results with experiment. The solid line is a Cabaret solution. The point line is experimental data.

Table 1. Material parameters.

Parameters/ material	ρ_0 [kg/m^3]	A [GPa]	n	E [MPa]	ν	K [MPa]	Y_0 [MPa]	γ	C_p [J/(kg·K)]	G [MPa]
Copper	8930	37.75	4	10^5	0.32	$1.39 \cdot 10^5$	$2.8 \cdot 10^2$	1.4	3790	$4.6 \cdot 10^6$
Aluminium	2785	$1.97 \cdot 10^5$	4.2							

5. Conclusion

In the course of the work, it turned out that the algorithm implemented by the Cabaret scheme describes the algorithm well the initial stage of uniaxial shock-wave loading of materials. The calculation according to the Cabaret scheme gives an excellent match with the analytical solution. With the experimental data and the results of calculations in LS-Dyna, the calculation according to the Cabaret scheme is qualitatively well matched at the initial moments of loading. In the future, the scheme will be tested on heterogeneous materials, a mathematical model of which was proposed in [13].

References

[1] Fomin V M, Gulidov A I, Sapozhnikov G A and others 1999 *Vysokoskorostnoye vzaimodeystviye tel* (Novosibirsk: Publisher of SB RAS) p 600

[2] Merzhievskiy L A 2015 Modeli deformirovaniya pri intensivnykh dinamicheskikh nagruzkakh *J. Physics of Combustion and Explosion* FGV 2 (2015) p 144

[3] Iserles A . Generalized leapfrog methods 1986 *IMA Journal of Numerical Analysis* V 6 pp 381–92 doi 10.1093/imanum/6.4.381

[4] Goloviznin V M and Samarskii A A Finite difference approximation of convective transport equation with space splitting time derivative 1998 *Matem. Mod.* 10:1 pp 86–100

[5] Goloviznin V M, Zaitsev M A, Karabasov S A and Korotkin I A 2013 Novel Algorithms of Computational Hydrodynamics for Multicore Computing (Moscow: Izd. Moskovskogo Univ.) p 472

[6] Kovyrkina O A and Ostapenko V V 2017 *Marchukovskiye nauchnyye chteniya* (Novosibirsk Izdatel'stvo IVMiMG SO RAN) pp 434-40
[7] Godunov S K and Romenskiy Ye I 1998 Elementy mekhaniki sploshnykh sred i zakony sokhraneniya (Novosibirsk: Nauchnaya kniga) 280 ISBN: 5-88119-012-2
[8] Samarskiy A A and Popov Yu P 1992 Raznostnyye metody resheniya zadach gazovoy dinamiki (Moskva: Nauka) 424 ISBN:5-02-014377-7
[9] Zyuzina N A, Ostapenko V V and Polunina E I 2018 Splitting method for CABARET scheme approximating the non-uniform scalar conservation law *Sib. Zh. Vychisl. Mat.*, 21:2 pp 185–200 *Num. Anal. Appl.,* 11:2 pp 146–157
[10] Gulidov A I and Shabalin I I 1987 Chislennaya realizatsiya granichnykh usloviy v dinamicheskikh kontaktnykh zadachakh Preprint № 12-87, SO AN SSSR, ITPM
[11] Kanel' G I, Razorenov S V, Utkin A V and Fortov V Ye 2008 Eksperimental'nyye profili udarnykh voln v kondensirovannykh veshchestvakh (Moskva: Fizmatlit) 248
[12] LS-DYNA Keyword User's Manual vol 2 Material Models. LSTC 2013
[13] Kolotilov V A and Fomin V M 2018 *AIP Conference Proceedings* 2027 pp 030144 doi 10.1063/1.5065238

Acknowledgments
The author is grateful to Academician V M Fomin for the formulation of the problem and V V Ostapenko for assistance in mastering the Cabaret scheme.

Comparative analysis of mechanical behavior of the tooth pair contacting with different mouthguard configurations

A A Kamenskikh, T N Ustjugova and A G Kuchumov

Department of Computational Mathematics, Mechanics and Biomechanics, Perm National Research Polytechnic University, Perm, Russia

anna_kamenskih@mail.ru, tat.ustugova@yandex.ru, kychymov@inbox.ru

Abstract. Sport mouthguards are an important piece of athletic equipment. They are used to prevent sport injuries of the dentofacial system. Paper presents comparative analysis of mechanical behavior of the tooth pair contacting with the mouthguard. Four contact types are considered: contact without mouthguard, contact with single-layer mouthguard, contact with three-layer mouthguard (with A-silicone thickness of 1.5 mm), contact with three-layer mouthguard (with A-silicone thickness of 3.5 mm). The axi-symmetric contact problem is solved with taking into account of the friction and all types of contact. We revealed a number of patterns in the results of computations with varying of indentation force in the range of 50-250 N. In the case of mouthguard contact (with A-silicone thickness of 3.5 mm), we obtained the dramatic decrease of maximal stress intensity. The maximal stress intensity vs indentation force dependence is close to linear. The silicone layer has an influence on the contact parameters magnitude and distribution. Dependence between strain intensity vs indentation force is non-linear.

1. Introduction

Due to the widespread occurrence of maxillofacial sport injuries [1] or professional injuries [2], there is a demand of studies related to the dentofacial system biomechanics. Application of new kinds of patient-specific devices made of contemporary materials is aimed at improving the quality of life of patients [2-5]. There is a wide range of research topics in the dental biomechanics: the center of rotation [6], modeling of the initial tooth movement under orthodontic stress [7], using 3D scanning technologies [8], numerical modeling of bite [9], shape memory alloy clamps applications [10], etc. A large number of studies are aimed at analyzing and solving the problems related to mouthguard usage during sports [1, 11-15]. Sports mouthguards and mouth protectors are different names for the same thing: a device worn over your teeth that protects them. Mouthguards are an important piece of athletic equipment. There are two types of sport mouthguards for contact sports and non-contact sports. Here is the mouthguard for non-contact sports is analyzed.

Recently, the new concept of individualized three-layer mouthguard (made of EVA and A-silicone) for the athletes in non-contact sports was developed at Perm State Medical University (Perm, Russia) [4, 14, 11].

Paper presents comparative analysis of mechanical behavior of the tooth pair contacting with the mouthguard. Four contact types are considered: contact without mouthguard, contact with single-layer mouthguard, contact with three-layer mouthguard (with A-silicone thickness of 1.5 mm), contact with three-layer mouthguard (with A-silicone thickness of 3.5 mm).

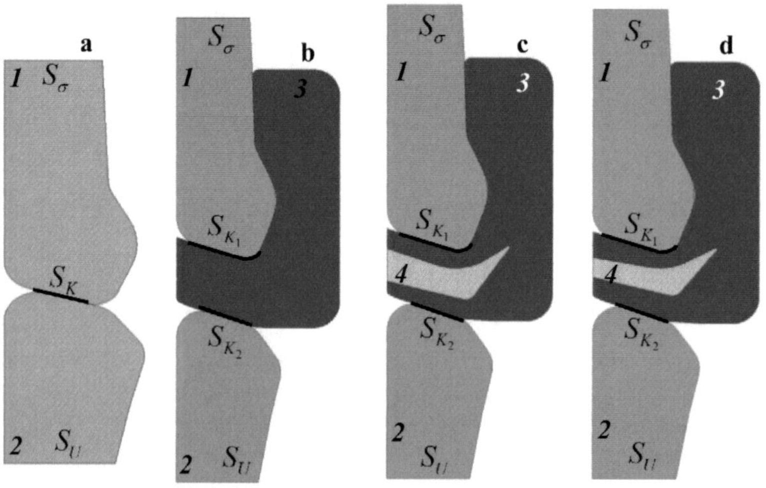

Figure 1. Computational schemes of contact interaction between a pair of teeth: a is without mouthguard; b is through single-layer mouthguard; c, d are through three-layer mouthguard (with A-silicone thickness of 1.5 and 3.5 mm, respectively).

2. Problem statement

Four cases of tooth contacting with mouthguard were considered (Figure 1): a is contact of upper tooth (1) and lower tooth (2) without mouthguard, b is contact of teeth and single-layer mouthguard (3), c and d are contact of tooth pair with three-layer mouthguard with A-silicone layer (4) with maximal thickness of 1.5 mm and 3.5 mm, respectively. Maximal dimensional parameters of upper tooth are height is 10 mm and width 6 mm, maximal dimensional parameters of lower tooth are height 8 mm and width 7 mm.

Mathematical problem statement includes equations of equilibrium, physical and geometrical relations [11]. Boundary and contact conditions are following: at the boundary S_σ, the constant functional load is applied varying from 50 to 250 N (indentation force) and bending is banned; at the boundary S_U, vertical displacements are prohibited; at the contact boundary S_{cont}, the contact interaction with friction (friction coefficient is equal to 0.3) and all types of contact (sticking, slipping, sticking-off) are considered. In computational schemes (c), (d), we did not take into account the A-silicone layer contact with other layers. Surfaces of teeth occlusion are S_K, S_{K_1}, S_{K_2}.

Mechanical properties of mouthguard materials was obtained in paper [4]. It was shown that the Young's modulus of EVA is 17.1 ± 1.58 MPa and Poisson's ratio is equal to $\nu = 0.46$; the material was exposed to demonstrate elastoplastic behavior. It was shown that EVA material exhibits elastoplastic properties, thus the dependence $\sigma - \varepsilon$ was plotted [11]. The Young's modulus of A-silicone is 0.34 ± 0.1 MPa and Poisson's ratio is equal to $\nu = 0.49$. The properties of the material of a pair of teeth are the properties of the tooth enamel. Young's modulus and Poisson's ratio were $E = 80.4 \cdot 10^3$ MPa and $\nu = 0.3$, respectively.

3. Results

In work an analysis of the convergence of the results of the numerical solution of the problem of the contact interaction of a pair of teeth was carried out, taking into account the degree of discretization of the system in the protective tire model. The influence of the geometric configuration of the protective prosthetic structure on the stress-strain state of the biomechanical assembly as a whole and the parameters of the contact zone, especially on the surface of the teeth closing, was considered. In the framework of a series of experiments, fields of stress and strain intensity distribution, regularities of change of contact parameters, as well as the dependence of maximum values of stress intensities and strains on the indentation force were obtained. For an example of the nature of the distribution of stress intensity σ_{int} fields, figure 2 shows the distribution for all considered models for the value of the indentation force 200 N.

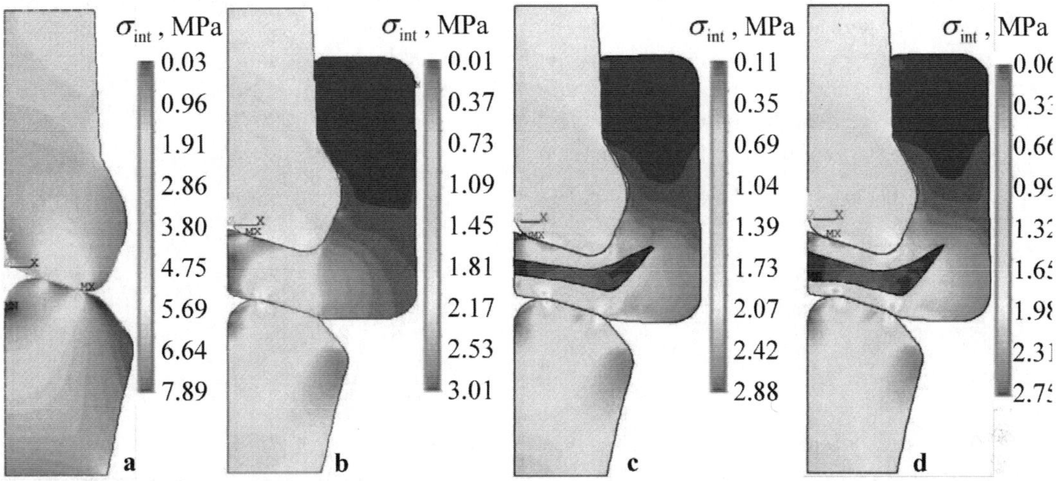

Figure 2. Stress intensity distribution in the teeth and mouthguard at indentation force 200 N: a is without mouthguard; b is through single-layer mouthguard; c, d are through three-layer mouthguard (with A-silicone thickness of 1.5 and 3.5 mm, respectively).

For cases b–d, a decrease of the maximum stress intensity is observed. The maximum stress intensity is observed on the surface of the closure of the teeth, while using the prosthetic structure, the nature of the stress intensity distribution does not contain local zones of maximum stress. Upon contact of a pair of teeth through a protective tire, a drop in the maximum level of stress intensity is observed: for a single-layer mouthguard (b), on average, by 61.85%; for three-layer mouthguards (c) and (d), on average, by 63.5% and 65.15%, respectively. It should be noted that the maximum stress intensity of the tooth in the lower dentition is approximately 5% lower than that of the tooth in the upper dentition. The dependences of the maximum stress intensity on the indentation force, which is shown using the example of the tooth of the upper dentition, are shown in figure 3.

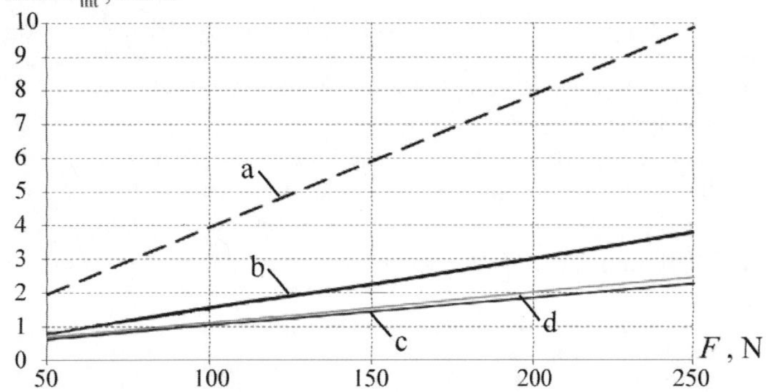

Figure 3. The $\max \sigma_{\text{int}}$ vs F in the upper tooth: lines a–d correspond to cases a–d.

The minimum stress intensity is observed when a pair of teeth contact through prosthetic structures with an A-silicone layer. As the indentation force increases, the difference between $\max \sigma_{\text{int}}$ the computational schemes, taking into account the single-layer mouthguard and three-layer protective prosthetic structures, increases, while the dependence on the indentation force is close to linear for all considered computational schemes. With an indentation force of 250 N, the maximum stress intensity in the hard tissues of the teeth decreased by more than 75% when using a three-layer prosthetic structure, and when using a single-layer mouthguard - by 60%.

Of particular interest is the nature of the distribution and the level of contact pressure on the surface of the closure of the teeth. Figures 3 and 4 show the contact pressure distribution fields for all the previously considered design schemes for an indentation force of 200 N.

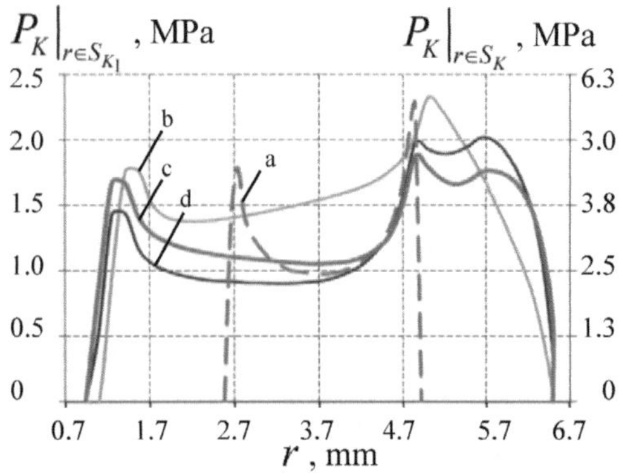

Figure 4. Contact pressure on S_{K_1} and S_K: lines a–d correspond to cases a–d.

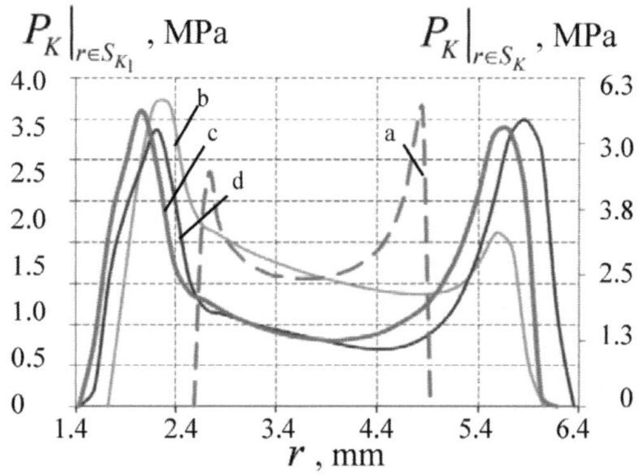

Figure 5. Contact pressure on S_{K_2} and S_K: lines a–d correspond to cases a–d.

The nature of the distribution of contact pressure is not uniform with peak levels p_K in the region of the maximum intensity of stress. The introduction of an A-silicone interlayer into the protective mouthguard had the most significant effect on the distribution of contact pressure: near the tooth closure surface S_{K_1}, the level of maximum contact pressure decreased by approximately 13% compared to the design diagram (b); on the contact surface S_{K_2} there is a more uniform distribution of contact pressure with two zones max p_K up to 3.5 MPa. For all three models of prosthetic structures, a decrease in the level of contact pressure is observed: on average S_{K_1} for a single-layer mouthguard (b) by 59.3%, for a three-layer mouthguard S_{K_2} (c) and (d), on average, by 64.56% and 66.7%, respectively; on a single-layer mouthguard (b) by an average of 35.8%, for a three-layer mouthguard (c) and (d) an average of 38.7% and 47.9%, respectively.

Since the main Eva material of the three considered variants of the geometrical configuration of prosthetic structures works within the framework of the deformation theory of small elastoplastic deformations, the nature of the distribution of plastic strain intensity in kappa is of interest. Figure 6 shows the nature of the distribution of strain intensity in the prosthetic structure at an indentation force level of 250 N

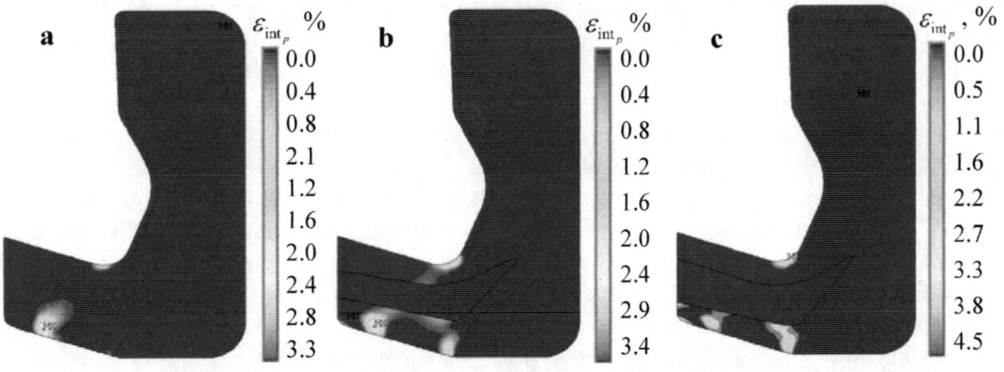

Figure 6. Plastic strain distribution in the mouthguard (case b).

It is worth noting that the maximum intensity of plastic deformations is observed near the tooth closure surface and does not exceed 5% at maximum indentation force. In the design schemes (b) and (c), the maximum $\varepsilon_{\mathrm{int}_p}$ is close to the zone of contact with the tooth of the lower dentition; for the design scheme (c), the maximum $\varepsilon_{\mathrm{int}_p}$ is observed near the zone of contact with the tooth of the upper dentition.

Figure 7 shows the dependence of the maximum level of plastic deformations in prosthetic structures of different geometrical configuration.

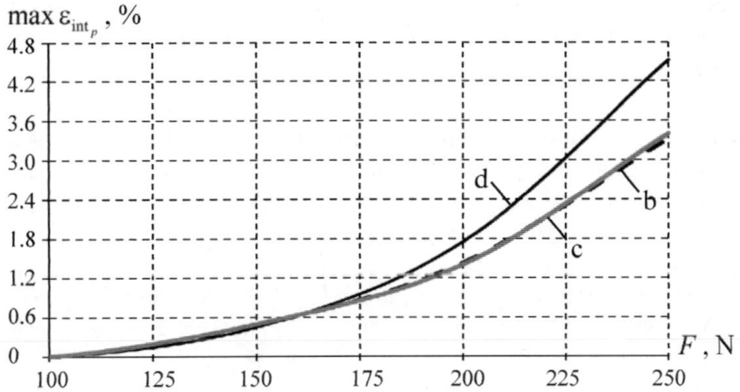

Figure 7. Dependence of $\max \varepsilon_{\mathrm{int}_p}$ on F : lines b–d correspond to cases b–d.

The nature of the dependence $\max \varepsilon_{\mathrm{int}_p}$ on the indentation force is not linear. For the three considered variants of the prosthetic design, the first plastic deformations appear when the indentation force is more than 100 N. For the design schemes (b) and (c), the maximum level of plastic deformations differs slightly. In the prosthetic design of the design scheme (d), the level of plastic deformations is higher than a maximum of more than 1%.

4. Conclusion

As part of the study, a numerical model of the problem of contact interaction of two teeth of the upper and lower dentition was constructed with and without consideration of the protective prosthetic structure of different geometric configuration (single-layer and three-layer). The analysis of the deformation behavior of the biomechanical contact node in the framework of the theory of deformation elastoplasticity under different functional loads is performed. The distribution of stress and strain intensities is obtained for all the models considered. When analyzing the results of a series of numerical experiments, it was established:

IOP Conf. Series: Materials Science and Engineering **511** (2019) 012003 doi:10.1088/1757-899X/511/1/012003

- Upon contact of a pair of teeth through a protective tire, a drop in the maximum level of stress intensity in the tooth of the upper dentition is observed: for a single-layer mouthguard, on average, by 61.85%; for a three-layer mouth guard with a layer of A-silicone 1.5 mm thick – by 63.5% and for a three-layer mouth guard with a layer of A-silicone 3.5 mm thick – by 65.15%. The maximum stress intensity of the tooth of the lower dentition is lower than that of the tooth of the upper dentition by approximately 5% in all variants of the design schemes.

- The maximum level of contact pressure when using a three-layer prosthetic design is reduced by approximately 13% in the zone of contact of the tooth of the upper dentition compared with the design scheme with a single-layer mouthguard. The maximum level of contact pressure in the vicinity of the tooth of the lower dentition has small differences for all variants of the dental splint.

- The maximum decrease in the level of contact pressure is observed in the construction of a three-layer mouthguard with an A-silicone interlayer with a thickness of 3.5 mm: in the zone of contact with the tooth of the upper dentition below by more than $\max P_K$ 66%; in the zone of contact with the tooth of the lower dentition - by 47%.

- The dependence of the level of intensity of plastic deformations on the indentation force is not linear, for all computational schemes it does not exceed 5%. In this case, the zone of localization of the maximum plastic deformations of a single-layer mouthguard and a three-layer mouthguard with an A-silicone interlayer with a thickness of 1.5 mm is located near the tooth of the lower dentition. In a three-layer mouthguard with a 3.5 mm A-silicone interlayer, the nature of the distribution of plastic deformation differs from the other two options considered, the zone of maximum plastic deformation is observed near the tooth of the upper dentition.

In general, prosthetic constructions with an A-silicone layer make it possible to reduce the level of stress intensity in the hard tissues of the teeth and the level of contact pressure by a larger amount than a single-layer prosthetic structure.

The work is performed under financial support of Russian Foundation for Basic Research (project № 17-48-590411 r_a). The article was published with the financial support of the Government of Perm Krai. AGK acknowledges Russian Government Assignment grant 19.7286.2017/8.9 for their financial support.

References

[1] Zakaryan G A 2016 *Actual problems of the liberal and nat. sci.* **10-2** 166-170
[2] Klimova T N, Shemonaev V I, Sargsyan K A and Borshcheva E S 2011 *Volgograd J. of Med. Res.* **3** 41-44
[3] Semenchishina V S and Zholudev S E 2016 *Actual problems in dentistry* **4** 34-39
[4] Lokhov V A, Kuchumov A G, Merzlykov A F, Astashina N B, Ozhgikhina E S and Tropin V A 2015 *Russ. J. of Biomech.* **4** pp 409-420
[5] Shemonaev V I, Klimova T N and Timacheva T B 2014 *Volgograd J. of Med. Res.* **2** 45-49
[6] Nyashin Y, Nyashin M, Osipenko M, Lokhov V, Dubinin A, Rammerstorfer F and Zhurov A 2016 *Comp. Methods in Biomech. and Biomed. Eng.* **3** 229-239
[7] Arutyunov S D, Gavryushin S S and Demishkevich E B 2014 *Herald of the BMSTU. Ser. Nat. Sci.* **3** 108-120
[8] Yumashev A V, Mikhailova M V, Kuderova I G and Kristal E A 2015 *J. of New Med. Technol.* **1** 2-6
[9] Tverier V M, Nikitin V N and Krotkikh A A 2017 *Russ. J. of Biomech.* 1 41-50
[10] Kuchumov, A G, Lokhov V A, Nyashin Y I, Mesnard M and Selyaninov A A 2009 *Russ. J. of Biomech.* **13** 18-27
[11] Kamenskih A, Astashina N B, Lesnikova Y, Sergeeva E and Kuchumov A G 2018 *Ser. on Biomech.* **1** 3-15
[12] Sevbitov A V, Achkasov E E, Kanukoeva E Y, Borisov V V and Sultanova O A 2014 *Sports med.: res. and pract.* **2** 42-46
[13] Ivanova A A, Slynko A Yu, Borisov V V and Shakaryants A A 2016 *Dental Forum* **4** 33

IOP Conf. Series: Materials Science and Engineering **511** (2019) 012003 doi:10.1088/1757-899X/511/1/012003

[14] Lokhov V A, Astashina N B, Sergeeva E S, Merzlyakov A F and Baydarov A A 2017 *Russ. J. of Biomech.* **2** 211-218
[15] Kopetsky I S and Vasilyev Yu L 2016 *Dental Forum* **3** 32-34

IOP Conf. Series: Materials Science and Engineering **511** (2019) 012004 doi:10.1088/1757-899X/511/1/012004

Destruction of LED heterostructures under high-current electron beam irradiation

Zixuan Li, Jiyao Xian and S G Sysoyeva

National Research Tomsk Polytechnic University, 30 Lenin Ave., Tomsk, 634050, Russia

E-mail: llzzxx0@163.com

Abstract. This paper presents the experimental results of the fracture morphology of InGaN / GaN heterostructures and epitaxial GaN layers deposited on sapphire substrates after irradiation with high current electron beams. A special type of damage - microspheres, ~1-5 μm was discovered and studied. The results show that the microspheres act as diffractive optical elements redistributing the stimulated emission of InGaN/GaN in space.

1. Introduction

The study [1] reveals the prospects of luminescent control of light-emitting InGaN/GaN heterostructures on sapphire substrates by high-energy high-current electron beams (HCEBs). In view of this, it is necessary to elucidate the mechanisms of breaking the light-emitting heterostructures during their HCEB diagnostics. As is known, the HCEB irradiation of dielectrics and semiconductors is accompanied by their fracture [2-4]. Threshold electron beam energy density H* required for breaking macrosamples with a size over the electron beam penetration depth in a material is determined by the type of a solid and increases at the transition from ionic crystal dielectrics to semiconductors. It was established that the main mechanisms of the HCEB initiated fracture of dielectrics and semiconductors are thermal shock [2, 3] and electrical breakdown [4]. Both mechanisms are related to the features of the HCEB effect on materials—specifically, high rates of charge and energy input (10^{10}–10^{11} C m^{-3} s^{-1} and 10^{11}–10^{13} Gy s^{-1}, respectively) [5]. The discharge mechanism of dielectric fracture is implemented, as a rule, in the regime of multipulse irradiation of samples by a low-density (H $\approx$ 0.1 J/cm^2) electron beam; the thermal shock is implemented in the regime of single irradiation by an electron beam of higher density (H $\geq$0.6 J/cm^2). These mechanisms can be experimentally distinguished by the characteristic morphology of fractures [4]. This article considers the morphology of electron-beam-initiated fracture of light-emitting diode heterostructures (HSs) based on InGaN/GaN quantum wells grown on sapphire substrates and determines possible mechanisms of their fracture.

The purpose of this work is to study the destruction patterns and features of GaN / InGaN LED heterostructures under irradiation by a high-current electron beam.

2. Methods

In this work, three LEDs with InGaN / GaN quantum wells, grown by various manufacturers using organometallic gas-phase epitaxy on sapphire with (0001) orientation, were investigated (figure1).The samples had a dislocation density of 10^7 - 10^9 cm^{-2}. The samples were irradiated in atmospheric air at room temperature.

IOP Conf. Series: Materials Science and Engineering **511** (2019) 012004 doi:10.1088/1757-899X/511/1/012004

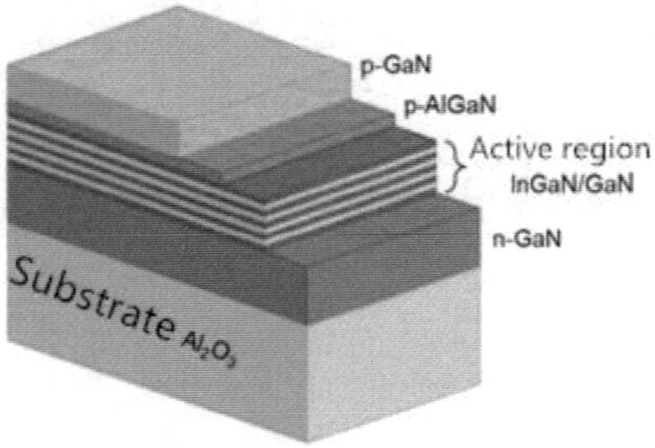

Figure 1. Sample LED heterostructure

The excitation source was an electron accelerator with a GIN-600 generator (350 keV, 2 kA, 15 ns). The electron beam energy density was varied in the range of 0.05–0.25 J / cm^2.

The spatial distribution of luminescence over the sample surface was photographed by a SONY DSLR-A500 digital camera in the "Bulb" mode during a single excitation pulse through an MBS-10 microscope (figure 2). The morphology of residual damage resulting from multipulse irradiation of HCEB was detected using a μVizo-101 transmission microvisor.

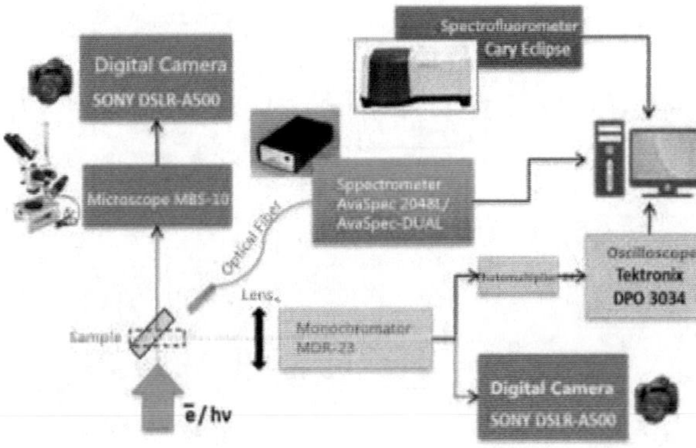

Figure 2. Source of excitation: HCEB (U = 350 keV; t = 15 ns; 0.02 - 1 J/cm^2);

3. Results and Discussion

Figures 3 and 4 show the morphology of damage, two mechanisms characteristic destruction of the HS by an electron beam. It can be seen that for thermal shock, the typical type of damage is cracks in both crystallographic planes and in other directions (figure 3). For the electrical breakdown, GaN's electrical breakdown channels and local micro-region damage are typical damage (figure 4).

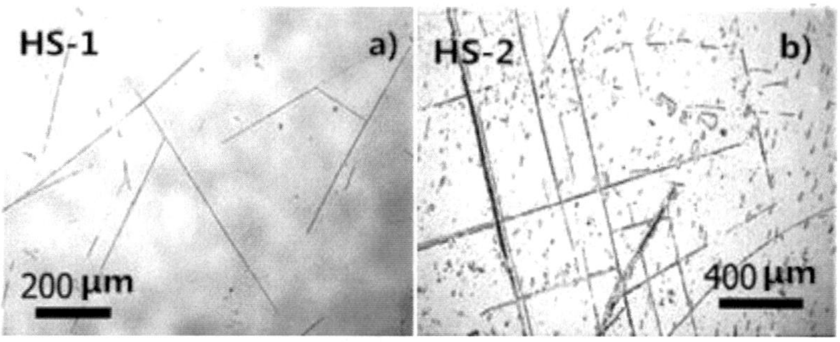

Figure 3. Cracks, formed in the samples after 10 pulses of the HCEB, H = 1 J/cm^2

Table 1 The angles between the cracks formed in the samples under the action of the HCEB

Sample	Angles between cracks, degrees
HS-1	2; 3; 4; 5; 6; 8; 9; 10; 12; 13; 15; 17; 22; 24; 31; 33; 37; 46; 62; 65; 71; 74; 75; 77; 80; 82; 83; 84; 86; 87; 88; 92; 93; 95; 108; 117; 119; 132
HS-2	1; 2; 9; 13; 28; 37; 41; 50; 92; 93; 94; 105; 106; 107; 133; 134; 135; 142; 143; 144

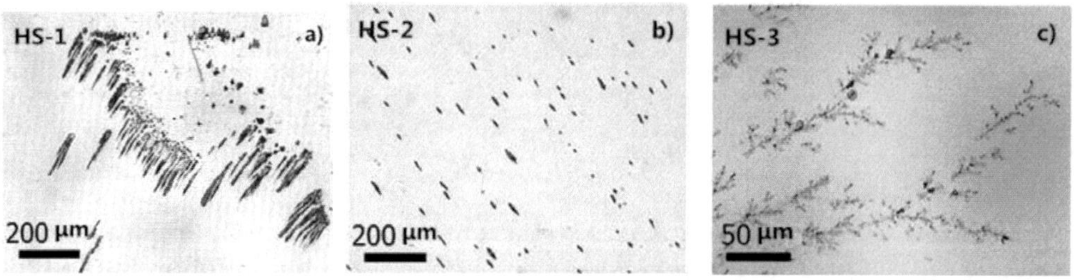

Figure 4. Photographs of micro destruction in the heterostructure samples after 100 irradiation pulses of HCEB (H = 0.25 J/cm^2)

A special type of degradation of the HS, recently discovered by us, is a spherical region with diameters of 1-10 microns (microspheres). The photograph of microspheres obtained on sample HS-3 is shown in Fig. 5. Similar structures were previously observed on the samples of Alkali Halide Crystals (AHC) irradiated by a HCEB. Studies have shown that these microspheres are formed as a result of the impact of a stretching wave on a crystal. As a result, spherical pores form in a solid. In our case, such pores are formed in HS-3.

IOP Conf. Series: Materials Science and Engineering **511** (2019) 012004 doi:10.1088/1757-899X/511/1/012004

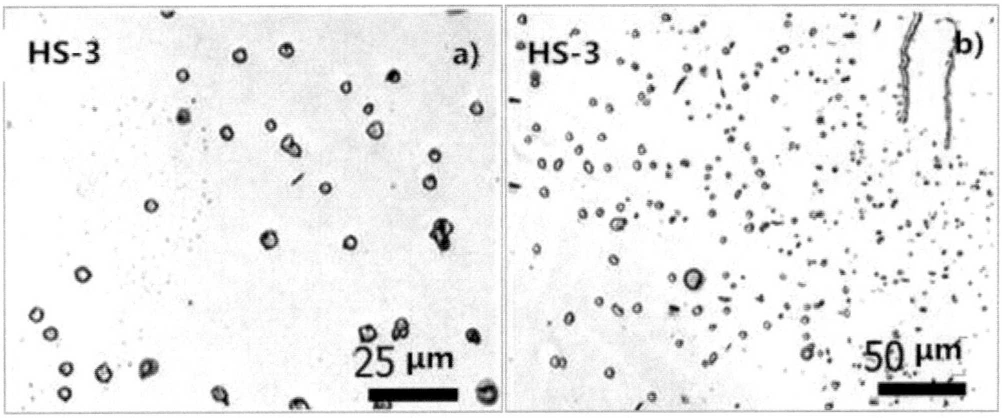

Figure 5. Destruction morphology of sample HS-3 after 100 pulses irradiation by HCEB at H = 0.6 J/cm^2

When photographing the spatial distribution of pulsed cathodoluminescence (PCL) over the surface HS-3, we observed stimulated radiation diffraction on the microspheres. Photographs of PCL diffraction on microspheres are shown in Figure 6.

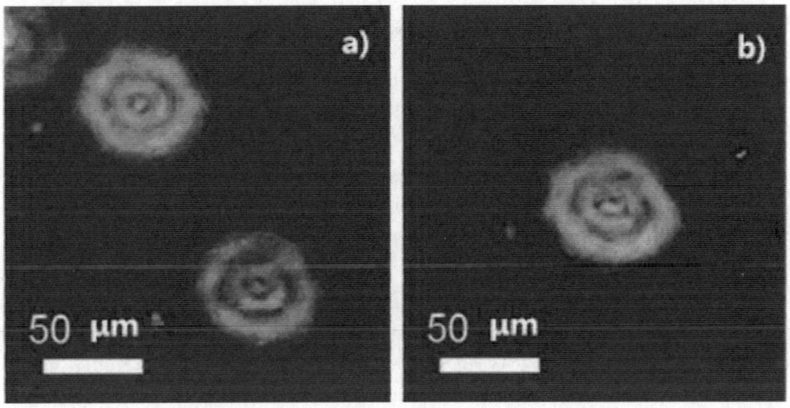

Figure 6. The pulsed cathodoluminescence photographs of the sample HS-3 with microspheres by excitation HCEB (H = 0.6 J/cm^2)

The diameter of the outer ring is 50 μm, and the diameter of the central spot is ≈ 5 μm.

The formation of microspheres (pores in a solid) during the irradiation of HCEB was observed by us in other samples: in polymethyl methacrylate (PMMA) and in glass with a protective film sprayed onto the surface (figures 7-8).

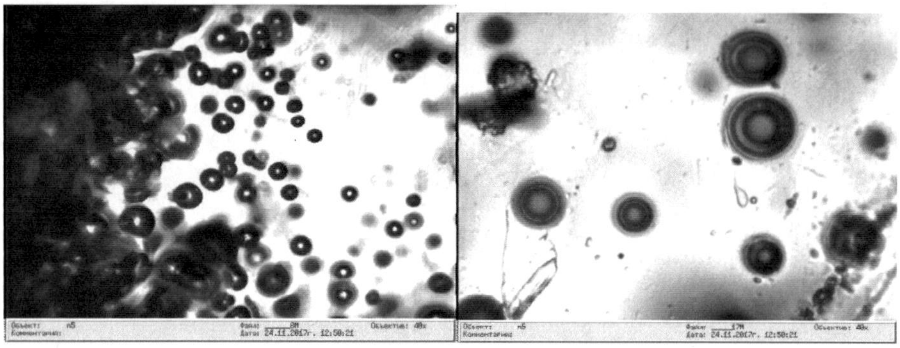

Figure 7. Diffraction of light on microspheres formed in PMMA when exposed to HCEB

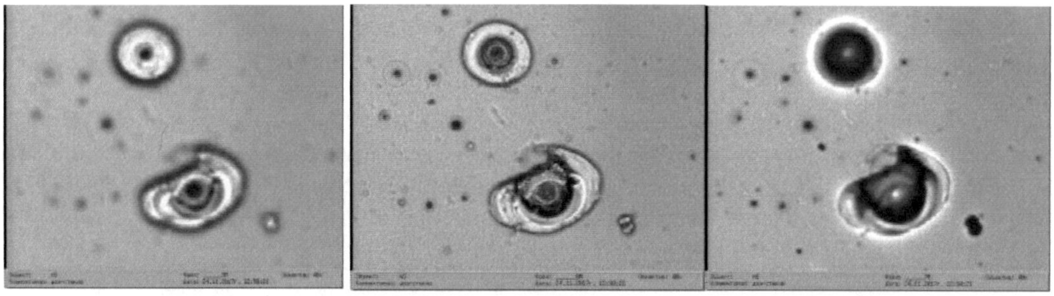

Figure 8. Diffraction of light on microspheres formed in glass with a protective film when exposed to laser radiation

An analysis of the experimental data obtained in this work suggests that the formation of microspheres is associated with the action of stretching acoustic wave on a solid. Acoustic waves with a high amplitude, sufficient for the destruction of solids, can be formed as a result of the HCEB, and laser radiation.

Microspheres play the role of local "mirrors" for stimulated InGaN quantum wells PCL: bright luminous microzones are characteristic of samples in whose luminescence spectra recorded stimulated InGaN quantum wells emission. It has been established that the spatial distribution of luminous microzones coincides with the localization regions of electron-beam microspheres. The spectral composition of the luminescence HS-3 before and after the electron-beam damage formation does not change.

4. Conclusion

As a result of the research, it was established that during the irradiation of the HS by an electron beam, three types of fracture structures are formed: cracks, electrical breakdown channels and microspheres. The electrical breakdown channels are formed as a result of the electrical discharges development in the HS. Cracks and microspheres are formed by the interaction of a bipolar acoustic wave with free surfaces of the irradiated samples. It is shown that microspheres can play the role of passive optical elements (local "mirrors" and diffractive optical elements) that do not affect the spectral composition of the luminescence, but only redistribute the stimulated emission of the quantum-well active InGaN / GaN region in space.

References

[1] Oleshko V, Gorina S G, Korepanov V I et al. 2013 *Izv. universities. Physics* **25 1** 55-58
[2] Weissburg D I 1982 *High-energy electronics of a solid* (Novosibirsk: Science) p 227

IOP Conf. Series: Materials Science and Engineering **511** (2019) 012004 doi:10.1088/1757-899X/511/1/012004

[3] Bogdankevich O B, Zverev M M, Ivanova T Yu et al. 1986 *KE* **113** **10** 2132-35
[4] Oleshko V and Shtanko VF 1987 *FTT* **29** **2** 320–324
[5] Oleshko V, Lisitsyna L, Malys D et al. 2010 *Nuclear Instruments and Methods in Physics Research B* **268** **19** 3265–68

IOP Conf. Series: Materials Science and Engineering **511** (2019) 012005 doi:10.1088/1757-899X/511/1/012005

Determination of residual stresses in multi-layer plasma coatings

V I Bogdanovich and M G Giorbelidze

Samara National Research University, 34, Moskovskoe Highway, Samara, 443086, Russian Federation

E-mail: bogdanovich@ssau.ru, m.giorbelidze@ssau.ru

Abstract. Method of determination of residual stresses in multi-layer coatings produced by plasma gas-thermal deposition is suggested. Residual stresses in a three-layer plasma coating Ni60Cr15 – NiCr80-20 + CaF_2 are determined. Evaluation of power effect of the plasma arc on the residual stresses distribution pattern in the coating, the sublayer, and the substrate is carried out.

1. Introduction

Residual stresses are among the primary reasons of destruction of plasma gas-thermal coatings. As a result, there are a lot of papers dealing with this problem, and the papers [1-18] should be distinguished. Nature of residual stresses development along with a number of reasons causing them constitutes some difficulties in modelling of these processes.

During plasma coating formation, powder material particles heated up to the melting temperature and accelerated by the stream, hit the part surface, and they are deformed on the surface. After crystallization of the melted particles, microscopic stresses balanced in the micro-volume of the whole particle material originate on the surface. According to the experience [1-3], adhesion and cohesive strength is not determined by the particle material strength, but by the attraction strength between the adjacent particles. Taking into account the fact, that the coating can consist of some layers of different materials, the macroscopic stresses attract the most interest. They are balanced in the volume that is essentially larger than the volume of a particle, and commensurable with the thickness of the whole coating or separate deposited layers. It allows to substitute modelling of microscopic stresses development process for modelling of macroscopic stresses development process and to consider them during the process of coating layer deposition.

2. Method of determination of residual stresses in multi-layer coatings

In some cases, necessity arises to determine residual stresses in multi-layer coatings. Using of well-known calculation methods in determination of residual stresses in multi-layer coatings for specific cases causes difficulties connected with cumbersome calculations. In this study, method of determination of residual stresses in multi-layer coatings produced by plasma gas-thermal deposition is suggested.

IOP Conf. Series: Materials Science and Engineering **511** (2019) 012005 doi:10.1088/1757-899X/511/1/012005

Let us consider the N-layer thin coating with thickness α_N with different moduli of elasticity E_i of layers, but with the same Poisson ratios μ_i. Let us assume that the layer of thickness η has been etched and further etching is taking place in the layer m. We apply stresses σ to the liberated surface (Figure 1) that existed in the etched layer of thickness η before its removal. Bending moment arises under these stresses, it can be determined from the following expression:

$$M(\eta) = -\int_{k-\eta-\delta(\eta)}^{h-\delta(\eta)} \sigma(z) z dz = \int_0^{\eta} \sigma(\xi)\left[\xi + \delta(\xi) - h\right] d\xi, \tag{1}$$

in this expression, transition to the variable ξ has been done according to the formula

$$\xi = h - \delta(\eta) - z;$$

$$\delta(\eta) = \frac{\int_0^{h-\eta} E(z_z) z_1 dz_1}{\int_0^{h-\eta} E(z_z) dz_1}.$$

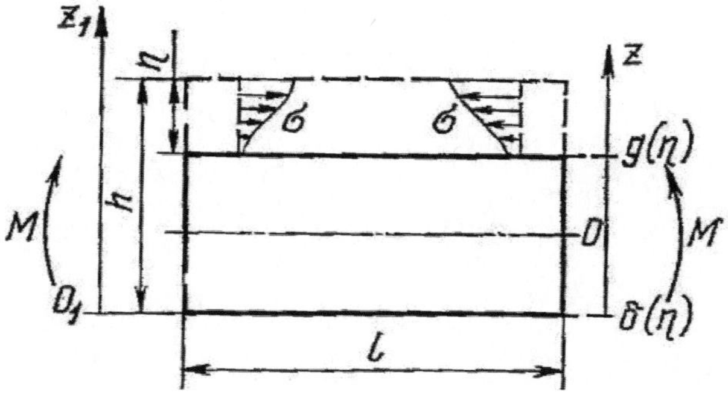

Figure 1. Scheme of determination of residual stresses after removal of the material of thickness η.

Bending moment $M(\eta)$ results in deflection (Figure 2) determined from the relation:

$$f(\eta) = \frac{l^2 M(\eta)}{2 I_E(\eta)}; \tag{2}$$

$$I_E(\eta) = \int_0^{h-\eta} E(z_1)\left[z_1 - \delta(\mu)\right]^2 dz_1,$$

where $I_E(\eta)$ – moment of inertia of the plate cross section in relation to its width.

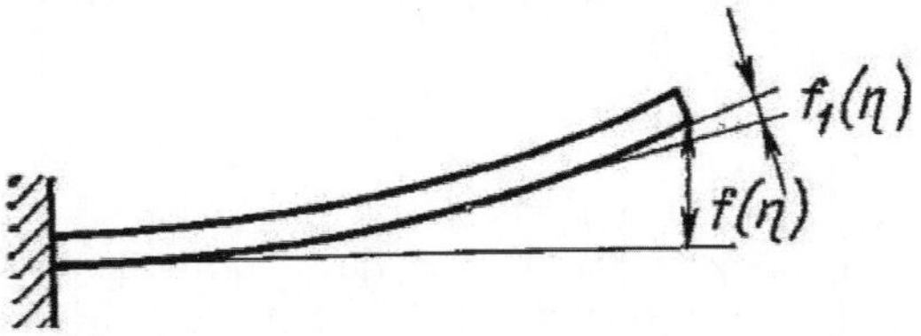

Figure 2. Scheme of deformation measurement.

Using (1-2), we get integral equation:

$$\frac{2f(\eta)I_E(\eta)}{l^2}=\int_0^\eta \sigma(\xi)\left[\xi+\delta(\eta)-h\right]d\xi. \tag{3}$$

Differentiating twice the equation (3) and taking into account that $\left[\frac{\eta+\delta(\eta)-h}{\delta'(\eta)}\right]'_\eta=-1$, we get

$$\frac{d\sigma}{d\eta}=\frac{2}{l^2}\frac{\left\{\frac{\left[f(\eta)I(\eta)\right]'_\eta}{\delta'(\eta)}\right\}'_\eta}{\frac{\eta+\delta(\eta)-h}{\delta'(\eta)}}. \tag{4}$$

Assume that etching is taking place in the layer m, i.e. $h-a_m<\eta<h-a_{m-1}$.
Integrating (4), we get

$$\sigma(\eta)=\sigma(h-a_m)+\frac{2}{l^2}\int_{h-a_m}^{\eta}\frac{\left\{\frac{\left[f(\xi)I_E(\xi)\right]'_\xi}{\delta'(\xi)}\right\}'_\xi}{\frac{\xi+\delta(\xi)-h}{\delta'(\xi)}}d\xi. \tag{5}$$

Taking the integral in parts twice in the equation (5) and determining $\sigma(h-a_m)$ after single differentiation of (3) we get

$$\sigma(\eta)=-\frac{2}{l^2}\left\{\frac{\left[f(\eta)I_E(\eta)\right]'_\eta}{\eta+\delta(\eta)-h}-f(\eta)I_E(\eta)\frac{\delta'(\eta)}{\left[\eta+\delta(\eta)-h\right]^2}-\right.$$

$$\left.-\int_{h-a_m}^{\eta}f(\xi)I_E(\xi)\frac{\delta'(\xi)}{\left[\xi+\delta(\xi)-h\right]^3}d\xi\right\}-$$

$$-\left[\frac{2}{l^2}f(\eta)I_E(\eta)\frac{\delta'(\eta)}{\left[\eta+\delta(\eta)-h\right]^2}+\frac{\delta'(\eta)}{\eta+\delta(\eta)-h}\int_0^\eta\sigma(\xi)d\xi\right]\Bigg|_{\eta=h-a_m}. \tag{6}$$

Formula (6) is right for a thin bar $\sigma=\sigma_x>>\sigma_y$. For an isotropic plate or a thin coating the derivation is the same, but the equation members from the first to the forth of the right part of the equation (6) are to be divided by $1-\mu$.

For the case $N=1$ equation (6) becomes (single layer plate):

$$\sigma(\eta)=-\frac{2}{l^2}\left\{\frac{\left[f(\eta)I_E(\eta)\right]'_\eta}{\eta+\delta(\eta)-h}-f(\eta)I_E(\eta)\frac{\delta'(\eta)}{\left[\eta+\delta(\eta)-h\right]^2}-\right.$$

$$-\int_{h}^{\eta} f(\xi) I_E(\xi) \frac{\delta'(\xi)}{\left[\xi+\delta(\xi)-h\right]^3} d\xi \Bigg\}. \qquad (7)$$

For the case $N=2$ when $h-a_1<\eta\le h$ equation (6) becomes:

$$\sigma(\eta)=-\frac{2}{l^2}\left\{\frac{\left[f(\eta)I_E(\eta)\right]'_{\eta}}{\eta+\delta(\eta)-h}-f(\eta)I_E(\eta)\frac{\delta'(\eta)}{\left[\eta+\delta(\eta)-h\right]^2}-\right.$$

$$\left.-\int_{h-a_1}^{\eta} f(\xi) I_E(\xi)\frac{\delta'(\xi)}{\left[\xi+\delta(\xi)-h\right]^3}d\xi\right\}-$$

$$-\left[\frac{2}{l^2}f(\eta)I_E(\eta)\frac{\delta'(\eta)}{\left[\eta+\delta(\eta)-h\right]^2}+\frac{\delta'(\eta)}{\eta+\delta(\eta)-h}\int_0^{\eta}\sigma(\xi)d\xi\right]\Bigg|_{\eta=h-a_1}. \qquad (8)$$

3. Results and discussion

This procedure has been used to determine residual stresses in the deposited layer NiCr80-20 + CaF_2 and the sublayer NiCr80-20 (Figure 3). Streaks of heat-resistant alloy 80 x 10 x 3 mm were used as substrate-samples. Deposition was carried by argon-hydrogen plasma at modernized installation UPU-8M using portable plasmatron of own design. Argon consumption was 3 m^3/h, hydrogen consumption 0.45 m^3/h, deposition distance 100 mm, rate of the burner travel 0.1 m/s, scanning pitch 7.5 mm. Before deposition, the samples were sand blasted.

When determining residual stresses, sample's deflection was measured during continuous electropolishing. For uniform electropolishing, special etchant, a cathode made of stainless steel, a barretter for keeping direct current, an automatic mixer for balancing of etching temperature, and special oil for impregnation of porous samples were used.

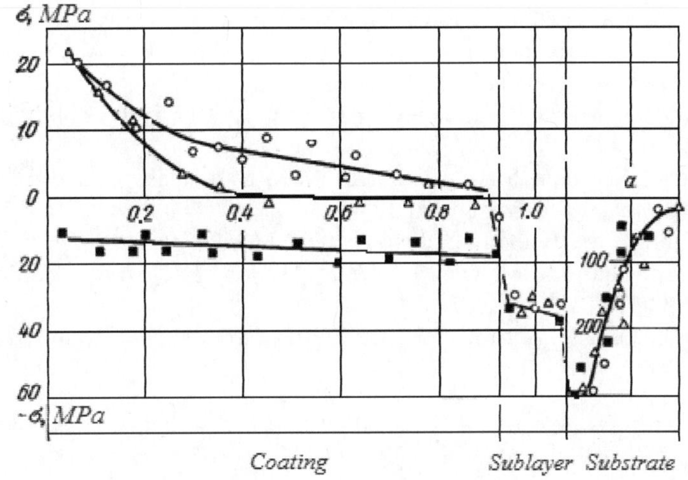

Figure 3. Distribution of residual stresses in the deposited layer, the sublayer and the substrate under the arc current: ■ 265 A; Δ 400A; ○ 535 A; – – – residual stresses in the transitional layers coating-sublayer, sublayer - substrate.

4. Conclusions

Performed work has allowed to develop method of determination of residual stresses in multi-layer plates, to determine residual stresses in three-layer plasma coatings, and to evaluate effect of plasma arc power on the distribution pattern in the coating, the sublayer, and the substrate. Study results of residual stresses are shown in the Figure 3.

Acknowledgments

The reported study was funded by the Ministry of Education and Science of the Russian Federation according to the research project No.11.2744.2017.

References

[1] Barvinok V A 2005 *Plasma in Technology: Reliability and Resource* (Moscow: Nauka i technologii Publishers)
[2] Barvinok V A 1990 *Strained state and properties of plasma coatings control* (Moscow: Mashinostroyeniye)
[3] Bobrov G V, Ilin A A and Spektor V S 2014 *Theory and technology of inorganic coatings formation* (Moscow: Alfa-M)
[4] Bobrov G V and Ilyin A A 2004 *The Application of inorganic coatings* (Moscow: Intermet-Inzhiniring)
[5] Ilyushchenko A F, Shevtsov A I and Okovity V A 2011 *The processes and modeling of thermal spray coatings formation* (Minsk: Belarus nauka)
[6] Puzryakov A F 2003 *Theoretical bases of plasma spraying technology* (Moscow: Bauman University Publishers)
[7] Kudinov V V, Pekshev P Yu and Belashchenko V E 1990 *Plasma coating application* (Moscow: Nauka)
[8] Kravchenko I N, Puzryakov A F, Gladkov V Yu, Pankratova E V and Glinskih M A 2011 *Repair. Recovery. Modernization* **10** 6-11
[9] Bondareva G I 2011 *Bulletin of mechanical engineering* **9** 32-35
[10] Kartsev S V and Shirshov V S 2012 Engineering technology **5** 37-38
[11] Kravchenko I N, Moskal O Ya, Pankratova E V and Shiyan A V 2012 *Repair. Recovery. Modernization* **10** 44-50
[12] Pankratova E V, Kravchenko I N and Moskal O Ya 2012 *Repair. Recovery. Modernization* **11** 34-40
[13] Kravchenko I N, Kolomeychenko A V, Pupavtsev I E, Puzryakov A A and Solovev R Y 2017 *Welding International* **31(10)** 809-813
[14] Mutter M, Mauer G, Mücke R, Guillon O and Vaßen R 2017 *Surface and Coatings Technology* **318** 157-169
[15] Nayebpashaee N, Seyedein S H, Aboutalebi M R, Sarpoolaky H and Hadavi S M M 2016 *Surface and Coatings Technology* **291** 103-114
[16] Croom B P, Bumgardner C and Li X 2016 *Extreme Mechanics Letter* **7** 126-135
[17] Capek J, Pala Z and Kovarik O 2015 *IOP Conference Series: Materials Science and Engineering* **82(1)** 012112
[18] Yang L, Yang F, Long Y, Zhao X and Xiao P 2014 *Surface and Coatings Technology* **251** 98-105

Development of personalized approach to the reconstruction of bone tissue defects using porous ceramic osteoimplants

D E Kulbakin[1,2], E L Choinzonov[1], M R Mukhamedov[1], A S Buyakov[2,3,4], S P Buyakova[2,3,4], S N Kulkov[2,3,4]

[1]Tomsk National Research Medical Center of the Russian Academy of Sciences, Tomsk, 634050 Russia
[2]Tomsk State University, Tomsk, 634050 Russia
[3]Institute of Strength Physics and Materials Science of Siberian Branch Russian Academy of Sciences, Tomsk, 634021, Russia
[4]Tomsk Polytechnic University, Tomsk, 634050 Russia

E-mail: alesbuyakov@gmail.com

Abstract. A fundamentally new complex technique for preoperative planning and creation of individual implants from ceramics for the reconstruction of the maxillofacial region is clinically applicable and in demand in modern trends of reconstructive surgery were developed and approved. The developed method implies using of alumina based ceramic material, which meets the requirements for medical materials used in reconstructive surgery.
It was shown that by varying the initial composition of the ceramic powder material, it is possible to achieve structural and mechanical affinity with the surrounding bone tissue. Building a 3D model of an osteoimplant on the basis of a computed tomography of the patient avoids the need for additional resection of the bone tissue and ensures complete geometric conformity of the osteoimplant to the prosthetic area.
The method was approved by the reconstruction of the patient's skull-orbital complex using a three-dimensional personalized ceramic implant. The complete absence of the inflammatory reaction after surgery was observed.

1. Introduction

Despite the small proportion of maxillofacial tumors (about 0.5%) in the overall structure of malignant tumors, their treatment is one of the most difficult problems in modern oncology [1]. The peculiarity of the maxillofacial region structure causes the rapid spread of the tumor process along the bone structures and soft tissues of the near regions. The local prevalence of tumors requires extensive resections leading to the formation of postoperative defects, but the possibility of reconstructing the defects with the body's own tissues is limited. Thus, the problem of finding new, safe, effective techniques for the reconstruction of the maxillofacial region defects is urgent task.

In modern reconstructive surgery research is aimed at developing an individual technology of endoprosthetics of the skull facial region. This direction contributes to the rapid development of 3D technology and 3D printing [2].

This may be the solution in developing a method for individual implants for reconstruction of maxillofacial area defects by using 3D prototyping technologies, ensuring the structural identity of the implant with bone tissue by organizing a hierarchical pore structure and identity of the implant mechanical properties to the properties of bone tissue.

At the present time, due to the combination of high chemical, corrosion and wear resistance and high strength there is a steady tendency to replace metals and polymers in artificial implants with ceramics [3, 4]. A special attention among the materials used for bones prosthetics deserve oxide ceramics, such as zirconia and alumina [5, 6]. These ceramics are close to the inorganic matrix of bone

IOP Conf. Series: Materials Science and Engineering **511** (2019) 012006 doi:10.1088/1757-899X/511/1/012006

tissue by the type of chemical bond and included in the ISO registry as materials for osteoprosthetics (ISO #6474-1:2010). Al_2O_3 and ZrO_2 are not perceived by the body as heterogenous and biochemical reactions at the interface with bone can lead to tissue invasion into the implant pore volume and provide conditions for osteogenesis [7-9]. Further, unlike metals and polymers, ceramic can be sterilized by all currently used methods: dry-heat cabinet, autoclaving, exposure to ozone, ultrasound and ionizing radiation.

The aim of this study is to develop a new approach to the personalized reconstruction of the bone defects of the maxillofacial region with implants based on alumina ceramic.

2. Materials and experiments

A virtual planning of the maxillofacial area defect reconstruction with 3D model of implant was used. Procedure of implant modeling was based on the data of the preoperative spiral computed tomography of the patient's maxillofacial region. The use of such technique and the close cooperation of the surgeon and the engineers at the stage of preoperative planning allow creating an implant as close as possible to a specific reconstructive task. [10, 11]

Ceramic osteoimplant was obtained using the method of precision additive prototyping with complex volumetric configuration. An inverse matrix of the implant 3D model was obtained, into which the fluid mass based on ceramic powder and organic plasticizers was injected into matrix with subsequent annealing and sintering, leading to the consolidation of ceramic powder into a 3D ceramic osteoimplant with a complex geometry and pore structure. The pores were obtained due to the burning out of the organic plasticizer and pore-forming particles during annealing.

Sample of an osteoimplant was obtained for the reconstruction of the cheek-orbital complex. The created implant replicate the shape and contours of the defected bone area of the facial part of the skull, figure 1.

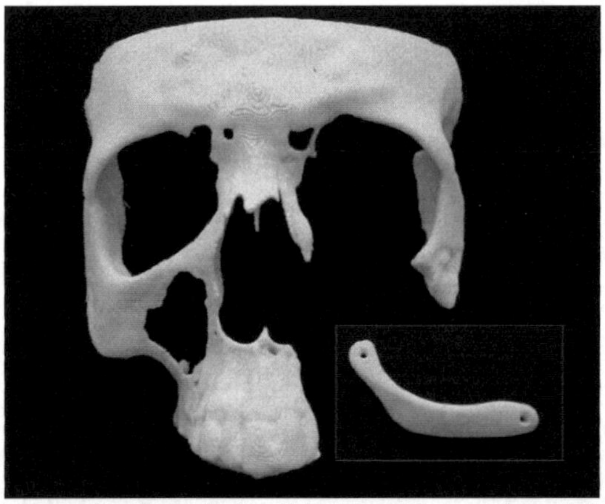

Figure 1. Ceramic osteoimplant skull-orbital complex and model of patient's skull.

Structural and mechanical studies of the material of the endoprosthesis were carried out: the microstructure was studied using scanning electron microscopy; flexural strength by three point bending method.

Obtained ceramic implant was sterilized by dry-heating and autoclaving. Implant fixation was carried out by mini-screws to the bone edges of the defect with subsequent control of fixation and covering with surrounding soft tissues and skin grafts.

3. Results and discussions

Studies of the ceramic have shown that by introducing an organic plasticizers and different rate of pore-forming particles in the initial powder composition it is possible to control the pore structure. This approach allow to obtain two types of pores: micropores, due to the presence of space between the particles of ceramic powder, and the macropores that inherit the configuration of pore-forming particles, figure 2.

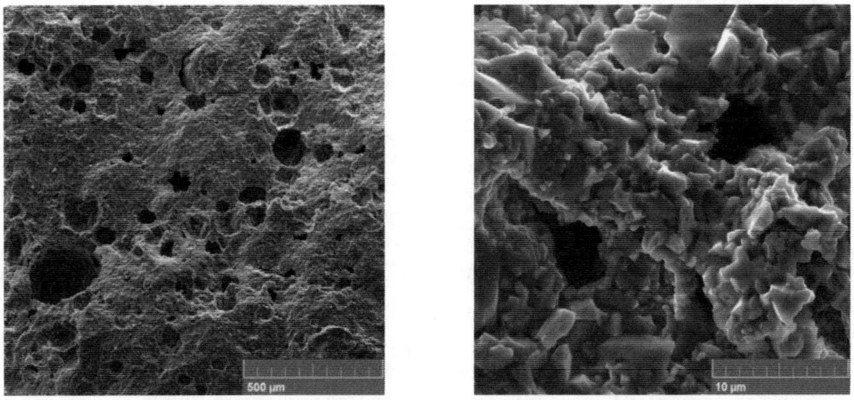

(a) (b)

Figure 2. Pores on the fracture surface of Al_2O_3 ceramic.

The strength of alumina ceramic significantly exceeds the strength of bone tissue, which undoubtedly leads to destruction of bone tissue in the contact area of the implant-bone system due to ubnormal internal stress distribution. Reducing the strength of the ceramic to the level of bone tissue strength is one of the main conditions for ensuring the biomechanical compatibility of the endoprosthesis. The organization of a developed pore structure reduces the strength of alumina ceramic to 97 MPa, figure 3, table 1. The increase of porosity leads to strength decreasing, which is consistent with the strength of various types of bone tissue [12].

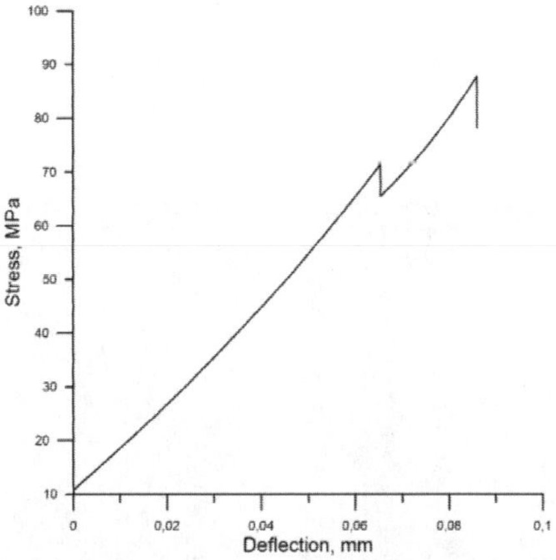

Figure 3. Stress-deflection dependence of Al_2O_3 ceramic with 53 % of porosity.

Table 1. Porosity and bending strength of Al_2O_3 ceramic

Porosity, %	Bending strength, MPa
4 %	284.5
11 %	246.3
39 %	117.2
53 %	96.6

The bones of the facial region of the skull have the most complex geometry among all the bones of the skeleton, which significantly complicates its prosthetics. The task of replacing postoperative bone defects in this case is not only the restoration of the mechanical functionality of the prosthetic area, but also more complex aesthetic aspects, such as the restoration of facial expressions, articulation and the facial contour [4]. The discrepancy between the geometry of the osteoimplant and the geometry of the resected fragment of the facial skeleton leads to limited mechanical functionality, impaired facial expression and articulation [12].

Modern reconstructive surgery of the maxillofacial area tends to single-stage procedure, reducing the number of postoperative complications, and choosing the method most safe for the patient, which will allow to achieve the earliest functional and social rehabilitation [13]. The use of generally accepted methods of postoperative bone defects of the maxillofacial region recovery (microsurgical reconstruction) have certain disadvantages: multistage, additional surgical trauma, high cost of treatment.

Full replacement of the body's own tissues with implants is possible only with full biochemically and biophysically compatibility with the surrounding tissues. From the point of view of reconstructive surgery, the implant should have a maximum affinity with the resected bone tissue of the maxillofacial region, and perform the function assigned to this region (protective, supporting, chewing, etc.) [14]. Ideal for mechanical stability of the implant is the bone tissue integration in the pore volume of implant. This leads to the stability of the implant with the functions it performs and the absence of inflammatory changes in the surrounding tissues.

By the method of creation a personalized of complex-shaped ceramic products, an osteoimplant of the skull-orbital complex was prepared and a surgical operation on its implantation was performed, figure 4 (a). In the area of reconstruction, there were no signs of active inflammation, figure 4 (b). The approach based on creating three-dimensional personalized osteoimplants from porous ceramic satisfies current trends in reducing the duration of surgery, reducing the area of bone resection and preventing the risk of complications.

It should be noted that the duration of the reconstructive surgery is reduced with using an individual ceramic implant compared to standard bone grafting from 540 to 360 minutes. An earlier restoration of the physical activity of patients in the postoperative period and a reduction of the postoperative period stay in the hospital compared to the bone grafting techniques.

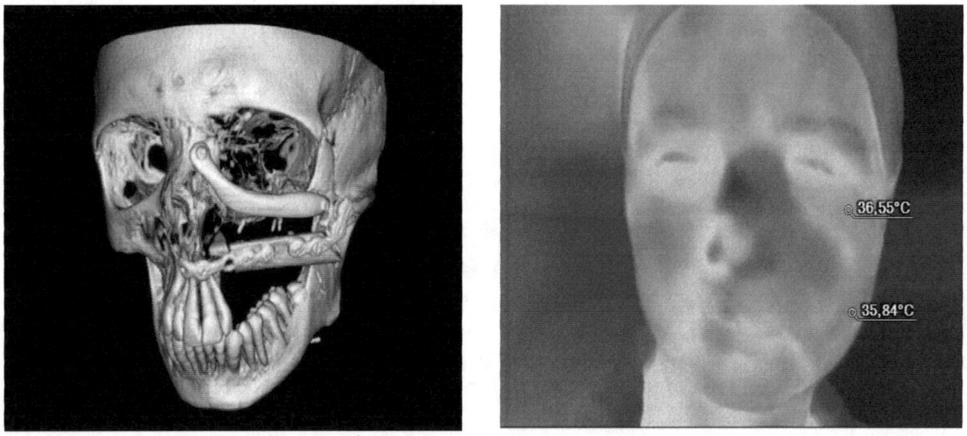

(a) (b)

Figure 4. Ceramic osteoimplant and skull model (a); Postoperative thermogram (b).

4. Conclusions

In this study were developed a fundamentally new approach for the reconstruction of the maxillofacial area using 3D-printing technologies for producing geometrically complex ceramic-based implants.

The developed complex technique for preoperative planning and creation of individual implants from ceramics for the reconstruction of the maxillofacial region is clinically applicable and in demand in modern trends of reconstructive surgery. Performing reconstructive surgeries using ceramics implants allows to achieve high functional and aesthetic results, as well as improving the life quality of patients with maxillofacial tumors without causing additional operating injury to the patient.

The developed ceramics material meets the requirements for medical materials used in reconstructive surgery. By varying the composition of the ceramic, it is possible to achieve structural and mechanical affinity with the surrounding bone tissue. Building a 3D model of an osteoimplant on the basis of a computed tomography of the patient avoids the need for additional resection of the bone tissue and ensures complete geometric conformity of the osteoimplant to the prosthetic area.

The clinical application of the developed technique for reconstruction of the maxillofacial area allows solving a complex problem that causes social isolation and functional disability of patients after surgical removal of maxillofacial area tumors – reducing the trauma, inefficiency and cost of reconstructive operations on the craniofacial area cancer patients.

References

[1] Caprin A D, Starinsky V V, Petrova G V 2016 *The state of oncological care for the population of Russia in 2015* (Moscow: MNIOI P.A. Gertsena) p 236
[2] Shin U S *J. of Tiss. Eng.* **2011** 674287
[3] Ginebra M P, Traykova T, Planell J A *J. Controlled Release* **113** 16740332
[4] Bing D, Weizhen L, Yue D *Int J Nanomedicine* **10** 4386782
[5] Subhadip B, Shekhar N, Bikramjit B J of Biomaterials Applications **23** 5
[6] Curran D J, Fleming T J, Towler M R *J Mater Sci Mater Med.* **21** 4
[7] Nair L S Laurencin C T *Progr. in Pol. Sc.* **32** 8
[8] Redondo A, LeSar R *Ann. Review of Mater Res.* **34** 279
[9] Jegoux F, Goyenvalle E, Cognet R *Arch. Otolaryngol Head Neck Surg.* **136** 10
[10] Centenero S A H., Hernández-Alfaro F *J. of Cranio-Maxillofacial Surg.* **40** 2
[11] Cavalcanti M G P, Haller J W, Vannier M W *J. of Oral and Maxillofacial Surg.* **57** 6
[12] Pal S 2014 *Mechanical properties of biological materials* (Springer) p 23
[13] Neligan P C. *Plastic and Reconstructive Surg.* **131** 2
[14] Wei W L, Han W P, Zhu M.G. *Laryngoscope.* **122** 11

IOP Conf. Series: Materials Science and Engineering **511** (2019) 012007 doi:10.1088/1757-899X/511/1/012007

Dissolution testing for prolonged-release solid dosage form containing drug substances degrading in test media. Application to polymeric delivery systems of paracetamol

A A Rakina, T I Spriridonova and S I Tverdokhlebov

National Research Tomsk Polytechnic University, 30, Lenin Ave., Tomsk, 634050, Russia

E-mail: tverd@tpu.ru

Abstract. Synthetic biodegradable polymers are widely used as materials for the targeted drug delivery and controlled release devices. In order to control the rate of drug release from polymer scaffolds surface ordinary dissolution testing is performed. It includes placing a device into testing media, taking probes at determined time and estimating the drug concentration. As for prolonged release the issue is the drug degradation in test media what leads to false interpretation of results. This study was focused on the investigation of paracetamol degradation in phosphate buffer saline media for three days and its effect on the results of polymeric drug delivery system dissolution test. It was shown, that relative error in the released drug concentration evaluation can rise up to 34% depending on the exposure time and the initial probe concentration. An approach of experimental results correction is proposed.

1. Introduction

Advanced drug delivery and controlled release are now one of the most important themes in modern pharmaceutics. Especially, solid dosage forms for prolonged release. Among all technologies biodegradable polymeric drug delivery systems are considered truly perspective because of their biocompatibility and non-invasive dissolution in the body what makes further surgical intervention for de-implantation not necessary. In order to study the drug release profile from any drug dosage form a dissolution testing should be performed.

Dissolution testing (DT) is required for most solid oral dosage forms [1]. The main aim of this analytical test is to detect the release of an active pharmaceutical ingredient (API) from the formulated product. For targeted drug delivery dosage forms to be efficacious, the API(s) must be absorbed in the space near the site of activity. This process characterizes the drug substance bioavailability and involves two stages: dissolution and absorption. Dissolution is the process of extracting the API out of the dosage form solid-state matrix into body fluids. Absorption is the process of transporting the drug substance to the target through diffusion. In the dissolution testing at set time points, aliquots of filtered medium are removed and analyzed for API content by UV-vis or HPLC methods. Common issue is the API degradation: if the drug degrades in the DT media, the results of the content analysis can represent smaller value of released API in comparison to real value. Detecting the API degradation and estimation of its kinetics is crucial in DT data interpretation [2].

Paracetamol is a commonly used anti-inflammatory drug. Moreover, it is a highly suitable model drug for testing new drug delivery systems. When incorporated, paracetamol can mimic other same average molecular weight molecules release. However most DT protocols of paracetamol dosage forms include not more than one-day testing. For prolonged release dosage forms DT has to be corrected due to paracetamol degradation to products such as hydroquinone, p-aminophenol, p-nitrophenol, 1,4-benzoquinone and, NAPQI (N-acetyl-benzoquinone imine) [3].

This study aims determining the degradation kinetics of paracetamol in phosphate buffer saline media during three days. Also the application of presented data to real DT of eletrospun polycarolatone nanofibers incorporated with paracetamol substance is presented.

2. Materials and methods

2.1. *Preparation of paracetamol solutions*
First, phosphate buffer saline (PBS, pH=7,2-7,4) was prepared by dissolving a tablet (Biolot, Russia) in distilled water according to the manufacturer instructions. Pure pharmaceutic substance of paracetamol was received from Shandong Xinhia Pharmaceutical (China). Substance was dissolved in PBS to get a highest in the experiment concentration – 875 μg/ml. Next fifteen points were prepared by double dissolution of the previous point. All experiments were performed in triplets. As a result, 16 solutions with different concentrations were prepared.

2.2. *Study of paracetamol degradation*
The amount of paracetamol in each probe was determined using UV-vis spectrometry (Specord 250 Plus, Analytik Jena AG, Germany), λ=245 nm (for concentations 0.21 – 27.4 μg/ml) and 300 nm (27.4 – 875 μg/ml). Analytical signal – absorption (peak height). In this study dependence of concentration drop in the solution on the exposure time (0, 1 and 3 days) and on the initial solution concentration (from 0.21 to 875 μg/ml) was studied. So, 16 solutions were prepared for every exposure time type. All experiments were performed three times.

2.3. *Preparation of PCL fibers by electrospinning*
Polycaprolactone (PCL) Mw~70–90 kDa (Sigma–Aldrich, Germany) was dissolved in hexafluoroisopropanole (HFIP) (Ekos-1, Russia) at a concentration of 7 wt.%. For the preparation of 2 wt./wt % and 32 wt./wt % (drug substance/dry polymer weight) paracetamol-loaded PCL solutions, previously dissolved in HFIP paracetamol powders were added to PCL granules and then refilled with the rest of the solvent. Mixtures were left for 30 hours at the room temperature in sealed glass containers until full homogenization. Before electrospinning solutions were stirred on the magnetic stirrer for 30 min.

Electrospinning of nanofibers was carried out on NANON-01 (MECC CO., Japan) with a 200 mm diameter drum collector. The process parameters used in the current study are shown in Table 1.

Table 1. The process parameters optimized for preparation of PCL nanofibers.

Characteristics	Value
Voltage, kV	20
Feed rate, ml/h	5
Collector rotation speed, rpm	50
Needle	G 21

After fibers formation scaffolds were removed and placed into a custom made vacuum camera for 24 hours (5×10^{-3} Pa) to remove residual solvents.

2.4. *Drug release study*
Untreated and e-beam treated electrospun PCL scaffolds, both with the incorporated drug, were immersed in Phosphate Buffer Saline (PBS, pH 7.4) at 25°C with three replicates for each type of scaffold. The experiments were run in 2 ml of PBS without stirring, and the drug release results were

found to be within one standard deviation to each other. At predetermined time points, a 1 mL aliquot was withdrawn for further analysis and replaced with an identical volume of the fresh medium. The amount of released paracetamol was determined using UV-vis spectrometry (Specord 250 Plus, Analytik Jena AG, Germany), λ=245 nm and 300 nm.

3. Results and discussion

It was shown in the experiment, that UV-vis absorption spectra of paracetamol in water solution has two characteristic peaks – at 245 nm and at 300 nm. Absorption at 245 nm is a suitable analytical signal for analysing solutions with paracetamol concentrations from 0.1 to 16 μg/ml, higher concentrations are better characterised by absoption at 300 nm. Figure 1 represents the difference in solutions' UV-vis spectra.

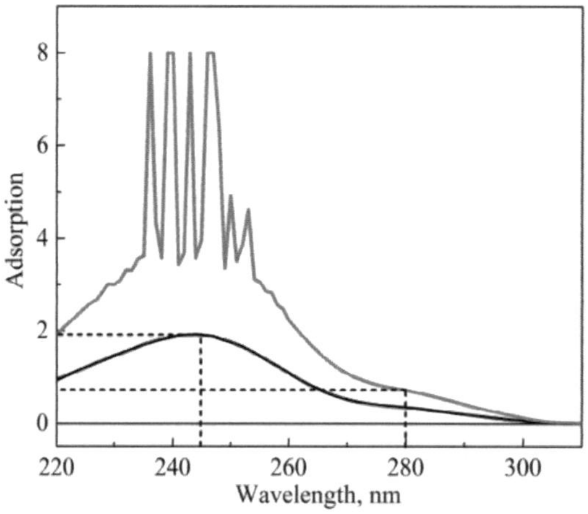

Figure 1. An example of absorption spectrum for 218 μg/ml and 14 μg/ml paracelamol solutions in PBS.

Thus, a study of the absorption value dependence on the initial concentration of paracetamol in solutions and on the exposure time were calculated to paracetamol concentration using two calibration curves (245 nm and 300 nm "0 day"). Results of all experiments are summarized in Figure 2.

It is shown that the observed concentration of drug significantly decreases after exposure for all initial concentrations and exposure times. Moreover, the drop of the absorption values are most noticeable for larger initial paracetamol concentrations and one day of drug exposure in the PBS. In some probes further growth of the absorption of the solutions is observed. We assume that the reason of this phenomenon may be due to the accumulation of the paracetamol degradation products in probes/ Literature data confirm that this products in water solutions have absorption area close to pure paracetamol solution [4].

Based on the obtained absorption data, calibration curves for wavelengths of 245 nm (0.21 - 13.67 μg/ml) and 300 nm 27.34 -875 μg/ml) were built. Results are presented in Figure 3.

All experimental points for each wavelength and storage time are approximated with high accuracy by a linear function. It means that paracetamol solutions with any initial concentration from studied range degrades with the same relative speed and it is possible to use the same tangent of calibration curve *k* for calculation of all paracetamol concentrations for probes taken in one day. As it was mentioned above, the values of absorption substantially decrease after the one day of the exposure and slightly increase after three days. However, this increase is almost absent for probes with lower concentrations, while the other curve is more sensitive to the storage time.

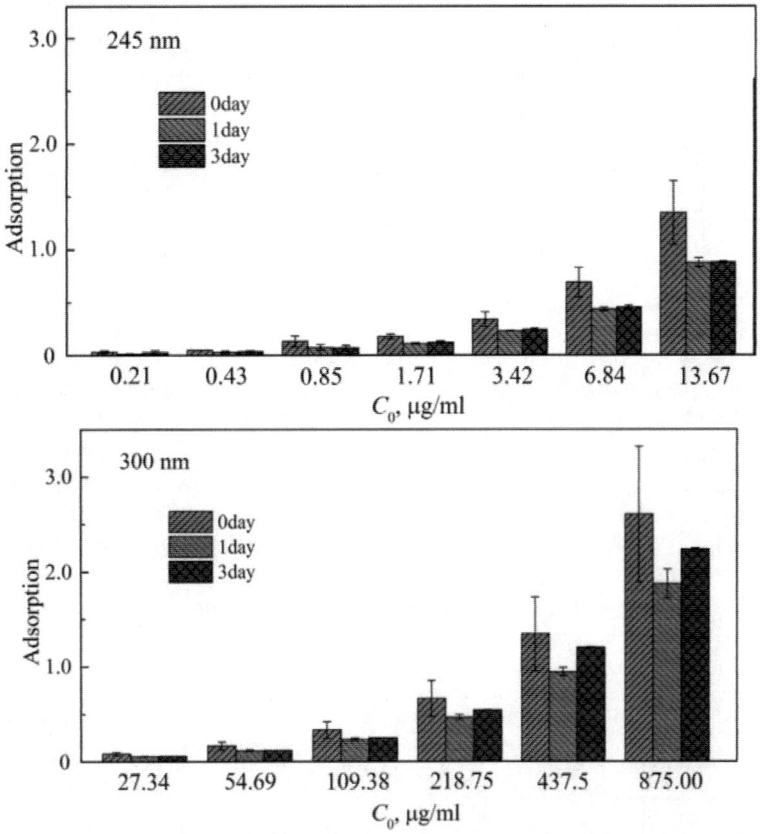

Figure 2. The dependence of absorption on the initial concentration of paracetamol and the time of storage in the PBS for 245 nm (at the top) and 300 nm (at the bottom).

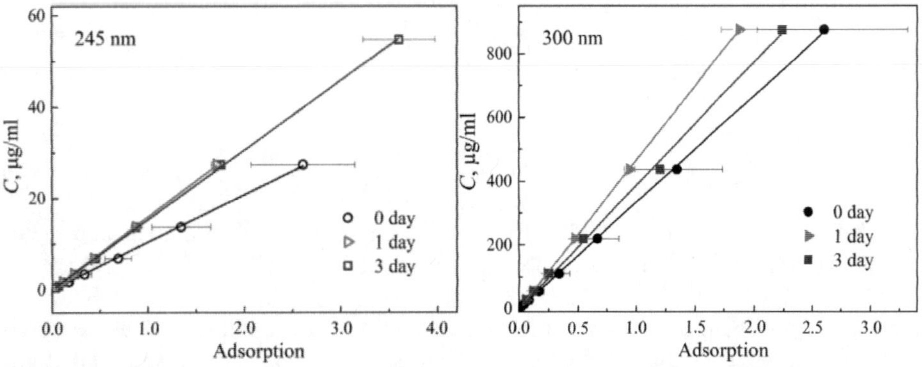

Figure 3. Calibration curves for the different storage time of paracetamol in the buffer.

In order to assess the effect of the paracetamol in-reactor degradation on the results of the drug release experiments, the relative difference in calculating the amount of the released drug (ΔC) was calculated as:

$$\Delta C = \frac{C(t)-C_0}{C(t)} \cdot 100\%, \qquad (1)$$

where C_0 is the initial paracetamol concentration, $C(t)$ is the concentration after the period of time (1 or 3 days). The relative differences for each time of storage and wavelength as well as parameters of calibration curves for experiment processing are presented in Table 2.

Table 2. The accuracy of the approximation of the experimental data for the calibration curves by a linear function (R^2), the tangent of the slope of the calibration lines (k) and the relative errors in the concentrations determination (ΔC) .

	245 nm			300 nm		
	0 day	1 day	3 day	0 day	1 day	3 day
R^2	0.99	0.99	0.99	0.99	0.99	0.99
k	10.37	15.84	15.24	333.23	465.25	386.21
ΔC, %	-	34.53	21.22	-	22.95	13.72

In order to show how the correction in concentration calculation helps to make a right interpretation of dissolution testing the results of drug release from three paracetamol-loaded polycaprolactone scaffolds were analyzed. One of scaffolds (2p) had a low concentration of paracetamol (2 wt./wt %) and probes taken from the reactor were exposed in the PBS for one day. The second sample contained a relatively high concentration of paracetamol (32 wt./wt %) and probes were exposed in the PBS for three days. The drug release profiles without taking into account paracetamol degradation are shown in Figure 4a. The largest amount of drug was released from the scaffold with the highest paracetamol concentration, while the amount of drug released from the sample with only 2 wt./wt % of paracetamol was around 10% smaller. Wherein, even the highest value of amount of released drug does not exceed 70%.

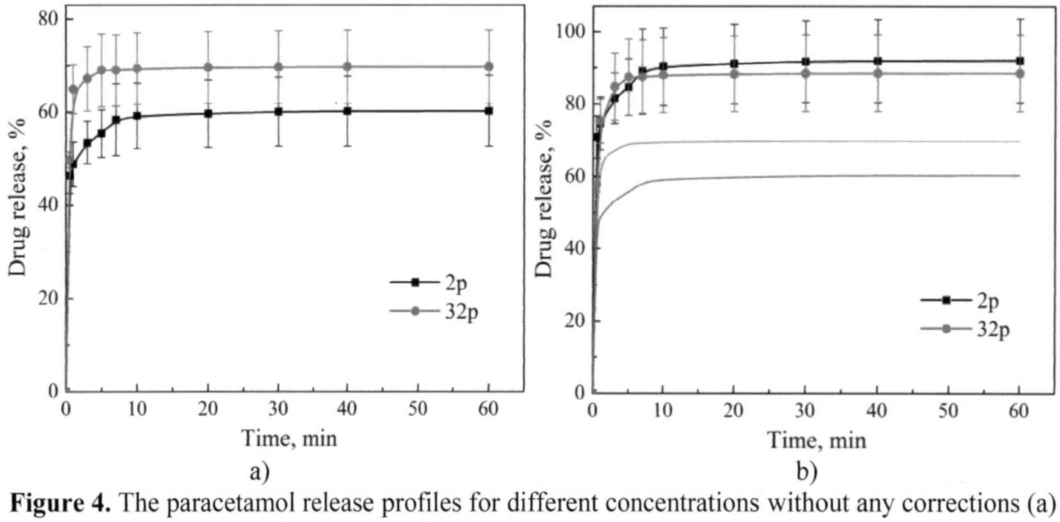

Figure 4. The paracetamol release profiles for different concentrations without any corrections (a) and after correction in comparison to the uncorrected ones (translucent lines) (b).

The paracetamol release profiles after taking into account the degradation process are presented in Figure 4b. Those profiles were obtained by using the corrected calibration curves instead of incorrected for 0 day. It is clearly seen that the corrected amount of released drug is much higher than uncorrected one. The relative difference for each time is already had been presented in the Table 2. In accordance with the previously obtained data, difference between profiles for sample 32 which was kept in the PBS for three days is smaller than for sample 2p which was kept only one day despite the higher drug concentration.

4. Conclusion

It was shown that for relatively long periods (one day or more) of conducting the dissolution testing and experiments of the release of paracetamol from polymers it is especially important to take into account the paracetamol degradation. The technique for the dissolution testing correction of paracetamol-loaded prolonged release drug dosage forms was proposed. By using the proposed corrected calibration curves results for tests on materials with concentration range between 0.21 and 875 μg/ml within three days of exposure can be corrected. This technique is easy to use and allows to significantly increase the accuracy of the experimental results. Also, proposed approach can be transferred to other systems with degrading drug.

Acknowledgements
This work was financially supported by the Ministry of Science and Higher Education of the Russian Federation, Federal Target Program (agreement #14.575.21.0140, unique identifier RFMEFI57517X0140).

References

[1] Azarmi S, Roa W, & Löbenberg R 2007 *International Journal of pharmaceutics* **328 (1)** 12
[2] Higuchi T 1963 *Journal of pharmaceutical sciences* **52 (12)** 1145
[3] Trovo A G, Nogueira R F P, Agüera A., Fernandez-Alba A R and Malato S 2012 *Water research* **46 (16)** 5374
[4] Moctezuma E et al. 2012 *Journal of hazardous materials* **243** 130-138

Effect of annealing on mechanical and morphological properties of Poly(L-lactic acid)/Hydroxyapatite composite as material for 3D printing of bone tissue growth stimulating implants

G E Dubinenko, A L Zinoviev, E N Bolbasov, V T Novikov and S I Tverdokhlebov

National Research Tomsk Polytechnic University, 30, Lenin Avenue, Tomsk, 634050, Russian Federation

E-mail address: tverd@tpu.ru

Abstract. In this work, effect of additional annealing on mechanical and morphological properties of 3D-printed PLLA/HAp composite scaffolds of three compositions (12.5, 25, and 50 wt.% of HAp) was investigated. Morphology and Young's modulus of 3D-printed scaffolds were investigated by scanning electron microscopy and nanoindentation. It has been shown that additional annealing does not have an effect on the homogeneous distribution of HAp powder in the PLLA-matrix. Results of nanoindentation showed growth of Young's modulus after annealing. The maximum value of 9393 ± 709 MPa Young's modulus was reached for the annealed composite with 50 wt.% of HAp.

1. Introduction

Reducing polytrauma and congenital diseases treatment duration is one of the most difficult and important objectives of orthopedics and traumatology. Unstable osteosynthesis and such related complications as delayed consolidation and nonunion of bone fragments often lead to a significant increase in the duration of treatment. One of the most effective approaches in the treatment of congenital and acquired diseases of the musculoskeletal system is the osteosynthesis with the use of metal implants. However, despite the effectiveness, metal implants have a number of drawbacks such as negative impact on the functioning of the musculoskeletal system and high probability of permanent implants rejection due to metallosis. Moreover, removing temporary metal implant from the body requires an additional surgical intervention.

One of the most promising alternative approaches in the treatment of musculoskeletal system diseases is the use of biodegradable implants made of thermoplastic polymers and polymer composites [1]. Advantages of such materials are their complete enzymatic bioresorbability in body fluids and non-toxicity [2]. However, poor mechanical properties and uncontrolled degradation of the polymer matrix restrict the use of such materials in orthopedics [3]. Used in orthopedics in the form of pins and screws for fixing bone fragments [4], poly(L-lactic acid) (PLLA) has mechanical characteristics that are significantly inferior to those of natural bone. The low elastic modulus of this polymer does not ensure the preservation of the implant morphology, thereby limiting its use as a bulk osteoconductive implant. It was shown that filling of the polylactic acid with bioactive hydroxyapatite (HAp) up to 50% by weight increases the stiffness of composite scaffold by 2 times [5]. Moreover, maintaining the melt flow rate at the level acceptable for extrusion allows creating of PLLA/HAp composite scaffolds by Fused Deposit Modeling (FDM) [6,7]. 3D printing of polymer composites allows implementing the personalized approach in traumatology and orthopedics [8], which, presumably, will have a positive effect on the reduction of treatment time.

In this research, the effect of annealing on the morphology and mechanical properties of the PLLA/HAp composite material with a HAp filling up to 50% by weight is investigated. Additional annealing of 3D-printed scaffolds allows obtaining PLLA/HAp composite with enhanced Young's modulus.

2. Materials and methods

2.1. Materials

In this work, poly(L-lactic acid) (USA, melting point ~ 175...180°C) was used as the polymer matrix. Biological hydroxyapatite (HAp) (Russia, fraction size 12.56 μm ± 5.28) was used as a reinforcing bioactive filler. Purchased from Ecros-Analytics (Russia) chloroform was used to dissolve PLLA pellets. Dimethyl sulfoxide (DMSO) (Russia,) was used as non-ionic surfactant to prevent aggregating of HAp particles. Composites were prepared in the weight ratio at 87.5% PLLA/12.5% HAp; 75% PLLA/25% HAp and 50% PLLA/50% HAp.

2.2 Preparation of filament and 3D printing

PLLA/HAp composites were prepared according to the following method. HAp was preliminarily dried at 150°C and mixed with DMSO for 3 hours in a ball mill. HAp weight fractions were 14.3 g, 33.3 g and 100 g for PLLA12.5, PLLA25 and PLLA50, respectively. Then, 15% solution of PLLA in chloroform was added into the mixture of HAp with DMSO. The composition was mixed for 8 hours on ball mill with zirconia balls. After mixing composite was granulated and dried in a vacuum oven at 60°C and $5\cdot10^{-3}$ mbar. Dried pellets were extruded on the horizontal single-screw extruder Filabot EX2 (Filabot, USA) at 175°C. Pristine PLLA filament for printing of reference scaffolds was also extruded. The diameter of the extruded filament was 1.75 mm.

Obtained PLLA and composite filaments were used to print scaffolds on FDM 3D printer PrintBox3D One (RGT, Russia). Printing was performed by extrusion of PLLA and composite melts on glass substrate through 0.2 mm nozzle. Temperature of the nozzle was 220°C, thickness of layers – 50 μm, infill – 30%.

After printing five samples of each composition were annealed in the air sterilizer GP-40-SPU (Smolensk SKTB SPU, Russia) for 12 hours at 110°C.

All scaffolds were divided into two groups: group 1 – 3D-printed scaffolds, group 2 – annealed 3D-printed scaffolds (table 1).

2.3 Scaffolds morphology

Scaffolds surface morphology was investigated by scanning electron microscopy (SEM). Preliminarily, all scaffolds were sputter coated with gold for 2 minutes on JEOL Smart Coater (JEOL USA, USA) magnetron sputter coater. SEM at 22x magnification was performed on JEOL JCM-6000 (JEOL USA, USA), at 4000× magnification – on Quanta 200 3D (FEI Company, USA). Sizes of HAp particles and scaffolds pores were estimated using ImageJ software (National Institute of Mental Health, USA).

Scaffolds deformation after annealing was evaluated as deviation from plane-parallelism using calipers.

2.4 Mechanical tests

Young's modulus of obtained materials was investigated by the method of nanoindentation on the Nanoindenter G200 (Agilent's Electronic Measurement, USA) according to the ISO 14577. Indentation was performed with triangular Berkovich pyramid at 10 mN.

Table 1. Groups of materials and compositions.

Group	Composition	PLLA, wt.%	HAp, wt.%
	PLA100	100	0
1	PLA87.5	87.5	12.5
	PLA75	75	25
	PLA50	50	50
	PLA100an	100	0
2	PLA87.5an	87.5	12.5
	PLA75an	75	25
	PLA50an	50	50

3. Results and discussion

3.1. Scaffolds morphology

Porous disk-shaped scaffolds of 20 ± 0.15 mm diameter and 2 ± 0.08 mm thickness were successfully prepared from pristine PLLA and composite filaments using FDM 3D printing. The average size of pores in scaffolds is 1.096 ± 0.181 mm. No influence of HAp weight fraction or additional annealing on the pores size was observed.

After annealing, bending and deviation of scaffolds from the original were observed. Presumably, the deformation was caused by volumetric changes in the crystal structure of the semicrystalline polymer matrix. The scaffold surface was rapidly cooled by airflow in the process of FDM 3D printing, which led to decrease of the size and increase of the density of crystallites in the surface layers of the extruded polymer. In the bulk of the polymer, layer cooling occurred with a lower degree of undercooling due to additional heating from the substrate and previous layers. As a result of the undercooling gradient, the difference in the densities and sizes of crystallites on the surface and in the bulk of polymer appeared. Heating above the glass transition temperature during annealing initiated diffusion processes leading to the recrystallization and redistribution of crystallite densities in the surface layers. Volumetric changes in the crystalline structure initiated tensile structural stresses on the surface of scaffolds and their deformation.

The deformation of scaffolds was evaluated as the deviation from flatness using the formula 1:

$$\Delta X = \frac{X_1 - X}{X} \times 100\% \quad (1)$$

where X – scaffold thickness after 3D printing, X_1 – the distance between most distant point in the cross section of a scaffold after annealing (figure 1).

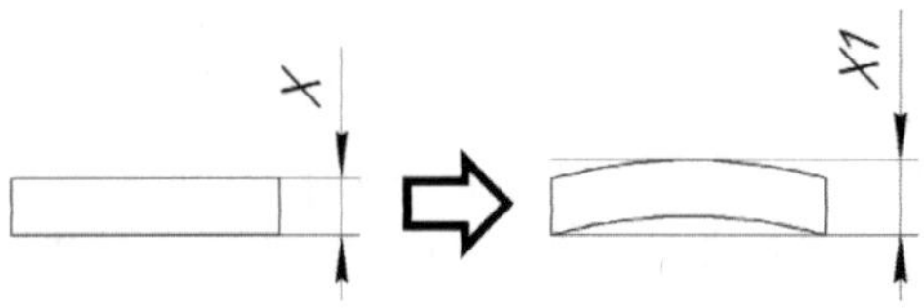

Figure 1. Method for determining deformation of scaffolds.

Deformation of reference scaffolds made of PLLA is 6.41 ± 0.11% (figure 2). Deformation increased up to 8.33 ± 0.08% for annealed composite filled with 12.5% by weight of HAp, but with further increase in the weight fraction of HAp, deformation decreased to 4.19 ± 0.06% and 1.86 ± 0.06% for 25 and 50 wt.% of HAp, respectively.

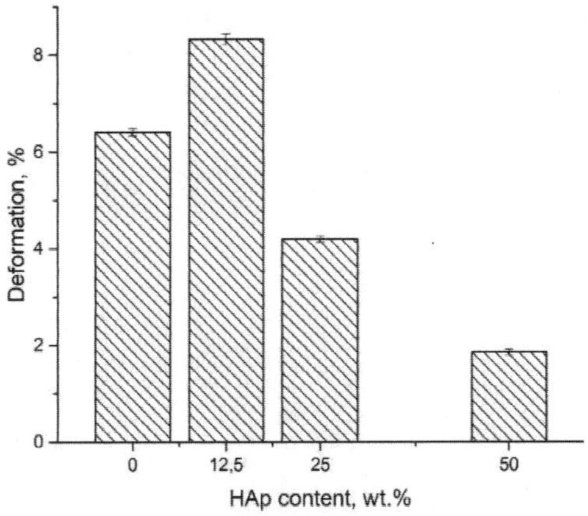

Figure 2. Scaffolds deformation after annealing.

SEM images (figure 3) demonstrated the presence of a large amount of homogeneously distributed hydrophilic HAp particles in the surface layer of scaffolds. The average particle size was 309 ± 71 nm, which indicates the successful grinding of HAp in the process of composites mixing.

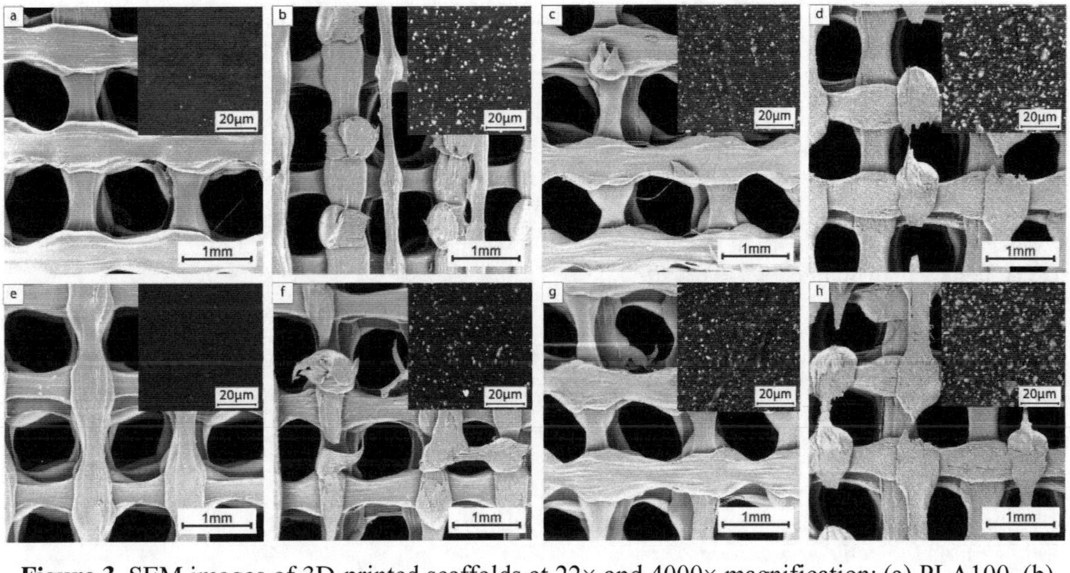

Figure 3. SEM images of 3D-printed scaffolds at 22× and 4000× magnification: (a) PLA100, (b) PLA87.5, (c) PLA75, (d) PLA50, (e) PLA100an, (f) PLA87.5an, (g) PLA75an, (h) PLA50an.

3.2. Mechanical tests

Young's modulus variation before and after annealing of obtained materials is shown in figure 4. As can be seen, Young's modulus of all compositions increases with increasing of HAp weight fraction. It is also seen that the module of all compositions, including reference pure PLLA, increased after annealing. The value of Young's modulus for pristine PLLA is 2468 ± 237 MPa. After annealing, the module of pure PLLA increased to 3725 ± 287 MPa. The addition of 12.5 and 25 wt.% of HAp into the polymer matrix led to a further increase in the modulus to 4774 ± 848 and 6138 ± 309 MPa, respectively. However, subsequent annealing did not show a significant increase in Young's modulus for the composite with 25 wt.% of HAp. The highest value of Young’s modulus 9393 ± 709 MPa was observed for the annealed composite with 50 wt.% of HAp. Presumably, the growth of Young's modulus with an increase in the HAp weight fraction means reinforcing effect in the polymer matrix as a result of the introduction of highly dispersed HAp powder. Additional annealing initiated heterogeneous recrystallization on the surface of HAp. Presumably, in the process of heterogeneous recrystallization, an increase in the degree of crystallinity of the polymer matrix occurred proportionally to the increase in the weight fraction of HAp, which led to the increase in the Young's modulus of the composites.

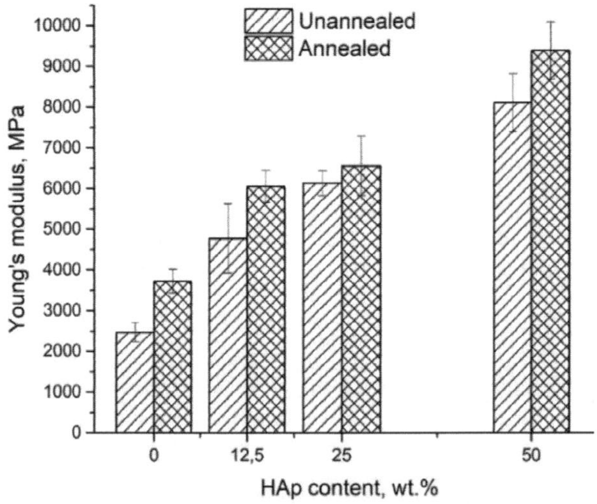

Figure 4. Young's modulus of obtained materials.

4. Conclusion

In this paper effect of annealing on morphology and elasticity of PLA/HAp composite investigated. It was shown that additional annealing does not influence the homogeneous distribution of HAp powder in the PLLA matrix. Deformation of 3D-printed scaffolds decreased with the increase of HAp weight fraction. Moreover, annealing led to the growth of Young's modulus up to 9393 ± 709 MPa for the annealed composite with 50 wt.% of HAp.

Acknowledgements

This work was financially supported by the Ministry of Science and Higher Education of the Russian Federation, Federal Target Program (agreement #14.575.21.0140, unique identifier RFMEFI57517X0140).

References

[1] Ebnesajjad S 2012 *Handbook of Biopolymers and Biodegradable Plastics: Properties, Processing and Applications*

[2] Saini P, Arora M and Kumar M N V R 2016 Poly(lactic acid) blends in biomedical applications *Adv. Drug Deliv. Rev.* **107** 47–59

[3] Farah S, Anderson D G and Langer R 2016 Physical and mechanical properties of PLA, and their functions in widespread applications — A comprehensive review *Adv. Drug Deliv. Rev.* **107** 367–92

[4] P. Pawar R, U. Tekale S, U. Shisodia S, T. Totre J and J. Domb A 2014 Biomedical Applications of Poly(Lactic Acid) *Rec. Pat. Regen. Med.* **4** 40–51

[5] Esposito Corcione C, Gervaso F, Scalera F, Padmanabhan S K, Madaghiele M, Montagna F, Sannino A, Licciulli A and Maffezzoli A 2018 Highly loaded hydroxyapatite microsphere/ PLA porous scaffolds obtained by fused deposition modelling *Ceram. Int.* 1–8

[6] Lin P L, Fang H W, Tseng T and Lee W H 2007 Effects of hydroxyapatite dosage on mechanical and biological behaviors of polylactic acid composite materials *Mater. Lett.* **61** 3009–13

[7] Russias J, Saiz E, Nalla R K, Gryn K, Ritchie R O, Tomsia A P, Division M S, Berkeley L and States U 2015 Fabrication and mechanical properties of PLA/HA composites: A study of in

IOP Conf. Series: Materials Science and Engineering **511** (2019) 012008 doi:10.1088/1757-899X/511/1/012008

vitro degradation Mater Sci Eng 1–15
[8] Bose S, Vahabzadeh S and Bandyopadhyay A 2013 Bone tissue engineering using 3D printing Mater. T*oday* **16** 496–504

IOP Conf. Series: Materials Science and Engineering **511** (2019) 012009 doi:10.1088/1757-899X/511/1/012009

Electrochemical hydrogenation after mechanical activated near equiatomic Ti-Ni powder

E Abdulmenova[1], O Vaulina[1] and S Kulkov[1,2]

[1]National Research Tomsk Polytechnic University, Tomsk, Russia
[2]Institute of Strength Physics and Materials Science SB RAS, Tomsk, Russia

E-mail: Ekaterina.V.Abdulmenova@yandex.ru

Abstract. It has been studied an effect of mechanical activation in a planetary ball milling on the electrochemical hydrogenation of Ti-Ni powder near equiatomic content. It has been shown that average particle sizes measured by scanning electron microscope decreased from 11.1 to 7.8 μm up to 30 s mechanical activation, then increased to 33.2 μm for 300 s activation. Beside, with an increase in the time of mechanical activation was formed state with small the coherently diffracting domain which correspond to quasi-amorphous state of material. It has been shown that after hydrogenation, the particle size distribution did not change. There is an incubation time hydrogenation for mechanical activation, when hydrogenation is not occurring. Mechanical activation for more than 30 s leads to the formation of the Ti_2NiH_x phase based on the Ti_2Ni and 50 s activation is sufficient to form the $Ti_2NiH_{0.5}$ stoichiometry. For a longer mechanical activation a phase with the stoichiometry $Ti_2NiH_{0.85}$ was formed and to obtain the Ti_2NiH stoichiometry it is necessary to carry out activation for 4.5 hours.

1. Introduction

Titanium – nickel alloys are perspective materials for applications in rechargeable nickel-metal hydride batteries due to their high volume density of hydrogen, small density, high electrochemical activity and corrosion resistance.

The system Ti-Ni near equiatomic composition as a rule has a complex phase content consisting with TiNi with BCC (B2), Ti_2Ni with FCC ($E9_3$) and $TiNi_3$ with HCP ($D0_{24}$). The intermetallic compound of Ni_3Ti is not capable for hydrogen storage [1-3], while the intermetallic compounds of TiNi [4] and Ti_2Ni [5] can absorb a great amounts of hydrogen. Ti_2Ni and TiNi absorbs a large amount of hydrogen by gas-solid reaction [6-8]. Besides this Ti_2Ni has four hydride phases ($Ti_2NiH_{2.5}$, Ti_2NiH_2, Ti_2NiH and $Ti_2NiH_{0.5}$), whilst TiNi has only one hydride phase, TiNiH [5].

According to the literature, the interaction of hydrogen with a metal may be accompanied by a change in the lattice parameter [9], e.g. Pelton, *et. al,* [10] showed that the presence of hydrogen increases the lattice parameter of the B2 phase from 0.3025 nm to 0.3047 nm (2.2 % volume increase). Buchner, et al. [5] have detected that the lattice parameter of the B2 structure increased from 0.301 nm to 0.310 nm, which corresponds to a volume increase of nearly 10 %, the lattice parameter increases due to the occupation of hydrogen an interstitial lattice site.

As it well known increasing of density of defects in materials ensured by the penetration of hydrogen via lattice defects [11]. A high-energy mechanical activation (MA) is used to increase a density of defects [12-14] and similar treatment is a very effective for powders. There are several methods of hydrogenation, for powders a simplest technique is electrochemical hydrogenation, when hydrogen is generated via electrochemical reaction in special solutions [15, 16].

The aim of this paper is to study the effect of mechanical activation on electrochemical hydrogen saturation of near equiatomic Ti-Ni powder.

2. Material and methods

The tests were performed using near equiatomic $Ti_{0,45}Ni_{0,55}$ powder, produced by the «Polema» plant. The mean size of particles 22 μm (standard deviation, sd=10 μm). Mechanical activation was carried out using a planetary ball mill «AGO-2», an acceleration during treatments was 60 g and the milling time was 5 – 300 second.

For the electrochemical hydrogenation of powders a special cell, figure 1, has been constructed. A 20 % aqueous solution of NaCl with the addition of 1.5 g/l of dextrin as surfactant served as electrolyte. The cathode current density during the experiment was 40 - 70 mA/cm^2. The temperature of the electrolyte was constant, approximately 20 °C. The absorption of hydrogen by metal occurs slowly in the first minutes (up to 30 minutes) of hydrogenation. This effect is associated with an oxide layer on the metal surface. An increase in time of 0.5 to 10 hours leads to a sharp increase in the rate of hydrogen absorption. In particular, the hydrogen content does not change significantly from 90 minutes to 10 hours with maximum absorption of hydrogen [17]. Thus, the hydrogenation time was 90 minutes.

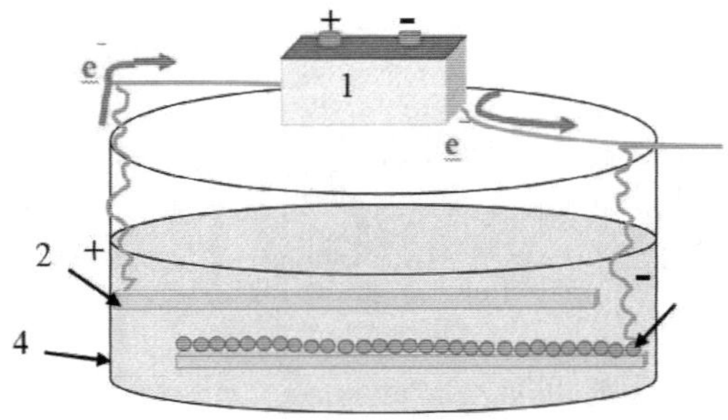

Figure 1. Basic outline of electrochemical hydrogenation : 1 – current source ; 2 – anode (graphite plate); 3 – cathode (TiNi powder immersed in a stainless steel cup); 4 – working vessel

Investigations of particle morphology and its size distribution were performed by scanning electron microscopy (SEM) TESCAN VEGA 3SBH and laser diffraction (LD) methods using a SALD-7101 nanoparticle diffraction analyzer by Shimadzu (Japan). The dispersion liquid was 0.2 wt. % aqueous solution of sodium citrate $Na_3C_6H_5O_7$.

Analysis of the phase composition and the parameters of the crystal structure of the powder were carried out by X-ray diffraction (XRD) on diffractometer with Cu-Kα filtered radiation with statistics accuracy better than 0.5 %.

The specific surface area (SS) of the powder was measured by the BET [18] method by Sorbi series instrument. From the obtained values of the specific surface it has been calculated the average particle size assuming their spheroids shapes.

3. Results

3.1. Morphology and particle size

The SEM pictures of powders are shown in figure 2. The particles mean size and size of the coherently diffracting domain (CDD) are shown in Table 1. As one can see, there are two stages during treatment: on the first step up to 30 s the fracture process of particles which changes agglomeration stage after this. The SEM and SS dat have shown that internal structure became more dispersive.

It was remarkable, that morphology and particle size distribution of powders do not change after hydrogenation.

IOP Conf. Series: Materials Science and Engineering **511** (2019) 012009 doi:10.1088/1757-899X/511/1/012009

(a) (b) (c)

Figure 2. SEM micrographs of the powder in the initial state (a) and after mechanical activation of 30 s (b) and 300 s (c)

Table 1. The particles mean size and size of the coherently diffracting domain (CDD)

Mechanical activation time, s	$<d_{SEM}>$ (µm), St.dev. (µm)	$<d_{LD}>$ (µm), St.dev. (µm)	SS ($m^2 \cdot g^{-1}$), Error ($m^2 \cdot g^1$) / $<d_{BET}>$ (µm), St.dev. (µm)	CDD (nm)
0	11.1±7.5	7.9±0.2	0.24±0.02 / 3.9±0.43	43
10	11.0±7.6	7,6±0.3	0.28±0.02 / 3.3±0.20	21
30	7.8±6.6	6.1±0.2	0.37±0.01 / 2.5±0.01	17
60	13.8±9.8	6.0±0.2	0.49±0.04 / 1.9±0.04	8
300	33.2±25.7	0.5±0.2	0.54±0.02 / 1.7±0.09	5

3.2. X-ray analysis

In figure 3a are shown X-ray patterns of powders in initial state and after mechanical activation. As one can see after mechanical activation, powder consist of TiNi (B2) and $NiTi_2$ ($E9_3$); TiNi (B19`) and Ni_3Ti ($D0_{24}$). In the process of mechanical activation, width of the peaks increase and after 200 s X-ray patterns looks like belongs to quasi-amorphous state of material, Figure 3a, and separate lattice parameter of phases could not be determined.

The phase content of the powder did not change after hydrogenation, figure 3b. After hydrogenation lattice parameters of all phases except of Ti_2Ni did not change, but for Ti_2Ni there is significant changing of lattice parameter, figure 4. From this figure one can see increasing this value from 1.132 nm to 1.163 nm, i.e. increasing by a factor of 3.1 %.From this plot one can conclude that after mechanical activation the lattice parameter does not changes and correspond to the table value, dashed black line [pdf2 №72-0442]. After hydrogenation parameter increase for samples activated more when 30 s reached $Ti_2NiH_{0.5}$ stoichiometry (blue dashed line) [pdf2 №27-0346] and more up to $Ti_2NiH_{0.85}$. Approximation of dependence of the lattice parameter of the $NiTi_2$ phase on the time of mechanical activation allowed us to obtain the equation $y = 0.1194\tau^{0.0052}$, which can be used to determine the hydrogen content achieved in the process of hydrogenation, figure 4b. Extrapolation of this dependence had shown that in order to obtain the Ti_2NiH_1 stoichiometry with the lattice parameter [19], it is necessary to carry out a mechanical activation for 4.5 hours. Unfortunately, this time is too long for such treatment. From figure 4 one can see that lattice parameter of Ti_2Ni after hydrogenation did not change (until 10 s mechanical activation). Accordingly the penetration of hydrogen via lattice defects does not observe. So, there is some «incubation period», for these experiments the value was a 10 s of mechanical activation.

IOP Conf. Series: Materials Science and Engineering **511** (2019) 012009 doi:10.1088/1757-899X/511/1/012009

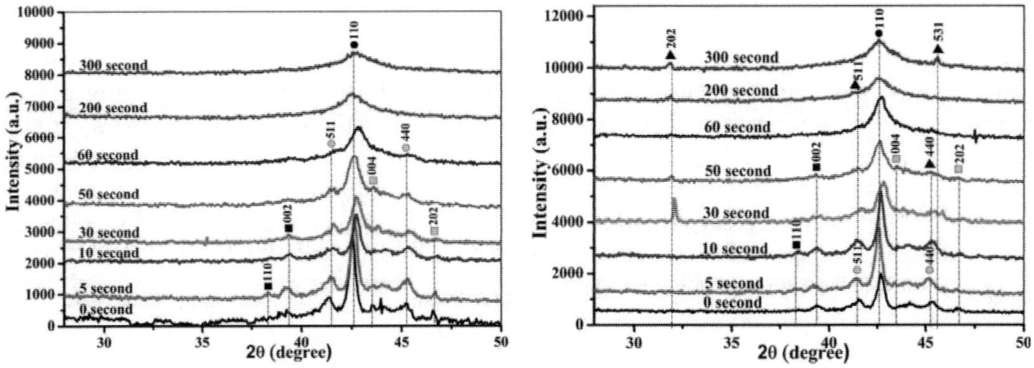

(a) (b)

Figure 3. XRD patterns of TiNi powders before (a) and after (b) hydrogenation with different mechanical activation time.

● – TiNi (B2); ■ – TiNi (B19`); ● – Ti_2Ni (E9$_3$); ■ – Ni_3Ti (D0$_{24}$); ▲ – Ti_2NiH_x (E9$_3$)

Probably it's connected to destruction of thin oxygen layers on the particles surfaces which removed during hydrogenation. So increasing of time hydrogenation (more 90 min) will allows reduce or remove "incubation period" of hydrogenation. But more likely hydrogen absorption are connected with defects accumulation during activation and such "incubation period" will not observe in all experiments, since, the hydrogen atoms may additionally be trapped via defects [20-23], which can provide access for the rapid diffusion of H-atoms [14], but it should be investigated in the following studies.

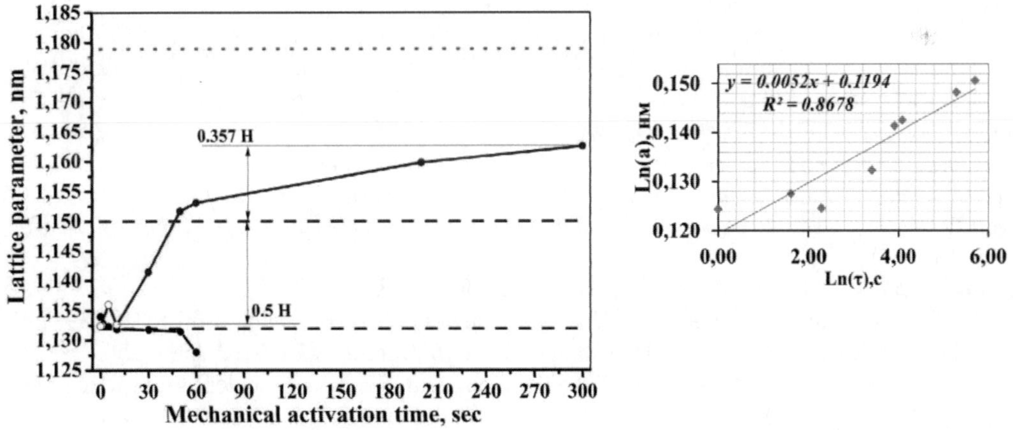

(a) (b)

Figure 4. Dependence of the lattice parameter of the $NiTi_2$ phase on the time of mechanical activation before and after hydrogenation (a); approximation of the dependence of the lattice parameter of the $NiTi_2$ phase on the time of mechanical activation of the powder after hydrogen absorption (b).

▬ ▬ - $NiTi_2$ pdf2 №72-0442; ▬ ▬ - $Ti_2NiH_{0.5}$pdf2 №27-0346; ---- Ti_2NiH_1 ASTM [8]; ▬●▬ - Ti_2Ni before hydrogenation; ▬O▬- Ti_2Ni after hydrogenation; ▬●▬ -Ti_2NiH_x after hydrogenation

Based on these results, one can assume that mechanical activation will allow obtaining a large capacity of hydrogen in this multiphase system.

4. Conclusion

With an increase in the time of mechanical activation, a state with a very small CDD is formed, which may correspond to quasi-amorphous state of material. CDD size is significantly less than <d> (LD and BET), that is, the powder particles consist of several crystallites, the number of which increases on average from 5×10^4 to 13×10^6 with increasing time of mechanical activation. There is an incubation time of hydrogenation when hydrogenation is not occurring. Mechanical activation for more than 30 s leads to the formation of the Ti_2NiH_x phase based on the Ti_2Ni and time 50 s is sufficient to form the $Ti_2NiH_{0.5}$ stoichiometry. Longer mechanical activation allows forming a phase with the stoichiometry $Ti_2NiH_{0.85}$.

Acknowledgement

This work was carried out according to the Program III.23 of Fundamental Scientific Research of the State Russian Academy of Sciences for 2013-2020. Authors are grateful for Leonov A. for his help in analyzing the particle size distribution in TPU Nano-Center.

References

[1] Justi E W, Ewe H H, Kalberlah A W et al. 1970 *Energy Conv.* **10** 183.
[2] Guiose B, Cuevas F, De´camps B et al. 2008 *Int. J. Hydrog.* Energy **33** 5795 – 5800.
[3] Saito T, Yokoyama T and Takasaki A2011 *J. Alloy. Compd.* **509** 779– 781.
[4] Van Rijswick M H J 1978 Metal hydride electrodes for electrochemical energy storage *Pergamon Press* ed. A F Andersen and A J. Maeland (London: Pergamon Press, Oxford). pp. 261-271.
[5] Buchner H, Gutjahr M A, BeccuK-D et al. *Zeitschrift Fur Metallkunde* 1972 **63** 497 – 500.
[6] Takeshita H T, Tanaka H, Kuriyama N et al. 2000 *J. Alloy. Compd.* **311** 188 – 193.
[7] Schlapbach L, Zuttel A 2001 *Nature* **414** 353 – 358.
[8] Saito T, Yokoyama T et al. 2011 *J. Alloy. Compd.* **509** 779 – 781.
[9] Skryabina N Y, Fruchartb D. Cagnon L and Shelyakov A V 2009 *J. Alloy. Compd.* **480** 91 – 93.
[10] Pelton A, Trépanier C Structural and diffusional effects of Hydrogenin TiNi 2009 *MenloPark CA: SMST Society* **480** p. 2 – 9.
[11] Yao X, Wu C, Du A et al. 2007 *Chem. Soc.* **129** 15650 – 15654.
[12] Bratanich T I, Getman O I et al. 2006 *Powder Metall. Metal Ceram.* **45** 11 – 12.
[13] Shan X, Payer J H and Jennings W D 2009 *Int. J. Hydrog. Energy* **34** 363-369.
[14] Burch R, Mason N B and Chem J 1979 *Soc. Faraday Trans. 1 Phys. Chem. Condens. Phases* **75** 561 – 577.
[15] Phillips I I, Poole P, Shreir L L 1972 *Corros Sci* **12** 855 – 866.
[16] Lunarska E, Chernyayeva O et al. 2010 *Materials Science and Engineering C* **30** 181 – 189.
[17] Burnyshev I N, Kalyuzhnyy D G 2014 *Khimicheskaya fizika i mezoskopiya* **2** 250 – 256.
[18] Greg C and Sing K 1982 *Adsorption, Surrface Area and Porosity* (London: Academic Press INC.) 313p.
[19] Luan B, Cui N, Zhao H et al. 1995 *Journal of Power Sources* **55** 101 – 106.
[20] Arantes D R, Huang X Y, MarteCet al. 1993 *Acta Metall Mater* **41** 3215 – 3222.
[21] Prochazka I, Cízek J, Havranek V et al. 2015 *J. Alloys Compd* **645** 69 – 71.
[22] Jambon F, Marchetti L, Sennour M et al. 2015*J. Nucl. Mater* **466** 120 – 133.
[23] Olsson P A T, Kese K et al. 2015*J. Nucl. Mater* **467** 311 – 319.

Electromagnetic characteristics of polymer dielectric matrixes obtained by additive technology at THz frequency range

S N Zhakupov, A V Badin and K V Dorozhkin

Faculty of Radiophysics, Tomsk State University, 36, Lenina av., Tomsk, 634050, Russia

E-mail: bdstsu@mail.ru

Abstract. There were presented the research results of electromagnetic characteristics of materials, which are widely used in 3D printing by fused deposition method. Using the spectrometer of continuous radiation, the frequency dependences of the real and imaginary parts of the dielectric constant of polystyrene, acrylonitrile sterol acrylate (ABS), polyethylene terephthalate glycol, thermoplastic elastomer, acrylonitrile butadiene styrene and nylon in frequency range 200-380 GHz at temperature 298 K were obtained. Applicability of researched materials for creating elements of the terahertz (THz) frequency range was considered.

1. Introduction

The creation of materials with given electrophysical properties is an urgent task in all spheres of human activity in which electromagnetic radiation is used. The rapid development of additive technology leads to the emergence of new materials and development of composites based on polymer matrixes [1-2]. The use of multi-extrusion layer-by-layer fusion systems is interesting in creation of interacting with THz radiation composites based on electrically conductive plastic and polymer dielectric matrices. [3-8]. The production of the EHF elements [9] using the 3D printing method opens up broad prospects for creating relatively cheap THz imaging systems in the food industry [10-11], medicine [12] and microelectronics[13]. An active research of matrixes with an artificial periodic structure consisting of conducting and dielectric elements is underway. These periodic structures allow controlling the laws of dispersion, refraction and reflection of electromagnetic waves in the material. Thus, resonators, filters, and phase shifters can be constructed on the basis of these matrixes. To create these THz devices, it is first necessary to research the electrophysical properties of the matrixes. This paper presents the results of the analysis of the electromagnetic properties of polymer dielectric plates created by additive technology in the THz frequency range. For research a number of polymeric materials, widely applicable in industrial and consumer 3D printers was selected.

2. Experimental work

In our research matrixes of popular materials that are used in 3D printing (polystyrene, acrylonitrile sterol acrylate, polyethylene terephthalate glycol, thermoplastic elastomer, ABS, nylon) were created using additive fused deposition modeling (FDM) technology. Matrixes are homogeneous square plates with transverse dimensions of 25 x 25 mm and 740-1970 μm thicknesses. Description and thickness of the fabricated samples of the matrixes are presented in Table 1. The diameter of the nozzle in

IOP Conf. Series: Materials Science and Engineering **511** (2019) 012010 doi:10.1088/1757-899X/511/1/012010

the manufacture of matrixes was 300 microns. The temperature of the table when printing samples was 343 K.

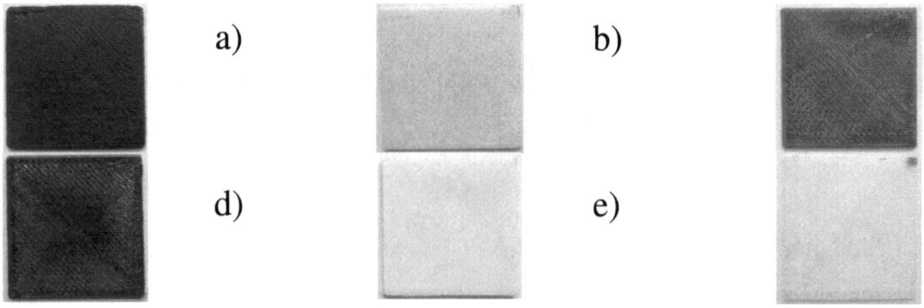

Figure 1. Optical photographs of dielectric plates produced by fused deposition modeling method: a) polystyrene; b) acrylonitrile sterol acrylate; c) polyethylene terephthalate glycol; d) thermoplastic elastomer; e) acrylonitrile butadiene styrene (ABS); f) nylon

Table 1. Description of researched materials.

№	Name	Thickness, mm	Description
1	HIPS	1.135	Polystyrene
2	ASA	0.800	Acrylonitrile sterol acrylate
3	PETAG	0.815	Polyethylene terephthalate glycol
4	Flex	0.890	Thermoplastic elastomer
5	ABS	1.970	Acrylonitrile butadiene styrene
6	Nylon	0.740	Nylon

To research the electromagnetic properties of polymer dielectric plates in the THz frequency range, we used a quasi-optical spectrometer STD-21 (Figure 2), which is based on the Mach-Zander interferometer at temperature 298 K. As a source of monochromatic electromagnetic radiation of 200-380 GHz frequency range, M-type backward wave oscillator was used. The THz wave was modulated by mechanically interrupting the quasi-optical beam with a chopper. Acousto-optic converter (Golay cell) performed detection of electromagnetic radiation transmitted through the researching sample. The matching of the THz radiation power with the allowable range for the detector was carried out with teflon attenuators in discrete steps. A mesh divider from microwires carried out the division of a linearly polarized wave into two arms of interferometer. Measurements of complex dielectric constant of each type of polymer dielectric matrix took place in two stages. At the first stage, the transmission coefficient of an electromagnetic wave through the matrix was measured as the ratio of the power at the detector with and without the researching matrix in quasi-optical path. At the second stage, the phase shift of electromagnetic wave as ratio of position of mirror in free arm of interferometer with and without the researching sample was measured. From the obtained results, the real and imaginary parts of the complex dielectric constant were calculated using equations relating the transmission and phase shift coefficients of electromagnetic wave with a complex magnetic and dielectric constant of the material [14]. The Figure 3 shows frequency dependences of complex permittivity of polymer dielectric matrixes calculated from the transmission and phase shift spectra of an electromagnetic wave.

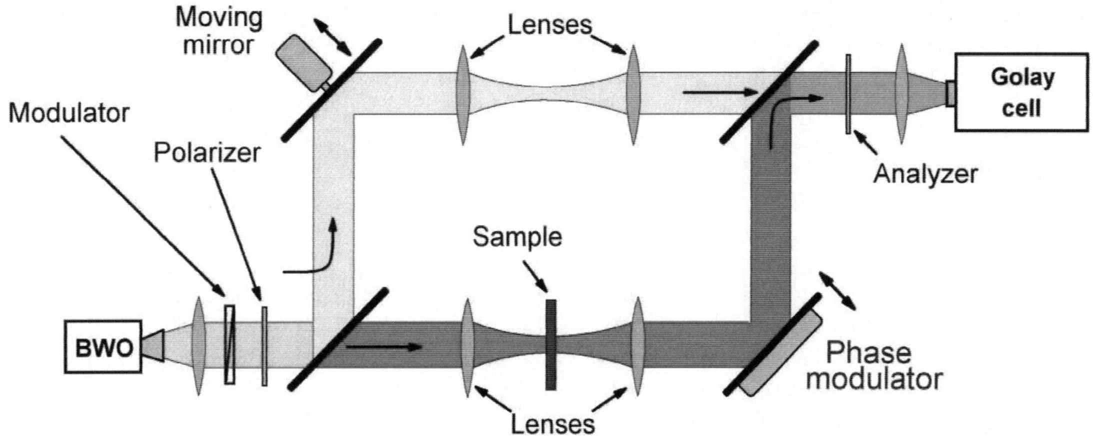

Figure 2. The experimental setup for research of the electrophysical properties of polymer dielectric matrixes obtained by additive technology.

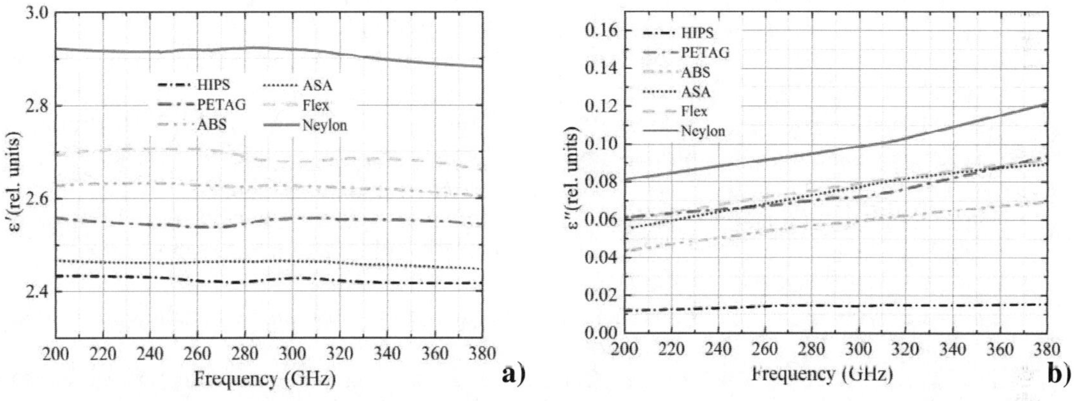

Figure 3. Frequency dependences of the real (a) and imaginary (b) part of the dielectric constant of polymer matrixes obtained by additive technology.

One can see from the graphs, the values of the real part of the dielectric constant of measured matrixes in the frequency range 200-380 GHz at temperature 298 K are in the range of 2.45-2.9 and remains constant with increasing frequency. The maximum value of the real part of dielectric constant is observed in nylon, the minimum in polystyrene. The imaginary part of the complex dielectric constant, which is responsible for the loss in the substance, has an almost linear increasing dependence on the frequency for all matrixes under research (Figure 3b). The exception is HIPS, for which the imaginary part of the dielectric constant remains unchanged in the range of 0.01-0.02.

Conclusion

Thus, based on this research, it can be said that the considered materials can be used to create focusing elements (lenses) of the quasi-optical terahertz tract. In this case, the best material for creating lenses by FDM 3D printing will be polystyrene due to low dielectric losses (< 0.02) in the THz range (Figure 3b). Another important direction of application of dielectric and conductive materials in additive technology can be the creation of frequency selective elements and metamaterials of extreme high frequencies. Besides the results obtained can be used in the simulation of the electromagnetic response of composite materials and objects in the THz frequency range. Based on the researched polymer matrixes and conductive plastic with carbon-containing additives, anisotropic materials can be manufactured by 3D printing method for THz frequency range technique. Thus, the knowledge gained about the electromagnetic characteristics of widely used materials in 3D printing will expand the use of additive technologies in the direction of the THz technology.

Acknowledgment
The reported study was funded by RFBR according to the research project № 18-32-00810.

References

[1] Eyers D R and Potter A T 2017 Industrial additive manufacturing: a manufacturing systems perspective *Computers in industry* **92** 208-18
[2] Bandyopadhyay A, Heer B 2018 Additive manufacturing of multi-material structures *Materials science and engineering: reports.* **129** (4) 1–16
[3] Badin A V, Kuleshov G E, Dorozhkin K V, Dunaevskii G E, Suslyaev V I and Zhuravlev V A 2018 Anisotropy of electrical properties of 3D-printing MWCNT composites at the THz frequency range, *proc. in 43th International conference on infrared, millimeter, and terahertz waves (IRMMW-THz 2018)* 1-2
[4] Busch S F, Weidenbach M, Fey M, Schäfer F, Probst T, Koch M 2014 Optical properties of 3D printable plastics in the THz regime and their application for 3D printed THz optics *Journal of infrared, millimeter, and terahertz waves* **35** (12) 993–7
[5] Sherrell P C, Mattevi C 2016 Mesoscale design of multifunctional 3D graphene networks *Materials today* **19** (8) 428-36
[6] Osa de la G, Pérez-Coll D, Miranzo P, Isabel Osendi M and Belmonte M 2016 Printing of graphene nanoplatelets into highly electrically conductive three-dimensional porous macrostructures *Chemistry of materials* **28** (17) 6321–2
[7] Gnanasekaran K, Heijmansa T, Bennekom S, Woldhuis H, Wijnia S, With de G and Friedrich H 2017 3D printing of CNT- and graphene- based conductive polymer nanocomposites by fused deposition modelling *Applied materials today* **9** 21–8
[8] Vendik I B, Vendik O G 2013 Metamaterials and their application in the technique of ultrahigh frequency (review) *Technical physics journal* **83** (1) 3 - 28
[9] Chen Q, Chen X, Xu K 2017 3-D Printed Fabry–Perot resonator antenna with paraboloid-shape superstrate for wide gain bandwidth *Applied sciences* **7** (11) 1134
[10] Wang K, Sun D W, Pu H 2017 Emerging non-destructive terahertz spectroscopic imaging technique: principle and applications in the agri-food industry *Trends in food science and technology* **67** 93-105
[11] Ok G, Park K, Kim H J, Chun H S, Choi S W 2014 High-speed terahertz imaging toward food quality inspection *Applied optics* **53** (7) 1406
[12] Yang X, Zhao X, Yang K, Liu Y, Liu Y, Fu W, Luo Y 2016 Biomedical applications of terahertz spectroscopy and imaging *Trends in biotechnology* **34** (10) 810-24
[13] Ahi K, Shahbazmohamadi S, Asadizanjani N 2018 Quality control and authentication of packaged integrated circuits using enhanced-spatial-resolution terahertz time-domain spectroscopy and imaging *Optics and lasers in engineering* **104** 274-84
[14] Lee Y-S. 2009 Principles of terahertz science and technology. 14-28

Electrophysical properties of ferrite-containing filaments for additive technology at EHF range

A A Kalugin, A V Badin and G E Kuleshov

Faculty of Radiophysics, Tomsk State University, 36, Lenina av., Tomsk, 634050, Russia

E-mail: thzlab@mail.ru

Abstract. The technology of producing ferrite-containing filament for 3D printer was described. The electrophysical properties of ferrite-containing filament sample at 115-258 GHz frequency range was investigated. The article presents the frequency dependence of complex permittivity and conductivity of sample obtained by fused deposition method (FDM) from filament based on acrylonitrile butadiene styrene (ABS) plastic with the addition of 20% ferrite.

Introduction

The increasing of power of microwave electronics devices leads to the fact that the high-frequency electromagnetic radiation formed during their operation creates strong interference to radio-electronic equipment and satellite communications. That is why, the development of new high-performance radar absorbing materials and products is very important for solution the problem of noise reduction and electromagnetic compatibility of radio-electronic devices [1]. The use of ferrites, carbon nanotubes and carbonyl iron additives in composites opens up great prospects for the development of the frequency-selective element in the field of hyper high and extremely high frequency ranges base up to 1 THz [2-4]. Today, various fillers are actively investigated for filaments of 3D printers based on carbon-containing structures, including graphene and MWCNT, as well as the addition of ferrites in order to obtain conductive radio-absorbing composites of complex geometric shape [5-14]. It is of interesting to research the ferrite-containing filaments for modern 3D printing technology, which have perspectives to manufacturing of radio-absorbing coatings, units and products in the field of extremely high frequencies.

Experimental work

In this research, ferrite of the structure of spinel Fe_3O_4, produced according to the classical ceramic technology, was used as a filler. The ABS plastic was used as a matrix. The manufacture of a composite based on ABS plastic and ferrite was carried out according to the following technology (block diagram in Figure 1). To the previously dissolved in dimethyl ketone ABS plastic, ferrite particles were added in a mass ratio of 80% ABS plastic and 20% ferrite. After ultrasonic treatment with a dispersant, the composite was dried. The mixture was polymerized at room temperature during the complete evaporation of dimethyl ketone. The dried mixture was grinded to small particles (less than 1 mm) in a mechanical mill and then the powder was sieved through a set of laboratory sieves for separation by size. Next, the raw material was used in an extruder for the production of composite filaments with an average diameter of 1.65 mm.

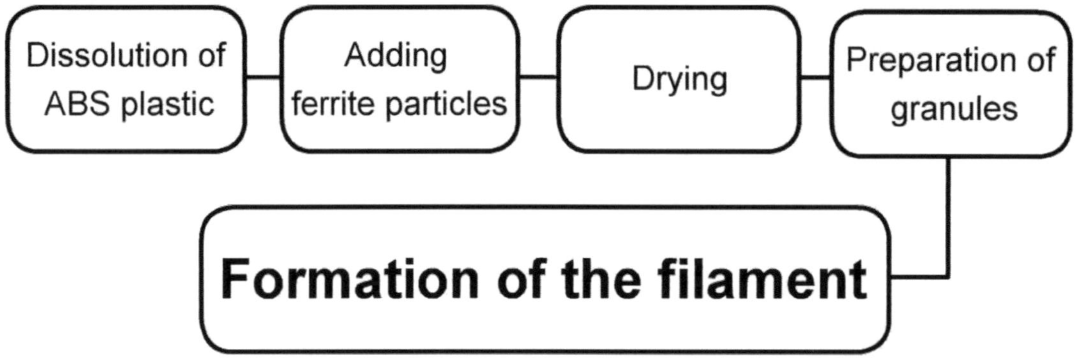

Figure 1. Block-diagram of the technology for producing ferrite containing filament.

The obtained filament was loaded into FDM 3D printer with a nozzle diameter of 500 μm. Plane-parallel round composite sample with diameter 48 mm and thickness of 700 μm was modelled for printing (Figure 2). A control sample was also made of pure ABS plastic for comparison of electrophysical properties in THz frequency range.

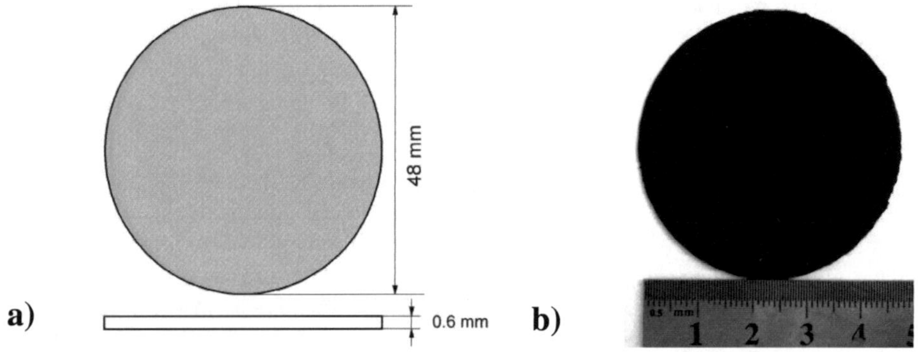

Figure 2. Model (a) and image (b) of sample of material printed by FDM technique from filament based on ABS plastic with additives of 20% ferrite.

The values of the complex dielectric constant and conductivity of a ferrite-containing composite sample were calculated from the transmission coefficient and phase shift spectra measured with a quasi-optical spectrometer STD-21 (Figure 3) in the frequency range of 115-258 GHz.

IOP Conf. Series: Materials Science and Engineering **511** (2019) 012011 doi:10.1088/1757-899X/511/1/012011

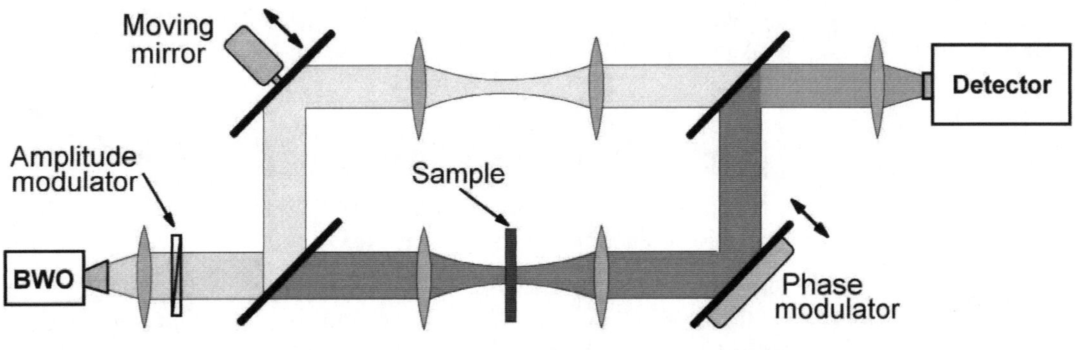

Figure 3. Scheme of THz spectrometer STD-21 for research frequency dependence of electromagnetic properties of ferrite-containing composites.

The generation of monochromatic radiation by a water-cooled backward wave oscillator was used. Sensitive acoustic-optical converter (Golay cell) carried out detection of THz radiation. Transmission coefficient was measured as the ratio of power with sample to without. Measurement of the phase shift of an electromagnetic wave was carried out by registering the position of the moving mirror when the sample was introduced into the working path. Figure 4 shows graphs of the frequency dependence of the complex dielectric constant of a material sample obtained by 3D printing from filament.

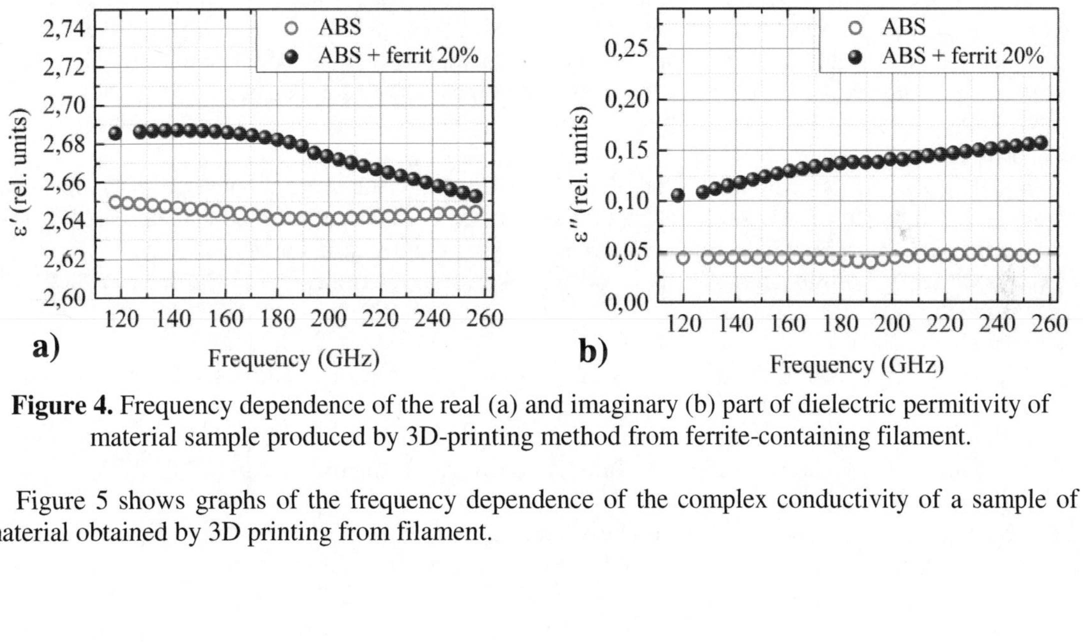

Figure 4. Frequency dependence of the real (a) and imaginary (b) part of dielectric permitivity of material sample produced by 3D-printing method from ferrite-containing filament.

Figure 5 shows graphs of the frequency dependence of the complex conductivity of a sample of material obtained by 3D printing from filament.

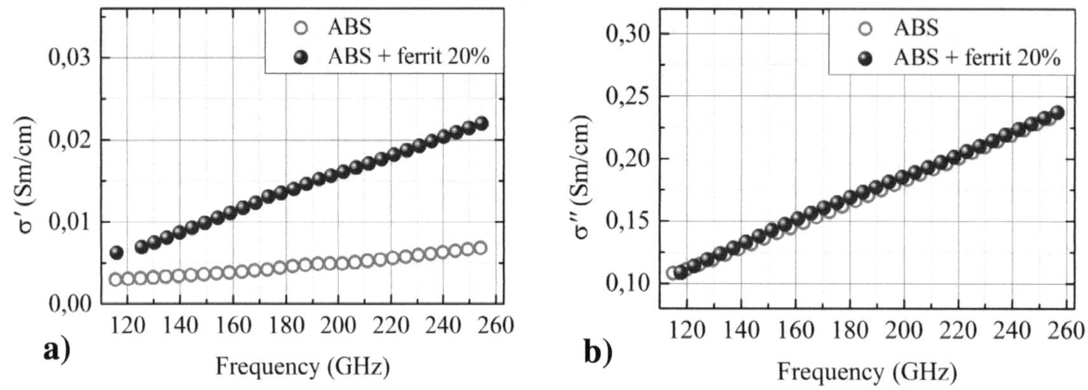

Figure 5. Frequency dependence of real (a) and imaginary (b) part of conductivity of material sample produced by 3D-printing method from ferrite-containing filament.

Summary

From the data obtained it follows that the addition of 20% ferrite to the ABS plastic does not make a significant contribution to the real part of the dielectric constant. At the same time, the imaginary part of the dielectric permittivity varies by 2 times. The real part of the conductivity also increases by 2 times, there is a sharper increase in conductivity with a frequency relative to pure ABS plastic. A sample of material printed on a 3D printer printed on a ferrite-containing filament interacts well with a permanent magnet and is capable of magnetization. The obtained composites can be used both as radio-absorbing coatings coatings located on a metal surface, as protective non-reflective coatings with small reflection and transmission coefficients. In addition, these composite materials can be applied for the manufacture of decorative products of complex geometric shape.

Acknowledgment

Authors are grateful to an engineer of the Shared Use Center "Centre of radiophysical measurements, diagnosis and research of parameters of natural and artificial materials" of National Research Tomsk State University, Dorozhkin K.V. for assistance in carrying out the experiments.
The reported study was funded by RFBR according to the research project № 18-32-00810.

References

[1] Latypova A F and Kalinin Yu E 2012 Analysis of promising radio absorbing materials *J. Vestnik VGTU* **6** 70-6

[2] Zhuravlev V A and Suslyaev V I and Korovin E Yu and Dotsenko O A 2010 Radar absorbing properties of carbonyl-iron-containing composites on microwave and EHF *J. Research in Russi* **35** 404-11

[3] Kachusova A, Dotsenko O and Dorozhkin K 2016 Influence of ultrasonic treatment on electromagnetic characteristics of composites based on multiwall carbon nanotubes at wide range of frequencies (100 Hz - 258 GHz) *J. Nano- Electron. Phys.* **8** 1–3

[4] Dunaevskii G E and Suslyaev V I and Zhuravlev V A and Badin A V and Dorozhkin K V 2016 Ferromagnetic resonance in hexagonal ferrite $Ba_3Co_2Fe_{24}O_{41}$ at the THz frequency range *proc. in. 41th International conference on infrared, millimeter, and terahertz waves* 1-2

[5] Gnanasekaran K, Heijmansa T, Bennekom S, Woldhuis H, Wijnia S, With de G and Friedrich H 2017 3D printing of CNT- and graphene- based conductive polymer nanocomposites by fused deposition modelling *Applied materials today* **9** 21–8

[6] García-Tuñon E, Barg S, Franco J, Bell R, Eslava S, D'Elia E, Maher R C, Guitian F, Saiz E 2015 Printing in three dimensions with graphene *Advanced materials* **27** (10) 1688–93

IOP Conf. Series: Materials Science and Engineering **511** (2019) 012011 doi:10.1088/1757-899X/511/1/012011

[7] Osa de la G, Pérez-Coll D, Miranzo P, Isabel Osendi M and Belmonte M 2016 Printing of graphene nanoplatelets into highly electrically conductive three-dimensional porous macrostructures *Chemistry of Materials* **28** (17) 6321–28

[8] Kwok S W, Goh K H H, Tan Z D, Tan S T M, Tjiu W W, Soh J Y, Ng Z J G, Chan Y Z, Hui H K and Goh K E J 2017 Electrically conductive filament for 3D-printed circuits and sensors *Appl. Mater. Today* **9** 167–75

[9] Wei X, Li D, Jiang W, Gu Z, Wang X, Zhang Z and Sun Z 2015 3D Printable Graphene Composite *Sci. Rep.* **5** 1–7

[10] Gardner J M, Sauti G, Kim J W, Cano R J, Wincheski R A, Stelter C J, Grimsley B W, Working D C and Siochi E J 2016 3-D printing of multifunctional carbon nanotube yarn reinforced components *Addit. Manuf.* **12** 38–44

[11] Yang F, Zhang X, Guo Z, Ye S, Sui Y and Volinsky A A 2019 3D printing of NdFeB bonded magnets with SrFe12O19 addition *J. Alloys Compd.* **779** 900–7

[12] Yang F, Zhang X, Guo Z and Volinsky A A 2018 3D gel-printing of Sr ferrite parts *Ceram. Int.* **44** 22370–7

[13] Huang D, Liao F, Molesa S, Redinger D and Subramanian V 2003 Plastic-Compatible Low Resistance Printable Gold Nanoparticle Conductors for Flexible Electronics *J. Electrochem. Soc.* **150** G412

[14] Zhang J, Yang B, Fu F, You F, Dong X and Dai M 2017 Resistivity and Its Anisotropy Characterization of 3D-Printed Acrylonitrile Butadiene Styrene Copolymer (ABS)/Carbon

IOP Conf. Series: Materials Science and Engineering **511** (2019) 012012 doi:10.1088/1757-899X/511/1/012012

Fatigue behaviour of CG and UFG titanium: DIC and fractography studies

A V Eremin[1], S V Panin[1,2] and Yu P Sharkeev[1]

[1]Institute of Strength Physics and Materials Science, Tomsk, Russia
[2]National Research Tomsk Polytechnic University, Tomsk, Russia

ave@ispms.tsc.ru

Abstract. The manuscript is devoted to the investigation of fatigue behaviour of course-grain (CG) and ultrafine-grain (UFG) titanium grade 2 by means of digital image correlation (DIC) and scanning electron microscopic fractography. The UFG state has been obtained utilizing severe plastic deformation technique. The results show that the high amount of defects (grain boundaries, dislocations, etc.) in UFG suppress plastic deformation and leads to earlier crack initiation, however at the macrocrack growth stage these impurities are obstacles for transgranular mechanism of fracture.

1. Introduction

The commercially pure titanium grade 1-4 is the most widely used material for dental and orthopaedic implants due to its biocompatibility, cost and relatively good processability [1]. However one of the very important limitations in a case of highly loaded prostheses is low mechanical properties which do not allow weight reduction of the components made from titanium. The prospective technique for enhancement of the mechanical properties is severe plastic deformation [2,3] which results in grain refinement and subsequent increase in yield and ultimate stresses [4,5] of the material, but due to lack of plasticity it could reduce of fatigue durability [6] which is also important for long-used implants.

Present study is aimed at investigation of fatigue properties and fracture kinetics of ultra-fine grain titanium (grade 2) in comparison with initial course-grain state.

2. Materials and fatigue test conditions

2.1. Grain refinement procedure

For the grain refinement the severe plastic deformation technique was implemented which consisted in three stages. At the first stage the blanks were subjected to abc-pressing and under elevated temperature (400-500°C) at the strain rate of 10^{-3}–10^{-1} s^{-1}. Strains at each pressing cycle were 40-45%. At the second stage the blanks were deformed using rolling with the total strains about 90%. The third stage was annealing at T=350°C for t=1 hour aimed at decreasing internal stresses caused by severe deformation. The final blanks had the shape of narrow thin bands with dimensions of 1.3×5×70 mm (thickness × width × length).

The average grain size of the course-grain titanium is 25μm, while it is 0.2 μm for ultrafine-grain titanium.

IOP Conf. Series: Materials Science and Engineering **511** (2019) 012012 doi:10.1088/1757-899X/511/1/012012

2.2. *Estimations of crack length and maximum strains at the crack tip*

The coordinates of crack tip position were estimated based on the images of the specimen surface at different operation time. Of course, the accuracy of such technique is lower than SEM based measurements, but comparable or higher than evaluations made by compliance method (COD gauge at the specimen edge).

The next level of measurement was performed via DIC – computation of displacement and strain fields based on the specimen images. The non-uniform surface of the polished titanium played the role of a speckle pattern which is obligatory for DIC. Then the extensometer tool was used in order to perform the evaluation of strain values directly at the crack tip which position has been found earlier. The extensometer length was 50 px which corresponds to ~30 μm. So the precise measurements of the strain response of the crack tip on the applied load during the fatigue tests were obtained.

2.3. *Fatigue test specimens and loads*

As far as the final titanium pieces after refinement were small therefore for the fatigue tests were cut the miniature specimens with single edge notches for crack initiation (figure 1). These specimens made of two types of materials – CG and UFG titanium – were subjected to cyclic loading under maximum load P_{max}=0.6 kN and loading ratio R=0.1 which corresponds to maximum stress intensity factor (SIF) K_{max}≈16 MPa·√m and SIF range ΔK≈14.7 MPa at the beginning of the test.

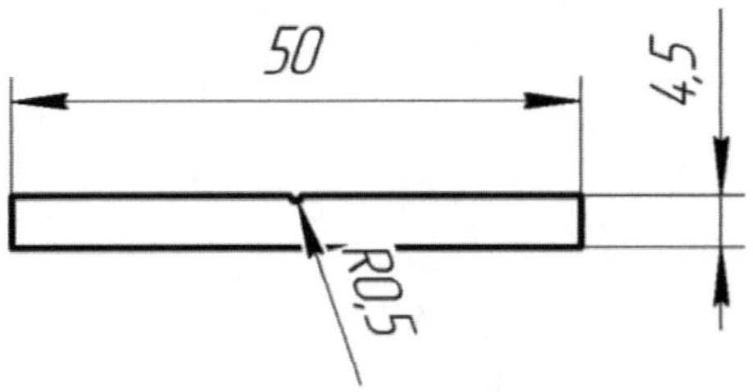

Figure 1. Fatigue specimen shape and dimensions.

The image acquisition of the crack tip at high magnification was performed under fixed maximum load (P_{max}) every 2000 cycles before crack initiation and every 500 cycles during the stage of macro crack growth. The digital camera follows the growing crack thus the tip was always in area of interest.

3. Experimental results and discussion

As a result of fatigue test the average fatigue durability was 210000±28000 cycles for CG titanium and 180000±38000 cycles for UFG titanium specimens.

3.1. *DIC study*

The crack growth process could be divided into two stages: (1) crack nucleation and (2) micro crack propagation. The durations (figure 2a) of the first stage were 186000 (88% of life time) and 135000 (77% of life time) cycles for CG and UFG correspondingly whilst the second one – 25000 (12%) and 41000 (23%). The crack growth stage for two specimen types is illustrated in figure 2b as a dependence on the life time. It could be noted that the crack appearance in UFG titanium occurs earlier than in CG Ti, whereas the total crack length achieved before fracture of UFG Ti is much higher for all tested specimens (approximately 2.5 mm for CG and 3 mm for UFG).

Figure 3a demonstrates fatigue fracture kinetic diagrams; it is seen that when crack is small the propagation rates are near the same, but when crack reaches ~0.7 mm (ΔK≈20 MPa·√m) it leads to the higher growth rate of the crack in CG titanium. However at the end of the test the rates became close to each other again, so the middle regions of the fracture diagrams of titanium are sensitive to the grain size.

Traditionally the processes of crack growth could be attributed to two main factors: threshold range of stress intensity factor (ΔK_{th}) and effective range of stress intensity factor (ΔK_{eff}) – both of them are different for presented types of specimen due to different yield stress which manages the near-tip stresses (σ*). The effect of each parameter needs to be studied.

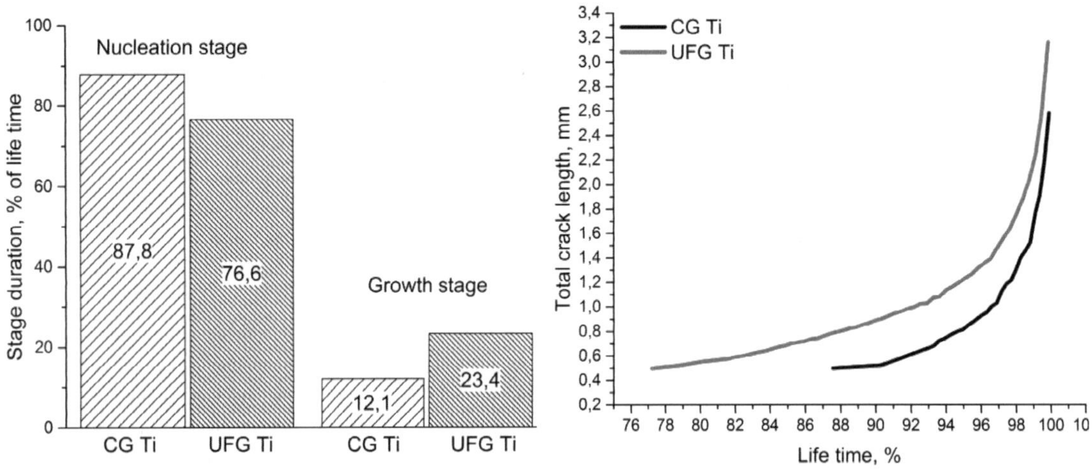

Figure 2. Bar diagram of crack nucleation and propagation stages (a) and the dependences of total crack length on the specimen life time (b) for course-grain and ultrafine-grain titanium.

Evolution of strains at the crack is shown in figure 3b and they demonstrate nonlinear behaviour. The last measurement point (prior to fracture) for UFG titanium is not presented in the plot due to high deformations and the loops of correlation between two adjacent images.

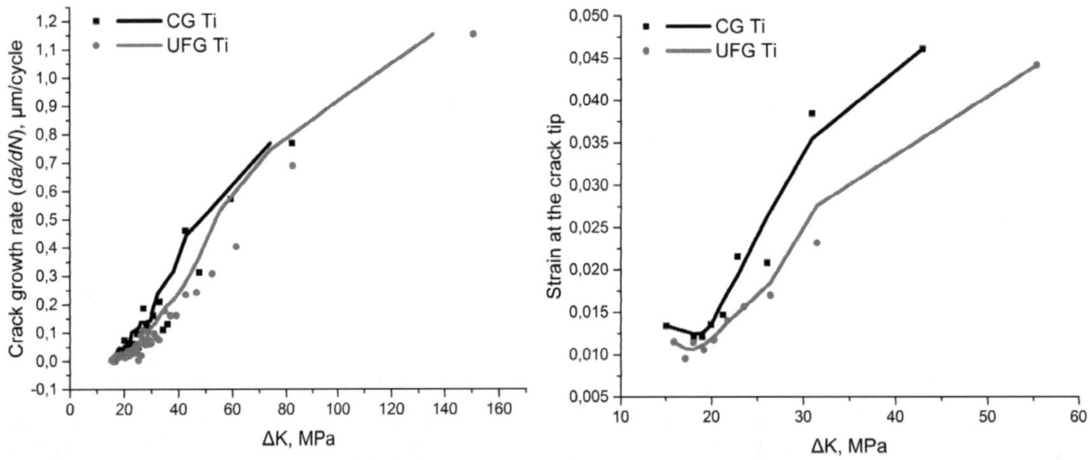

Figure 3. Fatigue fracture kinetic diagrams (a) and change of local strains at the crack tip during crack propagation (b) for course-grain and ultrafine-grain titanium.

3.2. SEM fractographic study

The SEM images of fracture surface of CG titanium are presented in figure 4. It demonstrates the panoramic view and high magnification images (2000x) of different fracture textures at different crack length. The fracture pattern consists of three texture types representing different crack propagation processes – origin area, where micro- and meso- cracks have been merged in one micro crack going thought the whole specimen thickness; area of stable crack growth and static fracture area at the right edge of the specimen. It should be noted that the during stable crack propagation period the texture becomes more rough and more cracks seen (figure 4b and 4c). Static fracture zone has a mostly uniform

IOP Conf. Series: Materials Science and Engineering **511** (2019) 012012 doi:10.1088/1757-899X/511/1/012012

pattern of ductile fracture. Moreover this area was significantly narrowed due to high level of plastic deformation.

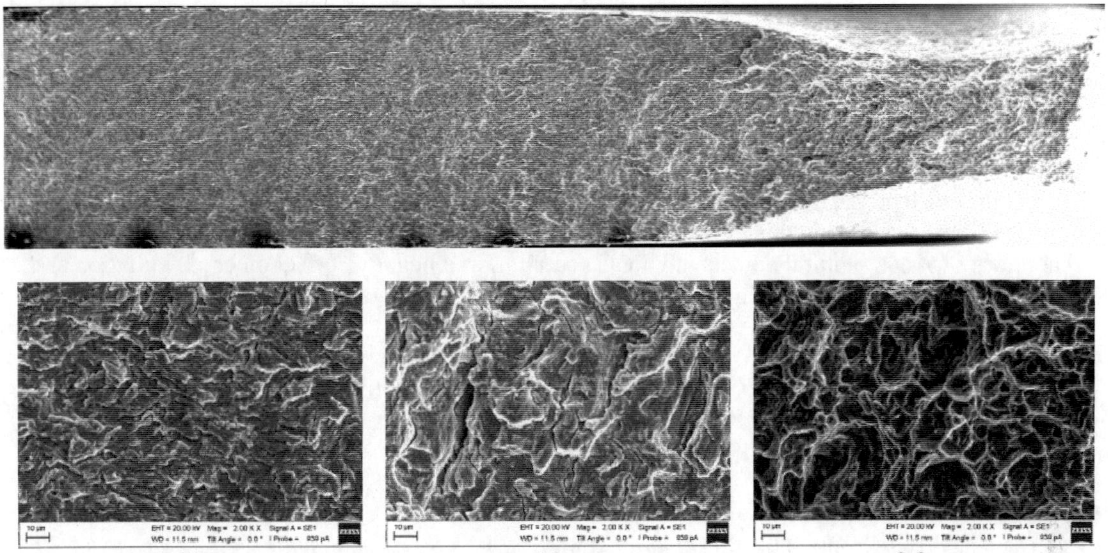

a≈0.37 mm a≈1,9 mm a≈2,9 mm

Figure 4. Panoramic image (a) of fatigue fracture surface for CG titanium specimens and surface texture at different crack length (b,c,d).

SEM images of UFG titanium fracture surfaces (figure 5) demonstrate more brittle fracture processes but still it consists of three main zones. The stable crack growth stage is represented by very smooth and fine microstructure with no river-like pattern (figure 5a). However, the micro- and mesocracking also occurs at longer crack length and it is even more pronounced as for CG titanium (figure 5b). Prior to the final fracture the crack extension at individual cycles is seen on the surface as a separate bands divided by striations (figure 5c).

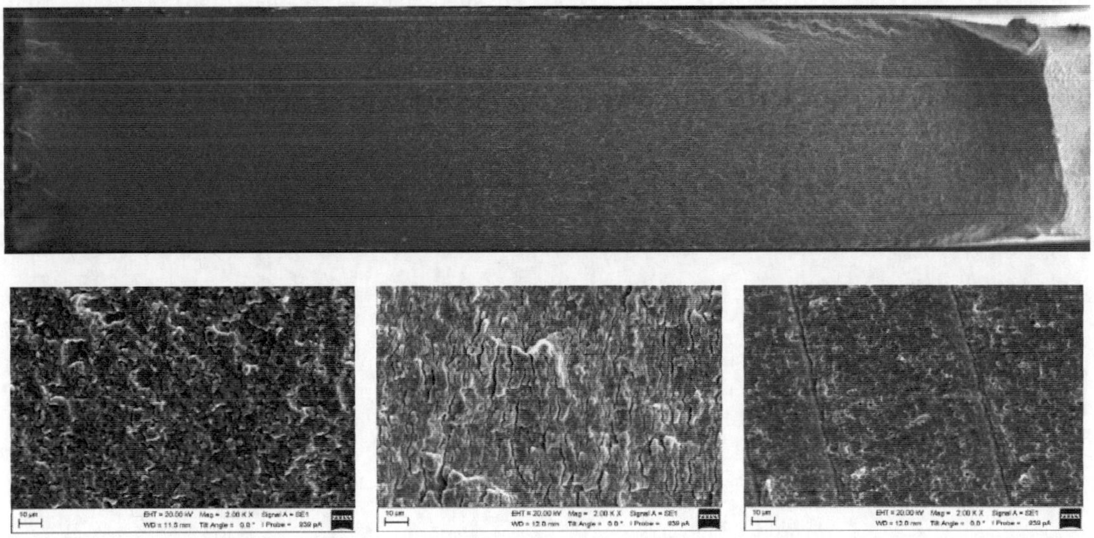

a≈0.34 mm a≈2,37 mm a≈3,45 mm

Figure 5. Panoramic image (a) of fatigue fracture surface for UFG titanium specimens and surface texture at different crack length (b,c,d).

4. Conclusion

The paper presents the comparison of fatigue fracture kinetics of CG and UFG titanium specimens by means of digital image correlation technique and SEM fractography. Analysis of the results obtained leads to the following conclusions:

- at the crack nucleation stage presence of many grain boundaries in UFG Ti prevent development of plastic deformations which would reduce local stresses and accumulated elastic energy in in the vicinity of appeared microcracks, thus crack emerges on the surface of UFG Ti earlier than for CG Ti;
- at the crack propagation stage UFG Ti has lower crack growth rates due to grain boundaries and dislocations which in this case are obstacles for transgranular type of fracture.

Therefore a lot of structural defects in UFG Ti could have a negative effect on crack initiation at the first stage and then have positive impact by suppression of crack growth processes at the second stage.

Acknowledgements

The work was supported by President Scholarship for young scientists and graduate students in 2018-2020 (reference number SP-198.2018.4). SEM fractography was performed by LEO EVO 50 microscope in “NANOTECH” of ISPMS SB RAS.

References

[1] The Freedonia Group 2006 *Biocompatible Materials. US Industry Study with Forecasts to 2010 & 2015 Study #2111* (Cleveland: The Freedonia Group) p 264
[2] Valiev R 2004 Nanostructuring of metals by severe plastic deformation for advanced properties *Nat. Mater.* **3** 511–6
[3] Ocid’Ko I A and Langdon T G 2012 Enhanced ductility of nanocrystalline and ultrafine-grained metals *Rev. Adv. Mater. Sci.* **30** 103–11
[4] Cavaliere P 2009 Fatigue properties and crack behavior of ultra-fine and nanocrystalline pure metals *Int. J. Fatigue* **31** 1476–89
[5] Fintová S, Arzaghi M, Kuběna I, Kunz L and Sarrazin-Baudoux C 2017 Fatigue crack propagation in UFG Ti grade 4 processed by severe plastic deformation *Int. J. Fatigue* **98** 187–94
[6] Panin V E and Egorushkin V E 2009 Physical mesomechanics and nonequilibrium thermodynamics as a methodological basis for nanomaterials science *Phys. Mesomech.* **12** 204–20

IOP Conf. Series: Materials Science and Engineering **511** (2019) 012013 doi:10.1088/1757-899X/511/1/012013

Finite element analysis of Al-Mg alloy elasto-plastic behavior and fracture during dynamic perforation

L R Akhmetshin[1, 2], E N Moskvichev[1, 2] and V V Skripnyak[1]

[1] National Research Tomsk State University, 36 Lenina ave., Tomsk, Russia
[2] Institute of Strength and Materials Science SB RAS, , 2/4 Academic ave., Tomsk, Russia

E-mail: em_tsu@mail.ru

Abstract. A mechanical behavior of Al-Mg alloy subjected to the dynamic perforation test at different strain rates was described by numerical modelling. Deformation - fracture process was investigated. Dynamic penetration tests were carried out; stress–strain curves were obtained at different rates. Johnson-Cook material and failure models were used. Model parameters have been determined experimentally. Numerical data were compared with experimental data, obtained by penetration tests with different strain rates. Finite element results are in good agreement with the experimentally obtained results under the domain of the strain rates considered in the present study.

1. Introduction

It is well known that an aluminum energy absorption per density unit is higher compared to steel. It is the reason why aerospace industries are interested in aluminum alloys for. Concerning that, the mechanical behavior and damage accumulation prediction tasks by numerical models are interesting for study. However, a numerical modeling of constructions made of Al-Mg alloys mechanical behavior at dynamical loading conditions is problematic due to lack of data. There is a request for dynamic testing and experimental results [1-4].

2. Materials and methods

Perforation testing is one of the necessary research methods, which allows to investigate material in a wide range of deformation rates. This method for determining plasticity of sheet alloys represents biaxial tension of test specimens; the scheme is shown in Figure 1a. A finite element mesh model for the numerical experiment is shown in Figure 1b. During the testing an axially symmetric indenter with a speed of 5 to 10 m/s, punches a rigidly fixed sample of sheet material. The sample material experiences large plastic deformations before fracture.

A widely used in aerospace industry Al-Mg 1560 alloy sheet in a delivery condition was chosen as a research subject. Aluminum alloy 1560 refers to the strain hardened aluminum alloys. Rolled sheets are delivered in annealed condition. Chemical composition: Mg – 6.124 %, Mn – 0.597 %, Fe – 0.351 %, Si – 0.310 %, Zn – 0.203 %, Ti – 0.0843 %, Cu – 0.086 %, Al – mass (Russian grade 1560 alloy). Test circle specimens with 60 mm diameter and 1.5 mm thickness were cut from an initial sheet.

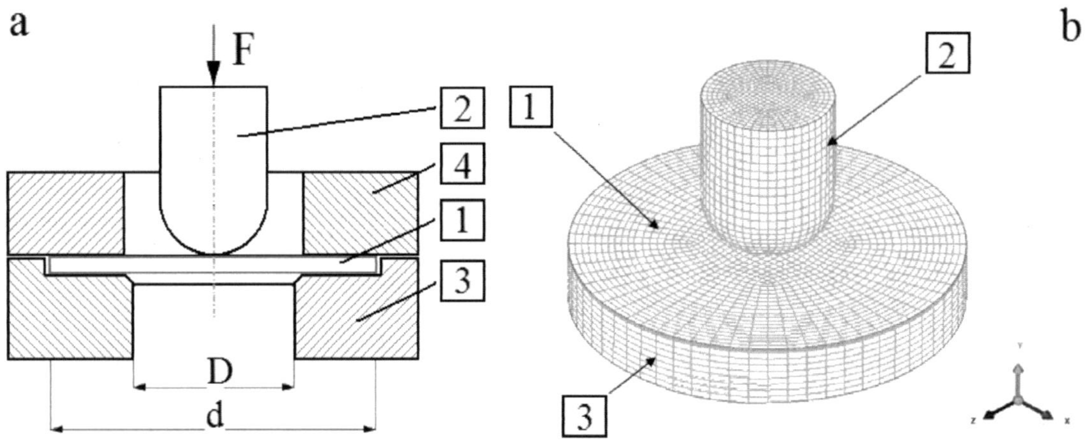

Figure 1. Experiment scheme (a) and grid model for calculations (b)
1 – specimen, 2 – indenter, 3, 4 – The lower and upper parts of the support matrix.

Numerical modeling complexity of such processes is the correct selection of the equation of state, strength and failure models. The high strain rate punch tests were performed in an Instron VHS 40 / 50-20 servohydraulic test bench for further comparison with the numerical modeling results of the deformation behavior of Al-Mg sheets under the large plastic deformations conditions. During the testing, changes of effort and displacement by time were recorded during punch tests of the 1.5 mm thick sheets. The forces dependencies up to the destruction at loading rates from 5 to 10 m/s were obtained.

The numerical task of the dynamic penetration modeling was solved by finite element method. A computational domain prepared by means of the ANSYS Workbench 15 software package is shown in Figure 1b. The numerical experiment was carried out by using the AUTODYN specialized for solving dynamic problems software. A round flat specimen rests on the lower parts support, which is rigidly fixed. A cylindrical indenter with a hemispherical end moves toward the sample at a constant speed. An indenter-sample interaction occurs before reaching the maximum permissible deformations and loads corresponding to the material destruction.

The Johnson-Cook material model was used to describe the Al-Mg alloy mechanical behavior [5]. The model considers the temperature and strain rate effect on the flow stress and the ultimate degree of material strain:

$$\sigma = \left[A + B\varepsilon_p^n \right] \left[1 + C \log \frac{\dot{\varepsilon}}{\varepsilon_0} \right] \left[1 - \left(\frac{T - T_r}{T_m - T_r} \right)^m \right] \tag{1}$$

where ε_p – is the effective plastic deformation, T_m – is the melting point, T_r – is the room temperature, $A, B, C, n, m, \varepsilon$ – are the model parameters.

3. Results and discussion

After analyzing the obtained numerical data (Figure 2), it was found that in the area before the maximum load was achieved, the results obtained using the Johnson-Cook model are in close agreement with the experimental data [6-7]. There is a need to select a failure numerical model, because of the sample destruction during the experiment. Description of the sample destruction process was carried out by the Johnson-Cook failure model:

$$D = \frac{1}{\varepsilon_f} \sum_i \Delta\varepsilon_p^i \qquad (2)$$

where ε_f – is the material ultimate strain, $\Delta\varepsilon_p^i$ – is the element effective plastic strain increment at the i-th integration step over time.

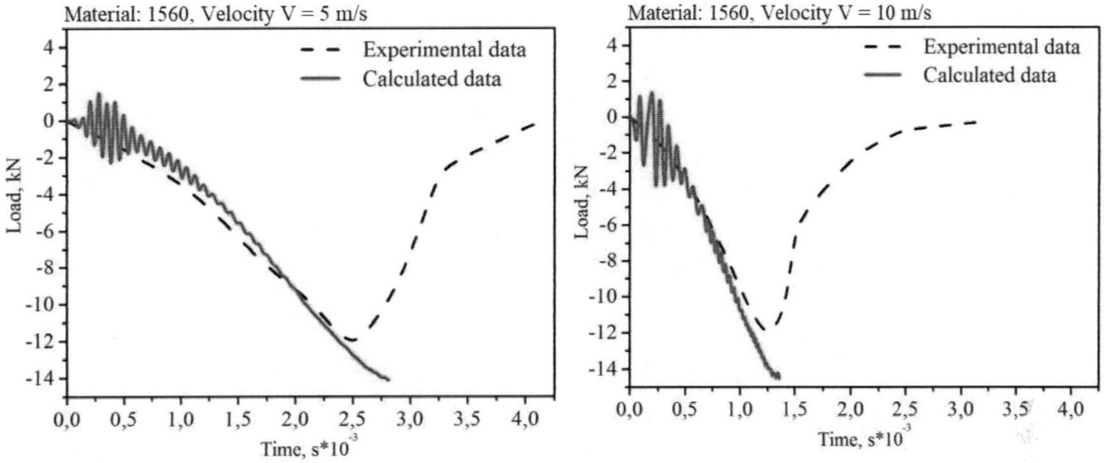

Figure 2. Time dependence of the deforming force during the experimental test of the sample (black dotted line) and the J-C material model simulation at loading speeds of 5 and 10 m/s (red line).

The $\Delta\varepsilon_p^i$ value is calculated by the formula:

$$\Delta\varepsilon_p^i = \left(D_1 + D_2 \exp\left(D_3 \frac{P}{\sigma_{ef}} \right) \right) \left(1 + D_4 \ln \frac{\varepsilon_p}{\varepsilon_0} \right) \left(1 + D_5 \frac{T - T_r}{T_m - T_r} \right) \qquad (3)$$

where $D_1,...,D_5$ – are the material parameters, σ_{ef} – is the effective stress, p – is the considered element pressure. According to the selected criterion, the final element destruction occurs if the damage parameter $D = 1$.

Constants values for the material and fracture models were determined experimentally. The experimental data were obtained earlier during tensile tests on flat specimens from the same material. Experimental data and numerical data obtained using the J-C deformation model and the J-C fracture model show qualitative and quantitative agreement throughout the all deformation region (Figure 3). This suggests the correct choice of the determining equation, the failure model, the models constants and the mesh density.

An erosion model was used to visualize the destruction process. The erosion model operation principle was as follows: the element was automatically excluded from the solution when the element's damage parameter D reached a value of 1, however, it continued to be present in the grid model, creating the illusion of integrity. When the erosion model was turned on, the elements in which the fracture criterion was achieved were removed. Thus, the erosion model inclusion to the calculation in conjunction with the Johnson – Cook failure model made it possible to obtain a fracture image as close to reality as possible.

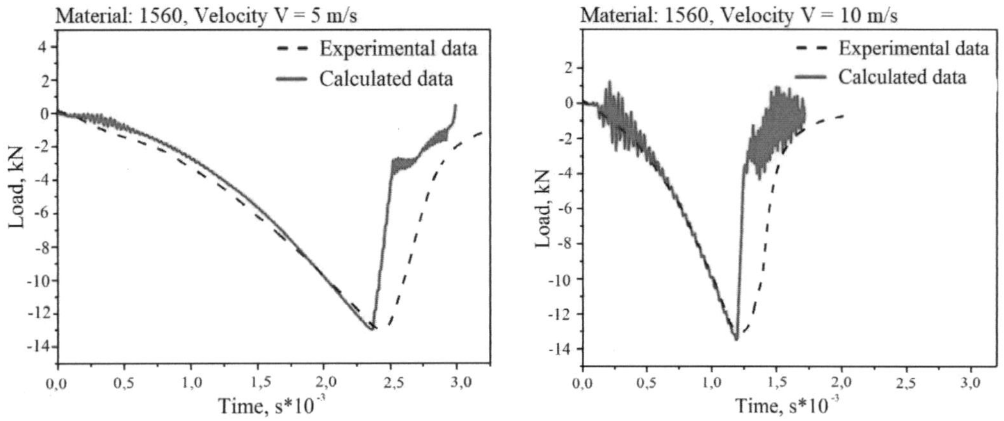

Figure 3. Time dependence of the deforming force during the experimental test of the sample (black dotted line) and the J-C material + J-C failure model simulation at loading speeds of 5 and 10 m/s (red line).

The numerical simulation results quality was estimated by comparing experimental and numerical results of the high strain rate punch tests at different strain rates. A good qualitative and quantitative agreement was obtained. The obtained results can be used in designing technological processes of sheet materials dynamic pressing, as well as in estimating strength and durability of constructions made of 1560 alloy under dynamic effects.

Acknowledgments

The practical part of the study was funded by RFBR according to the research project № 16-08-00037a. The simulation process was performed within the frame of the Fundamental Research Program of the State Academies of Sciences for 2013-2020, line of research III.23.

References

[1] Koubaa S, Mars J, Wali M and Dammak F 2017 Numerical study of anisotropic behavior of Aluminum alloy subjected to dynamic perforation. *International Journal of Impact Engineering* **101** 105–114

[2] Moskvichev E N, Skripnyak V A, Skripnyak V V, Kozulin A A and Lychagin D V 2016 Influence of structure to plastic deformation resistance of aluminium alloy 1560 after groove pressing treatment *Letters on materials* **6(2)** 141–145

[3] Skripnyak V A, Skripnyak V V and Skripnyak E G 2017 The mechanical behavior of metal alloys with grain size distribution in a wide range of strain rates *AIP Conf. Proc.* **1909** 020201-1-020201-4

[4] Xu M M, Huang G Y, Feng S S, Qin X Y, McShane G J and Stronge W J 2016 Perforation resistance of aluminum/polyethylene sandwich structure. *Materials and Design* **100** 92-101

[5] Johnson G R and Cook W 1985 Fracture characteristics of three metals subjected to various strains, strain rates, temperatures and pressures. *Eng. Fract. Mech.* **21(1)** 31–48

[6] Gambirasio L and Rizzi E 2014 On the calibration strategies of the Johnson–Cook strength model: Discussion and applications to experimental data. *Materials Science&Engineering A* **610** 370–413

[7] Buzyurkina A E, Gladkyb I L, Krausa E I 2015 Determination and verification of Johnson–Cook model parameters athigh-speed deformation of titanium alloys. *Aerospace Science and Technology* **45** 121–127

In situ estimation of fatigue crack parameters by digital image correlation under cyclic loading with single overload

A V Eremin[1], A V Byakov[1], P S Lyubutin[1], S V Panin[1,2] and V V Titkov[1]

[1]Institute of Strength Physics and Materials Science, Tomsk, Russia
[2]National Research Tomsk Polytechnic University, Tomsk, Russia

ave@ispms.tsc.ru

Abstract. The paper represents the investigation of crack propagation in aluminium alloy AA2024 in a case of constant cyclic loading with single overload. The parameters of the process were evaluated by means of optical and noncontact digital image correlation technique which allows precise in situ measurements and provides local strain data. There have been analysed strain fields, local strains at the crack tip and crack closure level in order to reveal the effect of overload on crack growth parameters. Results are consistent with published data and developed mathematical models – the highest impact is emerged when crack extended into overload plastic zone by 30%.

1. Introduction

Numerous structures and components are subjected to cyclic loading. Fatigue is the reason for the most of the fractures and accidents in engineering, however huge amount of experimental research and theoretical modelling cover mostly constant amplitude case while the real components undergo random spectrum loading. Such non-uniform loading attracts much interest for prognosis of residual life time and detailed behaviour of the crack. Most researchers utilize conventional technique for studying crack growth parameters via crack opening displacement gauge (COD) which is mounted at the edge of the specimen according to the standard ASTM E647 "Standard Test Method for Measurement of Fatigue Crack Growth Rates". This technique is based on the compliance principle which leads to low sensitivity and thus requires several loading-unloading cycle for more precise measurements of closure level and crack length. Such drawbacks limit the application of the gauge for evaluation of dynamic processes, particularly spectrum loading where the each next cycle overlap the effect of the previous one. According to the described aspects the COD gauge require complicated procedure for precise evaluation of loading history effects [1,2].

The other widely used method is optical or scanning electron microscopy which provides the most accurate results, but they might be obtained only after fracture, not during the test. Therefore it is not possible to perform detailed study of the crack parameters evolution which is critical in a case of random non-uniform loading fatigue tests.

The new approach which has prospects to replace the COD is digital image correlation technique, which allow one to determine the displacement and strain fields on the specimen surface. By post-processing of these data it is possible to evaluate the required characteristics of crack propagation process. The fundamental capabilities of the DIC was founded by M. Sutton [3,4].

The successive results for evaluation of crack-tip fields are published in [5,6]. Such technique was successfully implemented for crack growth characterization [7] including in presence of overload [8].

The current study is continuation of DIC enhancement for fatigue crack characterization and its implementation for different materials and loading conditions. The purpose of the work is measurements of crack growth parameters (propagation rate, strains at the tip and crack closure) in aluminium alloy AA 2024 under single overload by means of digital image correlation.

2. Laboratory equipment and evaluation of crack parameters

The fatigue tests were performed using servo-hydraulic testing machine BISS Nano 15 kN. As was mentioned above higher sensitivity of measurements of crack parameters in comparison with conventional COD method could be achieved with optical measurements at high magnification. Such a technique was implemented in present study by using optical microscope equipped with c-mount CCD camera Basler piA2400-17gc. The observed region was ~2.5x2.0 mm that provides the image scale of 1.04 μm/px. But such technique has difficulties for automatization due to crack growth out of the observed region. Thus the set up contains high-precise linear actuators with step motors for smooth microscope motion in X, Y and Z directions.

In this study several major crack growth parameters have been evaluated by means of digital image correlation technique by post-processing of displacement/strain fields.

2.1. Strains at the crack tip

The crack length and coordinates of the crack tip were obtained manually directly from acquired microscopic images and on this basis crack growth rate (CGR) was evaluated. In order to evaluate the mechanical response of the materials to the applied load the program extensometer tool was used. This extensometer locates at the crack tip and provides strain values based on the displacements obtained via DIC (see figure 1) and in doing this way local strains were evaluated.

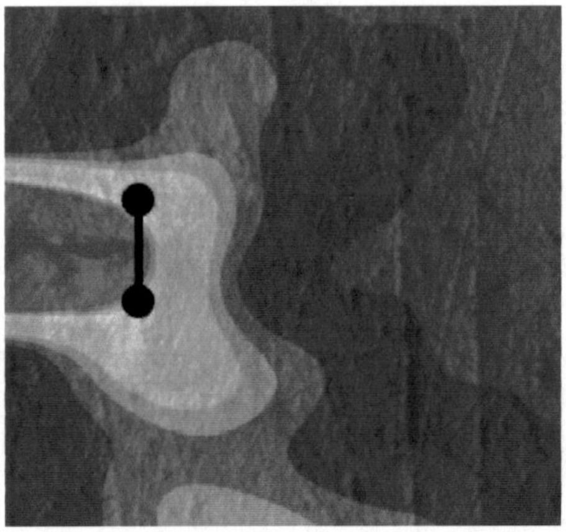

Figure 1. Position of the program extensometer on the ε_{yy} (y axis is vertical) strain field at the crack tip.

2.2. Hysteresis loops

As far as the surface images were acquired during the whole loading cycle, thus it is possible to evaluate the characteristic of deformation processes in the vicinity of the crack tip in the form of hysteresis loop presented in figure 2.

The shape of the loop characterizes nonlinear behaviour of the material at the crack tip due to local plastic deformations and crack closure – the wider ascending and descending parts, the higher level of plastic deformations. The plasticity could be numerically characterized by interior loop area which has a meaning of plastic deformation energy, while crack closure requires more complicated evaluation technique described in the next subsection.

2.3. Effective loading ratio

The value of effective loading ratio numerically describes the crack closure process. The computation procedure contains the following steps (illustrated in figure 2):

- calculation of the total loop area S;
- evaluation of loading value P_{mean} which cut the loop on two equal parts ($S_1=S_2$);
- in a case of no crack closure the loop will be symmetric and the relation
$$P_{max} - P_{mean} = P_{mean} - P_{min}$$
will be efficient, while in presence of closure $P_{max} - 2\Delta P = P_{closure} > P_{min}$ and therefore the effective loading ratio could be found from the following equation:
$$R_{eff} = 2\frac{P_{mean}}{P_{max}} - 1$$

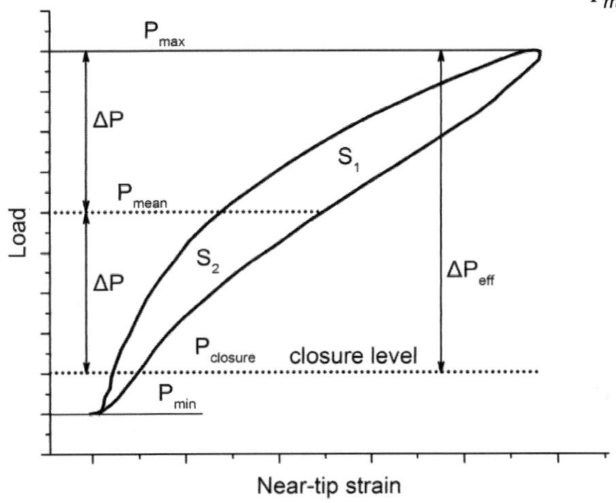

Figure 2. Mechanical response of the material in the vicinity of the crack tip to the applied loading.

3. Schematic of cyclic loading with single overload

The fatigue tests were performed for the notched specimen cut from 3 mm aluminium plate. The schematic of the loading procedure is presented in figure 3 and consists of measurement cycle before overload, overload cycle and cycles after overload at different intervals.

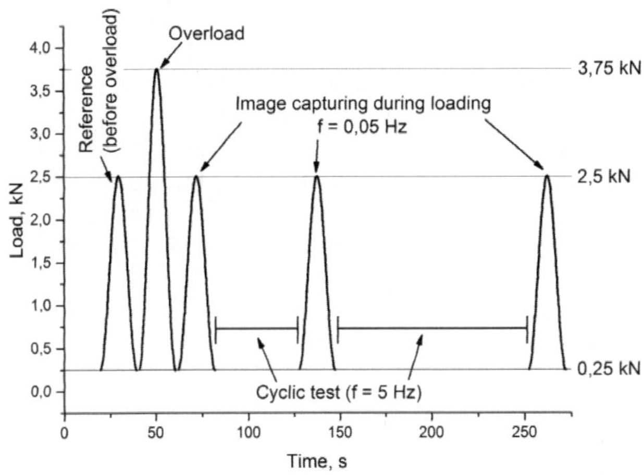

Figure 3. Schematic of the fatigue loading procedure with single overload (50% of nominal load). It consists of basic cycles (0.25-2.5 kN) – some of them had reduced cyclic frequency during which image acquisition has been performed and single overload (0.25-3.75 kN).

According to linear-elastic fracture mechanics the stress intensity factors at the beginning of the fatigue tests were the following: K_{max}=12.9 MPa·√m, while K_{OvL}=19.3 MPa·√m that corresponds to the plastic zone size for plane stress state is R_{OvL}=580 μm. Nominal load level provides the plastic zone ahead of the crack tip from 295 μm at the beginning of the test to 350 μm at the end whilst the crack extended from ~15.5 to ~16.6 mm.

4. Experimental results and discussion

The graph in figure 4 illustrates the changes of CGR due to single overload. At the beginning of the test the average growth rate was ~0.01 μm/cycle. Single overload cycle results in crack extension for 35 μm, but it could be attributed to better visualisation of the crack tip due to blunting rather than real propagation. Crack tip blunting and residual compressive stresses after overload retarded the crack growth for first several cycles, but there were removed very fast, because after 10 cycles the rate was 0.6 μm/cycle. The subsequent crack tip re-sharpness and extension in the overload plastic zone leads to the pronounced retardation of the growth rate to 0.004 μm/cycle at 30% of R_{OvL}. 30% is the critical value where the effect of overload is a maximum; further propagation leads to accelerated CGR up to 0.05 μm/cycle when crack tip completely exits the overload plastic zone.

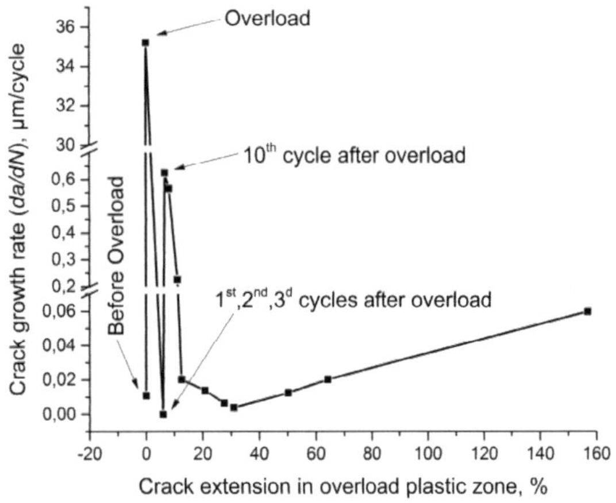

Figure 4. Crack growth rate before and after overload depending on the crack extension in the overload plastic zone.

Figure 5 shows the evolution of maximum strain value measured at the crack tip by program extensometer. Before the overload maximum strain was $4.7 \cdot 10^{-3}$, then overload increased ε_{max} to $13 \cdot 10^{-3}$ which means that crack tip was blunted due to intensive plastic flow. After overload crack tip re-sharpening occurred rather fast – after 500-1000 nominal cycles it recovers to the original values, but the sharp tip in presence of residual stresses caused by plastic deformations leads to development of crack closure and thus ε_{max} continue to reduce. When the crack extended to the overload plastic zone by 20-30% the strains at the tip start increasing.

Very important characteristic of fatigue crack propagation process is closure level, named in the manuscript efficient loading ratio. The changing of Reff is shown in figure 6. Fatigue tests were performed at the ratio of 0.1, however at the beginning of the fatigue test the closure was existed and the efficient ratio was slightly higher – Reff=0.12. Higher ratio means that real stress range at the tip is reduced. Overload results in crack blunting which keeps the tip opened, but re-sharping process occurring over the several cycles restores positive values. When the crack tip reaches 20% of overload plastic zone the closure level rises very fast to its highest value and leads to significant crack retardation. Recovering of closure level ends at 50-60% of the overload plastic zone size and remains constant; however the final value is slightly higher than the initial one.

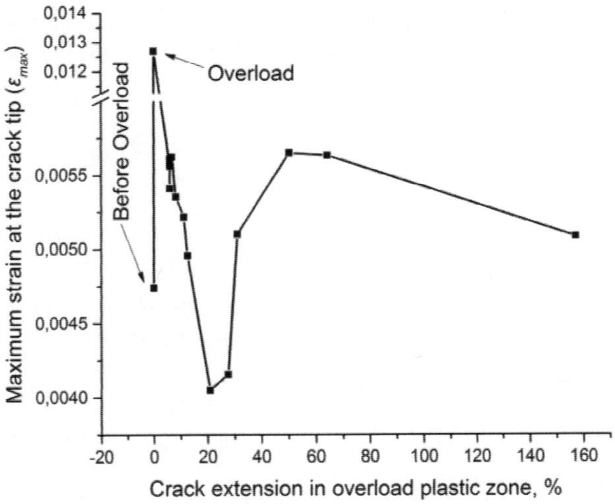

Figure 5. Maximum strain (ε_{max}) at the crack tip for measurements cycles before and after overload depending on the crack extension in the overload plastic zone.

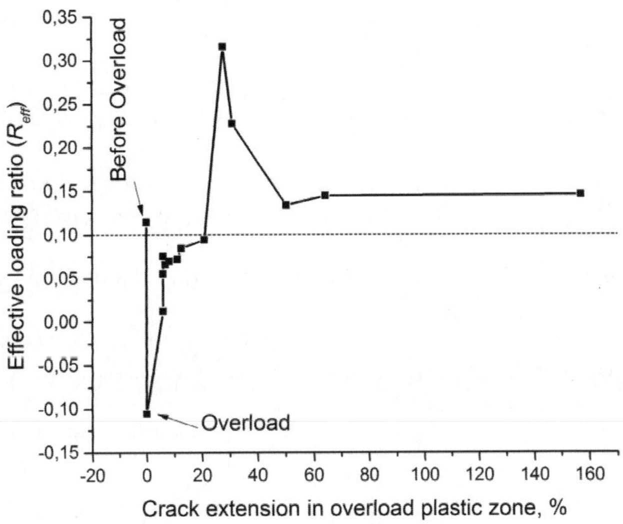

Figure 6. Effective loading ratio (R_{eff}) at measurements cycles before and after overload depending on the crack extension in the overload plastic zone.

5. Conclusion

The study demonstrates the procedure for evaluation and changes of crack growth parameters (crack growth rate *da/dN*, maximum strain at the crack tip ε_{max} and effective loading ratio R_{eff}) evaluated by means of post-processing of strain fields obtained via digital image correlation technique. It is seen that proposed technique is sensitive to the local changes in strains and closure levels.

The fatigue test of AA2024 specimens demonstrates that after overload there occur several processes – crack tip blunting which remains at the first few cycles and crack re-sharping which accelerates crack growth rate and increasing of closure level due to influence of plastically deformed area around the crack tip. Blunting was removed after several cycles and leads to increase of closure level. In order to be increased and play a significant role the latter needs extension into the overload plastic region where residual compressive stresses will keep the tip partially closed and thereby reduces stress concentration level. Thus when the crack tip extended into the overload plastic zone at 30% it leads to minimum of crack growth rate (*da/dN*=0.004 μm/cycle) and strains at the crack tip

(ε_{max}=4·10^{-3}) while loading ratio reaches its maximum (R_{eff}=0.32, when initial ratio was R=0.1). When crack achieves 60% of overload plastic zone the influence of overload is almost removes and the parameters returns to the values close to the initial.

Acknowledgements

The work was partially supported by RF President Council Grants for the support of young researchers № MK-6762.2018.9 and by RFBR 18-38-00659 mol_a.

References

[1] Salvati E, Zhang H, Fong K S, Song X and Korsunsky A M 2017 Separating plasticity-induced closure and residual stress contributions to fatigue crack retardation following an overload *J. Mech. Phys. Solids* **98** 222–35
[2] Newman J C, Walker K F and Liao M 2015 Fatigue crack growth in 7249-T76511 aluminium alloy under constant-amplitude and spectrum loading *Fatigue Fract. Eng. Mater. Struct.* **38** 528–39
[3] Sutton M A, McNeill S R, Helm J D and Chao Y J Advances in Two-Dimensional and Three-Dimensional Computer Vision *Photomechanics* (Berlin, Heidelberg: Springer Berlin Heidelberg) pp 323–72
[4] Sutton M A, Matta F, Rizos D, Ghorbani R, Rajan S, Mollenhauer D H, Schreier H W and Lasprilla A O 2017 Recent Progress in Digital Image Correlation: Background and Developments since the 2013 W M Murray Lecture *Exp. Mech.* **57** 1–30
[5] Carroll J D, Abuzaid W, Lambros J and Sehitoglu H 2013 High resolution digital image correlation measurements of strain accumulation in fatigue crack growth *Int. J. Fatigue* **57** 140–50
[6] Lopez-Crespo P, Moreno B, Lopez-Moreno A and Zapatero J 2015 Characterisation of crack-tip fields in biaxial fatigue based on high-magnification image correlation and electro-spray technique *Int. J. Fatigue* **71** 17–25
[7] Sgambitterra E, Maletta C, Furgiuele F and Sehitoglu H 2018 Fatigue crack propagation in [0 1 2] NiTi single crystal alloy *Int. J. Fatigue* **112** 9–20
[8] Wang D Q, Zhu M L, Xuan F Z and Tong J 2017 Crack tip strain evolution and crack closure during overload of a growing fatigue crack *Frat. ed Integrita Strutt.* **11** 143–8

Influence of annealing on corrosion properties of electroplated coatings

A A Kulemina, I M Kovenskiy and S S Michiy

Industrial University of Tyumen, 38 Volodarskogo st., Tyumen, 625000, Russia

E-mail: kuleminaaa@tyuiu.ru

Abstract. The paper discusses the features of the conditions of electroplating and heat treatment of traditional electroplated coatings to ensure the best anti-corrosion protection in the conditions of oil fields in Western and Eastern Siberia. The patterns of formation of the structures of the obtained coatings are given. Recommendations on the use of protective coatings are given, taking into account the climatic conditions of their operation.

1. Introduction

Modern electroplated coatings have a wide range of properties and are actively used as protective ones, including protection of oilfield equipment operating under harsh climatic conditions. Traditional methods of coating allow obtaining various types of structures after electrodeposition, which certainly affects their properties. However, the formation of the structure of coatings can be considered more widely by conducting heat treatment after electroplating.

As the object of study, coatings traditionally used as protective materials were taken: nickel, chromium, cadmium and zinc [1].

2. Material and methods of research

Coatings were obtained in factory conditions in industrial baths and from traditional electrolytes. The composition of electrolytes and the parameters of electrodeposition are shown in Table 1.

Annealing of the coatings was carried out in the argon shield at temperatures not exceeding the recrystallization temperature.

The corrosion rate was evaluated in gravimetric tests in media recommended by GOST, as well as in media consisting of produced water from various oil fields in Western and Eastern Siberia. The compositions of corrosive media are shown in Table 2.

Hardness was determined on a PMT-3M microhardness meter by the reconstructed imprint method.

The structure was investigated using optical and electron microscopy on Olympus GX51F and JEOL JSM-6510A microscopes.

X-ray diffraction analysis was performed using a DRON-7 X-ray diffractometer.

Table 1. The composition of electrolytes and modes of electrodeposition of the studied coatings.

Coating material	Electrolyte composition, g/l	Deposition modes
Cd	CdO – 20-40 NaCN - 80-130 NaOH – 20-30 $(C_6H_{10}O_5)n$ – 1-2	Temperature– 15-30 °C Cathode current density – 0.5-2 A/dm^2 + chromating
Zn	ZnO – 20-45 NaCN – 40-70 NaOH – 50-100 Na_2S – 0.5-5 $C_3H_5(OH)_3$ – 3-5	Temperature – 15-40 °C Cathode current density – 0.5-3 A/ dm^2 + chromating
Cr	CrO_3– 220-250 H_2SO_4– 2-2.5	Temperature – 40 – 60 °C Cathode current density – 20-70 A/ dm^2
Ni	$NiSO_4$ – 200-220 H_3BO_3 – 25-30 NaCl – 8-10	Temperature – 50-60 °C Cathode current density – 1.2-2.5 A/ dm^2

Table 2. Composition of corrosive media

Medium / field	pH	The content of ions, mg/l							Total mineralization	
		Cl^-	HCO_3^-	SO_4^{2-}	Ca^{2+}	Mg^{2+}	Na^++K^+	Fe_{total}	мg/l	мg-equiv/l
Kalchinskoye	8.03	5751	3416	5.6	72	7.8	4918.92	0.08	14171	419.1
Ety-Purovskoye	7	4468	549	8	276	12	2725	0	8038	258.8
Kuyumbinskoye	4.26	188150	28.06	170	42000	8640	57244.7	1.32	296234	10413.8

3. Results and discussion

The type of structure obtained after electrodeposition of coatings is significantly different. So, for example, during electrodeposition of refractory metals (chromium and nickel) with a high overvoltage of the cathode a cellular structure is formed, and with decreasing current density it may be subgrain or even monoblock.

During the deposition of coatings of relatively low-melting-point metals (zinc and cadmium), only subgrain or monoblock structures are formed, and a cellular one is not.

The corrosion rate during the "tightening" of electrodeposition modes increases. For example, when moving from a monoblock structure to a subgrain, the corrosion rate increases on average by 20 %.

Annealing of the obtained coatings leads to a transformation of the structure and a significant change in the properties of the coatings [2]. When conducting corrosion tests, the dependences of the corrosion rate on the annealing temperature were obtained (Figure 1).

It can be noted that during annealing of electroplated low-melting-point materials, such as zinc and cadmium, at temperatures not exceeding 150 °C, there are no significant changes in the corrosion rate. However, annealing at these temperatures makes it possible to stabilize the properties of the coatings and reduce the internal stresses arising during electrodeposition. Heating to temperatures higher than the homologous temperature of recrystallization significantly increases the corrosion rate, which can be explained by the appearance of secondary porosity in the coatings.

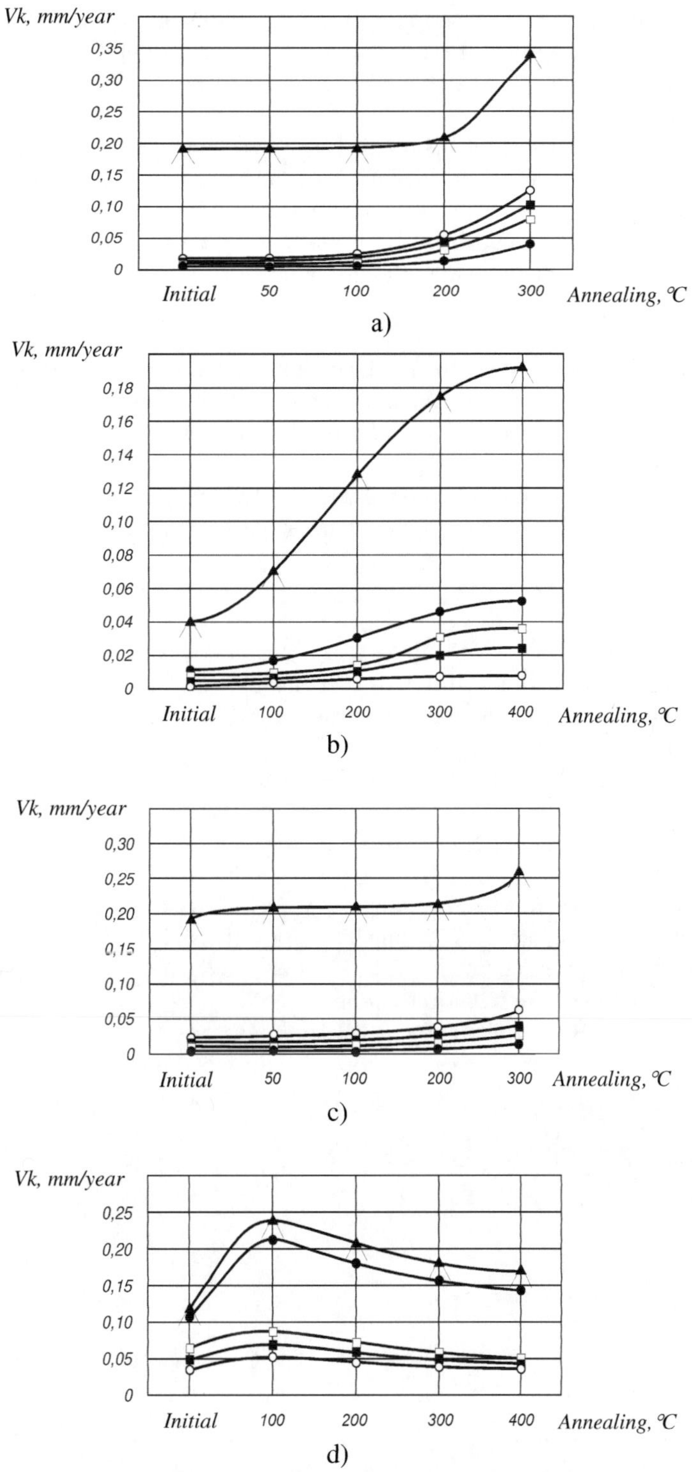

Figure 1. Change of the corrosion rate from annealing temperature
a) zinc; b) chromium; c) cadmium; d) nickel. (corrosive media: ▲ – 3%NaCl+ CH_3COOH; O – 3 % *NaCl*; ■ – Kalchinskoye field; □ – Ety-Purovskoye field; ● – Kuyumbinskoye field).

IOP Conf. Series: Materials Science and Engineering **511** (2019) 012015 doi:10.1088/1757-899X/511/1/012015

When chromium is annealed, an increase in the corrosion rate is observed at temperatures close to the recrystallization temperatures. When nickel is annealed, the maximum increase in the corrosion rate occurs at a temperature of 100 °C, which corresponds to the polygonization temperature and can be explained by the redistribution of atomic-crystal defects [3].

It should also be noted that the best results for corrosion resistance in produced water with a pH of 4.6 and total mineralization of 296.234 g / l are observed for coatings of zinc and cadmium, for coatings of nickel and chromium in these media the worst result is observed. With an increase in pH and a decrease in water salinity, the opposite result is observed: the worst value of corrosion resistance is in low-melting-point coatings, the best is in refractory ones.

4. Conclusion

A key factor affecting the corrosion rate in the produced waters of oil fields is the pH of the medium.

Taking into account the natural and climatic features of the fields, their water cut and the composition of produced water, it is possible to increase the corrosion resistance of oilfield equipment using protective coatings.

By analyzing the data obtained, it can be concluded that by changing the annealing parameters of galvanic coatings, it is possible to choose the most suitable parameters for specific operating conditions.

References

[1] Kovenskiy I M, Venediktov A N, Venediktova I A, Neupokoyeva A A and Obukhov A G 2012 Proceedings of Higher Education Institutions. Oil and Gas № 2 pp 101-107 (in Russian)
[2] Kovenskiy I M and Povetkin V V 2003 Electrolytic Alloys (Intermet Engineering, Moscow), pp 9–24 (in Russian)
[3] Kovenskiy I M, Kulemina A A 2016 IOP Conference Series: Materials Science and Engineering vol 12 1 012041
[4] Kovenskiy I M, Kulemina A A 2017 AIP Conference Proceedings vol 1886 26 September 020101
[5] Kovensky I M, Neupokoeva A A, Venediktova I A, Obukhov A G 2015 Proceedings of Higher Educational Institutions. Oil and Gas № 2 pp 97-100 (in Russian)
[6] Kovenskiy I M 2016 Heat Treatment of Electrodeposited Metals and Alloys (Science and Technology, Moscow) pp 14, 62–96 (in Russian)
[7] Abrahams M S, Rao S T, Buiocchi C J 1986 J. Electrochem. Soc. vol 133 №9 pp 1786-1791
[8] Katz J D, Pickering H W, Bitler W R 1980 Plating and Surface Finishing vol 67 №11 pp 45-49

Acknowledgments
This work was financially supported by Transneft.

Investigation structure and properties of heterogeneous materials based on submicron powders of boron carbide, produced by hot-pressing

A A Filippov[1], V A Kuznecov[2], E E Kornienko[2], V A Bezrukova[2]

[1]Khristianovich Institute of Theoretical and Applied Mechanics of the Siberian Branch of the Russian Academy of Sciences, Institutskaya 4/1, Novosibirsk 630090, Russia
[2]Novosibirsk State Technical University, Karl Marx Avenue 20, Novosibirsk 630092, Russia

E-mail: filippov@itam.nsc.ru

Abstract. This paper discussed the microstructure and mechanical properties of ceramic tablets obtained from mixtures of boron carbide-based powders with various submicron additives. As the main powder for the study was used crystalline powder of boron carbide brand F400 (PJSC "Zaporizhia Abrasive Plant") with an average size of 20 microns. Submicron powders of boron carbide, boron nitride and carbon nanotubes were used as additives. It is shown that with the hot pressing parameters used, active sintering does not occur. Sintered tablets based on boron carbide F400 with additions of 10% by weight have the same hardness as tablets obtained from submicron powders. These additives could be perspective for further applications.

1. Introduction

For ceramic powders with a high melting point, the use of cold pressing is practically not feasible due to the extremely low rate of diffusion flowing in the substance. The process of hot pressing allows producing sintered compacts from the powder. Mechanical properties of these compacts can be described by the behavior of a continuous heterogeneous deformable solid. With a decrease in the dispersion of the powder, diffusion processes are significantly accelerated, which leads to an increase in both the sintering rate and the strength of the sintered material [1, 2]. At the same time, reducing the dispersion makes it difficult to predict the form and size of the resulting product, since the amount of shrinkage becomes significant. Production of dense products from boron carbide-based materials with high values of hardness and strength is possible when sintering temperature is about 0.9 of melting temperature and pressure reached above 60 MPa. These technical requirements significantly limit the product range and the volume of production of superhard ceramics. Researchers are trying to decrease the temperature and sintering pressure by applying novel methods of compaction, mechanical activation, alloying additives [3-5]. The purpose of this work is to obtain the dependence of the hardness of hot-pressed tablets based on boron carbide on the composition of the alloying additives. Selection of the additive and the method of consolidation are generally dictated by the end use of the product and the properties that are required. The additive by itself or due to in situ reaction with boron carbide would form a non-volatile second phase aiding in densification and property enhancement.

2. Materials

As the base powder for this study were used crystalline powder of boron carbide brand F400 and F1200 (PJSC "Zaporizhia Abrasive Plant") with an average size of 20 and 3 microns respectively. Submicron powders of boron carbide, boron nitride and carbon nanotubes were used as additives (table 1). Obtaining submicron powders of boron carbide was carried out by reducing boric acid with carbon. Carbon nanotubes (CNT) participated in the interaction with boric acid, and boron oxides are remnants of unreacted boric acid. Grinding and mixing of the powders after preparation of the mixtures was carried out in a ball mill for one hour. Mixing with additives of boron nitride was carried out after the synthesis of boron carbide by weight.

Table1. Powder mixtures for hot pressing.

№	Powder name	Mechanical activation	Boron nitride	Submicron boron carbide	Submicron boron nitride	Carbon nanotube
1	F400	–	–	–	–	–
2	F400_MA	+	–	–	–	–
3	F400+10%BN(K)_MA	+	+	–	+	–
4	F400+10%BN(K)+1%CNT	–	+	–	+	+
5	F400+10%(B4C,B2O,BO,C)	–	–	+	–	–
6	F1200	+	–	+	–	–

3. Methods

For uniaxial hot pressing of tablets, a compact automatic laboratory hot press was used. This is periodic action set-up with indirect resistance heating, bottom loading elevator-type and autonomous water cooling system of closed type. The internal camera in the hot zone filled an inert gas (argon). Heater and insulation made of carbon materials and the possibility of manual and automated controls. As the mold was used to press has the cylindrical form with outer and an inner diameter of 40 mm and 16 mm respectively. As a result, samples were obtained in the form of tablets (cylinders) diameter 15.8-16.0 mm and a height of 2 to 3 mm. The walls were built of matrix graphite paper to extend the life of the graphite mold. The sintering parameters were chosen as low as possible to produce solid tablets. Heating of powder mixtures in a graphite mold was carried out at a rate of 50 ° C/min in argon atmosphere. A powder weighing 2 g was poured into a graphite matrix with a diameter of 16 mm. The maximum heating temperature was 1700° C, the pressure was –25 MPa. Hot-pressing at maximum parameters was carried out during 15 min. These parameters are reliable to early stage of sintering (solid phase sintering) of the powder when solid particles of boron carbide powder are sintered at the points of contact between the particles. Low rates of powder sintering are necessary to control the potential reactions of additives with the powder and the material of the mold, as well as to extend the service life of the mold. Low rates of powder sintering are necessary to control the potential exothermic reactions of additives with the powder and the material of the mold, ensure durability of the mold. The resulting tablets are easily removed from the graphite mold, able to retain their shape.

4. Microstructure

The cross-section of produced tablets was polished for further analysis using a Zeiss EVO MA 15 scanning electron microscope (SEM) and an Oxford Instruments INCA X-Max energy dispersive analyzer (EDX) (Table 2). It is shown there are many pores and cracks in all tablets. If there are impurities in the tablet, the pores move to the impurities; for materials without impurities (F400 and F1200), the pores begin to dissolve in the tablet material. The sizes of cracks and pores for samples with additives are approximately the same and range from 1 to 20 microns. Grains of boron carbide do not completely sinter and there are iron additives between grains (figure 2a, c, d, and e). EDX-analysis evidenced presence of iron in all tablets and nitrides in tablets with boron nitrides (table 2). Besides in samples with boron nitride high oxygen concentration detected. This is due to the additives are obtained by reducing oxygen from boron oxides and nitrogen. Mixing various boron carbide powders and mechanical activation in a ball mill leads to an elevated iron content in sintered tablets of up to 3 percent by weight. Other impurities were less than 0.5% wt.

Table 2. EDX-analysis of hot-pressed tablets.

Mixture name	F400	F400MA	F400+10%B_2O_5,B_4C, BO,C	F400+10%BN	F400+10%BN+1%CNT	F1200
Elements						
B	73,95	68,22	67,82	70,75	73,77	77,62
C	22,97	29,12	26,44	18,12	18,81	20,73
O	0,82	0,63	2,73	3,52	2,28	0,74
Fe	0,06	1,69	2,51	3,05	1,98	0,10
Al	0,11	0,15	0,11	0,21	0,10	0,05
Ca	0,30	0,14	–	0,26	0,30	–
Cr	0,07	–	0,13	0,14	0,09	–
Si	0,09	0,05	0,27	0,32	0,26	0,76
N	–	–	–	3,62	2,40	–

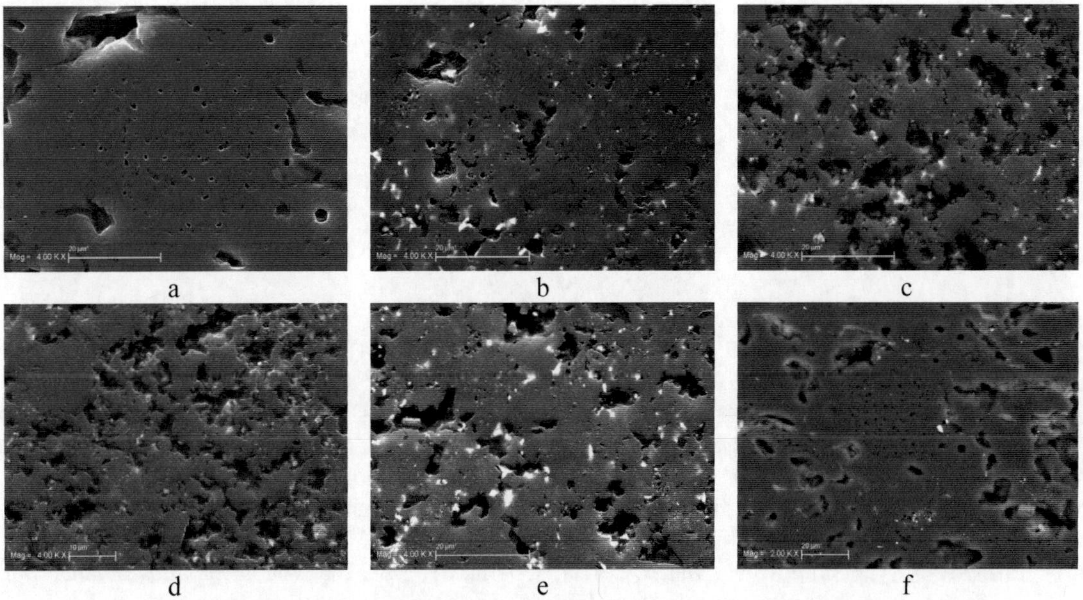

Figure 1. SEM-images of cross-sections hot-pressed tablets with magnification X4000: a - F-400; b - F-400+MA; c - F400+B_2O_5; d - F400+BN; e - F400+BN+CNT; f - F1200.

5. Microhardness

The study of microhardness was carried out on a Wilson Hardness Group Tukon1102 microhardness tester with a load of 1 kg (figure 2). The last column is the hardness of boron carbide sintered at pressure 90 MPa and temperature 2350 °C. Powders F400 and F400+MA crashed under indenter loading. Boron carbide grains split off from sample therefore hardness values could not be set correctly. Other tablets demonstrate similar hardness values slightly smaller as compared to fine powder (F1200). Submicron additives could work as sintering process accelerators at the given hot-pressing parameters. The rise of microhardness is attributed to the presence of oxide layer which helps in precipitation of B_4C through liquid B_2O_3.These data are confirmed by other researchers [5].

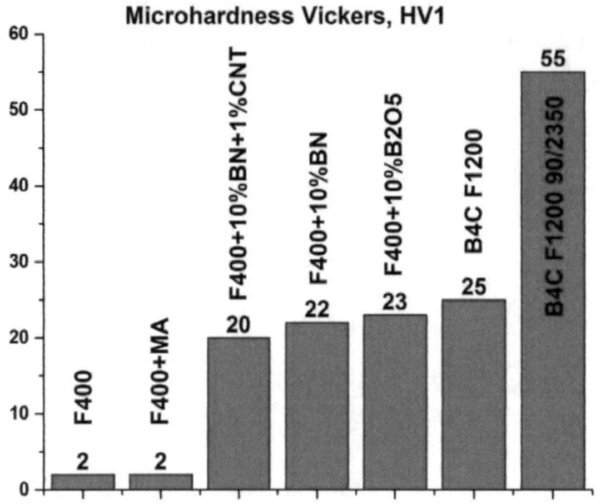

Figure 2. Micro hardness of tablets.

The obtained tablets exhibit loose structure, high porosity and a large amount of iron in the samples, which is directly related to insufficient parameters of temperature and pressure. The hardness of the obtained tablets corresponds to the hardness of boron carbide with small impurities (up to 4% by weight) of pure boron or carbon [5]. Nevertheless, the mechanical activation and the addition of carbon in the form of nanotubes ensure the activation of sintering processes, and could be perspective in feather applications such as selective laser sintering [6].

Acknowledgments
This work is supported by basic research program of the Russian Foundation of Basic Research (grant No. 18-41-543013)

References

[1] Gursoy A, Kalemtas A 2009 *J European Ceramic Society* **29** 473-480
[2] Zhang G J, Ando M, Yang J F, Ohji T, Kanzaki S 2004 *J European Ceramic Society* **24** 171-178
[3] Khasanov O L, Dvilis E S, Khasanov A O, Bikbaeva Z G, Polisadova V V, Milovanova T V 2014 *Advanced Materials Research* **872** 45-51
[4] Roy K, Subramanian C, Suri A K 2006 *Ceramics International* **32** 227-233
[5] Suri A K A. K. Suri, Subramanian C, Sonber J K and Murthy T S R Ch 2010 *International Materials Reviews* **55** 4-40
[6] Filippov A A, Fomin V M, Malikov A G, Orishich A M *2017 Nanomechanics Science and Technology: An International Journal* **8** 1–12

Investigation of the resource potential of the aluminum oxide waste formed during testing and disposal of energy-saturated materials

A N Ponik[1], N Y Karpova[1] and G M Batrakova[2]

[1]Joint-stock company "Research Institute of Polymeric Materials", JSC "RIPM", Russia, 614101, Perm, Chistopolskaya street, 16,
[2]Perm National Research Polytechnic University, "PNRPU", Russia, 614990, Perm Region, Perm, Komsomolsky Prospect, 29

E-mail: karpowa.nadezhda2010@yandex.ru

Abstract. Theoretical and experimental investigations of aluminum-containing waste formed during testing and disposal of energy-saturated materials are carried out. Conversion directions of using aluminum oxide are considered. The research results allow us to recommend cleaning products for dust and gas emissions of thermal destruction of energy-saturated materials for use as additives in various types of civilian products.

Introduction

Environmental safety control in the field of waste handling generated in the process of testing and disposal of products of energy-saturated materials is based on the principle of hazardous properties identification and introducing a set of measures aimed at reducing the amount of non-recyclable wastes and placing in the environment only non-recyclable waste residues, identifying the resource potential of waste and its resource-saving.

The priority directions of the state policy in the field of waste management are the maximum use of raw materials; prevention and reduction of waste formation; reduction of the hazard class of waste in the sources of their formation, as well as the introduction of the best available technologies for treatment, disposal and neutralization of waste. The stated directions (changes from December 29, 2014 458-FZ "On Production and Consumption Wastes") are intended to solve technological, economic, social and organizational issues of low-waste technologies creation in a complex, to combine environmental and economic interests of society in order to ensure sustainable development.

The known mechanisms for reducing the formation and hazard of industrial waste are:

- improvement of technological processes and minimization of waste formation;
- changing the properties of the waste, above all, toxicity, reducing the hazard of subsequent stages of waste management - the processes of temporary storage, transportation, processing, location in the environment (disposal);
- usage of the resource potential of waste components, processing and preparation of market products [1-2].

During one of the technological processes associated with the development, testing and destruction of high-energy systems, large quantities of waste are formed with a high content of aluminum oxide

(Al_2O_3). Waste formed during the process (Figure 1) - alkali metal salts and aluminum oxide can be partially disposed. At present, up to 200 t/year of waste is accumulated at production sites.

The biotesting method was used to investigate waste formed during testing and disposal of energy-saturated materials, and environmental hazard class 4 (low hazardous waste) was established. A certificate was received for this waste and the waste was included in FKKO No. 76511411203 [3, 4].

Figure 1. Waste from the gas cleaning system of the test stand.

Industrial wastes with a high content of aluminum oxide are considered as high-quality technogenic mineral raw materials with potential for recycling. Taking into account the possibility of release from impurities, dispersed and phase compositions, the directions of use and preparation of market raw materials and products were determined.

To determine the method of using aluminum-containing waste, the theoretical and experimental investigations are carried out.

The low toxic hazard of waste formed on the stand makes it possible to assess its applicability to the production of civilian products.

The following conversion directions for the use of aluminum-containing waste are proposed:

I. As an additive to materials for the soil stabilization in the construction of the subgrade and layers of pavement.

II. As a filler mixed with fireclay sand for the manufacture of refractory materials;

III. As the strengthening additive in cement mixtures, heat-resistant solutions (concrete) and in fired rock materials;

IV. As an additive for fired rock materials (technical ceramics).

Experiment

In the first direction, investigations were conducted on the use of aluminum oxide as an additive to materials for soil stabilization in the construction of the subgrade and layers of pavement, and the optimal mixing proportions were determined to give the greatest strength in the following ratio:

- portland cement (grade 300) - 1 part;
- sand - 3 parts;
- aluminum oxide - 2 parts;
- water - 1 part.

This cement-soil mixture was introduced and used in pilot production for the repair of the pavement base. However, the use of such valuable raw materials as a filler in pavements is unpractical.

Therefore, together with the «Orgtekhstroy» trust, experimental research work was carried out on the use of recyclable Al_2O_3 as an additive in building materials in order to increase their density, water impermeability and frost resistance.

According to the second direction of aluminum oxide usage, the effect of Al_2O_3 on the strength of the hardened composition was evaluated according to the following attributes:

- according to the amount of Al_2O_3 used, depending on the cement consumption;
- according to the influence of the water-cement ratio (W/C) and the heat-humidity conditions (HHC) on the sample structure.

The compositions were formed from a mixture of portland cement mark 300 and river quartz sand with the addition of Al_2O_3. The consumption of aluminum oxide ranged from 0.125 to 30.0 % by weight of the binder (water / cement (hereinafter referred to as W / C) - 0.5 and 1.0). (Figure 2).

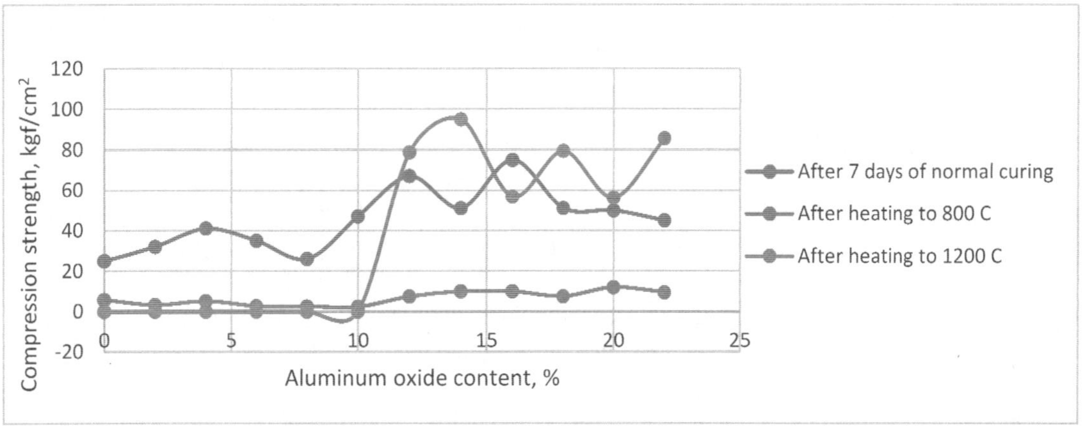

Figure 2. The test results of heat-resistant solutions with the addition from 5 to 22 % of aluminum oxide at the different firing temperatures.

The test results of the investigated compositions showed the following:

1. Cement mixtures with natural hardening for 28 days (standard term) and the W/C ratio about 0.5 (dry mixtures compacted by pressing) gave the strength increase from 37 to 64 %;
2. In the compositions with the W/C ratio about 1.0, a stable increase in strength was not obtained;
3. With the natural keeping of the samples for more than 28 days with the ratio W / C = 0.5, the strength increase is 29 % (with the consumption of aluminum oxide - 10%).

In the third direction, the effectiveness of using aluminum oxide as the refractory additive was studied on the formed concrete mixture:

- fireclay powder - 350-450 kg/m^3;
- portland cement (grade 300) - 350-450 kg/m^3;
- aluminum oxide - 5-55 % by weight of cement;
- water - 40-60 % by weight of cement.

The formed mixtures were subjected to firing at the temperature of 1200 °C; at tests of samples, the following results are received:

- the best performance achieved with the content of 35 % Al_2O_3;
- strength increase with respect to the control sample during firing up to 800 °C was 68.9%.

In the fourth direction of the aluminum oxide application for fired rock materials (technical ceramics) as additives were used: spent catalyst IM 2201, caustic magnesite, perlite sand, and fireclay sand. (Figure 3).

The total amount of refractory additive together with aluminum oxide in all compositions was 16.66 %, while the content of the other components was 83.34%. This ratio was maintained in all compositions. Before firing, all the compositions pressed in a punch with diameter of 50 mm under loads up to 411 kgf/cm^2 and fired. The firing mode was carried out in two stages: firing in a muffle furnace with the temperature rise to 800 and 1200 °C.

A comparative analysis of the results shows that the optimal amount of aluminum oxide content is from 8.3 to 13.3 %. At the same time, on the fireclay sand with the aluminum oxide content of 10.83 %, the maximum compression strength is 629.0 kgf/cm^2, and the strength increase to the control sample was 201 %. The strength increase with respect to the control sample during firing to 800 °C was 68.9 % [5].

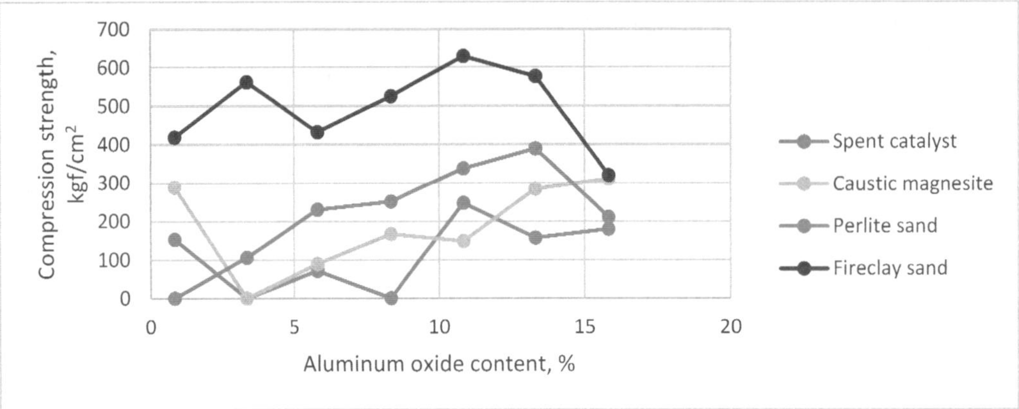

Figure 3. The test results of artificial fired rock materials based on the Chusovaya's slag, aluminum oxide and additives.

Summary

According to the results of exploratory research, we can distinguish rational areas of use for cleaning products of dust and gas emissions of thermal destruction of energy-saturated materials containing aluminum oxide in the production of building materials, namely, their use as a plasticizing additive to cement systems in the production of ready-mixed concrete and mortar mixes, as well as filler for dry construction mixtures. Also, taking into account the experience of handling this kind of wastes, we can assume a positive effect from their use in the production of high-temperature composites - in ceramic and refractory materials of general construction and special purposes.

Thus, the results of investigations in the above four directions allow us to recommend the cleaning products for dust and gas emissions of thermal destruction of EM for use as additives in the following types of civilian products:

- Refractory materials;
- Cement mixtures;
- Heat-resistant solutions (concretes) and fired rock materials (technical ceramics).

References.

[1] Sokolovsky M I and Vaisman Y I 2009 *Technical and environmental aspects of the disposal of intercontinental ballistic missiles* (Perm: Perm. St. Tech. University) p 636

[2] Gorbachev V A, Naumov S P, Ubey-Volk E Yu 2014 *Ecological safety at disposal (liquidation) of solid-propellant rocket motor* (Krasnoarmeysk: FKP 2 NII Geodesy Collected works of the scientific-practical conference Ensuring industrial and environmental safety in explosive productions of the military-industrial complex *Industrial Safety - 2014*) pp 351-358

[3] Criteria for classifying waste to the I-V hazard classes according to the degree of negative impact on the environment (Russia: Order of the Ministry of Natural Resources of Russia of 04.12.2014. No 536)

[4] Methods for determining the toxicity of water extracts from the soils, sewage sludge and waste, drinking, waste and natural water from the mortality of the test object Daphnia Magna Straus 2005 (Federal Service for Environmental, Technological and Nuclear Supervision) pp 1-46.

[5] The report Introduction of cement-saving technologies, including the work on the use of powdered production waste - aluminum oxide in building materials 1990 (Perm: MINURALSTROY RSFSR TSO Zapaduralstroy Orgtekhstroy Trust) pp 1-17

Measurement Affecting Errors in Digital Image Correlation

V V Titkov[1,a], **S V Panin**[1,2,b]

[1]Institute of Strength Physics and Material Science SB RAS, Tomsk, Russia, ispms.ru
[2]National research Tomsk polytechnic university, Tomsk, Russia, tpu.ru

[a]titkov.vladimir@gmail.com, [b]svp@ispms.ru

Abstract. Optical full-field measurement methods such as Digital Image Correlation (DIC) are increasingly used in the area of experimental mechanics. The reliability for each measurement technique depends on the knowledge of its uncertainty and the sources of errors of the results. The aim of this work is to systemize the sources of errors related to digital image correlation (DIC) technique applied to strain measurements. The paper is concluded by some suggestions proposed in order to minimize the errors.

1. Introduction

Nowadays Digital Image Correlation (DIC) is widely spread optical technique for strain measurement. The technique was proposed by Peters et al. in the 1980s [1, 2]. It allows to carry our investigation of deformation and fracture processes of heterogeneous materials (metals, alloys, ceramics, polymers etc.) as well as makes it possible estimating mechanical state of loaded machine parts and structure components. Application of this method benefits on high resolution and sensitivity, and provides reducing manufacturing and operational costs of the instrumentation. The strain measurement in the DIC technique consists of two principal stages: determination of optical flow and subsequent computation of deformation components.

As a measurement method, the accuracy of DIC at strain calculation is of crucial importance. There are many error sources involved in the DIC measurement procedure, either originating from the experimental set-up or from the employed correlation algorithm, which bring in difficulties for the uncertainty quantification [3].

2. Errors affecting DIC measurements

It is of importance to distinguish between systematic error (displacement of the average, resulting in lack of accuracy) and random error (large standard deviation, resulting in lack of correctness/robustness) [4]. In fact, accuracy and robustness of the DIC measurements cannot be taken for granted if the measuring system and the numerical processing have not been optimized and validated [4]. The displacement and strain affecting errors are originally induced by the overall quality of the raw experimentally gained images. A DIC-computed displacement field is less sensitive to variation of calculation parameters (specified by the user), while their effect is larger computation the strain [5, 6]. Palanca M. showed [6] the importance of a careful optimization of the DIC software and hardware settings to minimize random and systematic errors. The settings that allowed minimizing the random errors were also associated with averaging over a larger area, and were therefore associated with poorer spatial resolution. Accuracy and robustness of the DIC-system in computing the displacements are in the order of 0.01 pixel [5, 7, 8]; however with some optimizations errors can be further reduced [9]. The discrepancies between DIC evaluated displacements and specified ones have been statistically analyzed [8] in terms of random errors and systematic ones; in doing so they were correlated with the fractional part of the displacement component expressed in pixels. Main results are as follows: calculated displacement amplitude is almost insensitive to subset size, standard deviation of random error increases with enlarging noise level and decreases with enhancing subset size. However, the DIC formulations can be split up into two main families regarding displacement sensitivity to noise. For the first one, the amplitude of the latter increases with noising while it remains nearly constant against the subset size. In addition, for the first family, a strong dependence of random error with τ is observed for noisy images. DIC-computed strain values are generally quite accurate (systematic errors is of the order of few microstrains). Conversely, large noise usually largely affects DIC-computed strains: an accuracy of some hundreds of unit can be achieved only under optimal conditions [4].

The DIC analysis relies on the presence of a suitable pattern on the specimen surface [4]. In order to evaluate the errors related to its morphology, digital images of the speckle patterns were varied [10], numerically deformed [11], correlated in a zero-strain condition [12]. Inappropriately chosen speckle pattern is likely to make the correlation impossible at some patches (square–shape image regions), reducing the number of measurements points [10]. It is shown that the size of the speckles combined with the size of the used pixel subset clearly influences the accuracy of the measured displacements [11]. An optimal ratio exists between the patch size and the mean speckle size to reduce errors affecting DIC-computed displacements [11, 13]. They also showed that a limited scatter of speckle sizes yields more accurate displacement measurements, and that larger size dots ensures larger random errors in the constructed displacement field. The differences between black-on-white and white-on-black speckle patterns are negligible in terms of measurement quality [9]. A clear relationship exists between the measurement error and the uniqueness of the pattern, which depends on the speckle size and shape, and on the patch size [14]. Errors reduced with increasing number of speckles in the pattern, with enlarging speckle size providing a greater variation of shape, resulting in a lower error than patterns with smaller speckles. The airbrush airgun method provides a better detectability of the dots dimension as compared to the use of powder speckle [15]. Even if at a limited extent an airbrush airgun can be adjusted to produce the desired speckle dots [14], the performance of DIC stands quite robust and stable [16].

Images acquired by the digital camera are subjected to random errors affect, being induced by thermal noise (or dark noise), excess noise due to the CCD-sensor and electromagnetic noise of the relative measurement channel [17]. Moreover, a source of systematic error in 2D-DIC derives from out-of-plane displacements of the specimen during loading. That is why 2D-DIC is often chosen in investigations at the microscopic level [18, 19]. In practical applications, various unavoidable disadvantageous factors, such as small out-of plane displacement of the test object surface occurred during loading, small out-of-plane motion of the sensor target due to the self-heating or temperature variation of a camera, and geometric distortion of the imaging lens, may seriously impair or slightly change the originally assumed linear correspondence. In certain cases, these disadvantages may lead to significant errors in displacements and strains measured by 2D-DIC [19]. Suboptimal choice of the calculation parameters (specified by the user) can result in higher noise, or, conversely, could hide existing strain gradients [20]. The optimal parameters can be identified through virtually imposed displacement [10]. Numerically "deformed" images were prepared to evaluate the accuracy and robustness at constructing the displacement field, and identify the optimal parameters [16, 21, 22, 23].

3. Conclusions and recommendations to minimize measurement errors

It is required to validate DIC measurements by their comparison against independent measurements; in [7, 24, 25] the DIC-computed strains were compared against single strain gauges. A more extensive validation may include the use of specimens with known material properties, subjected to well-defined loading conditions [7, 24, 25]. Moreover, preliminary tests to identify the spatial displacements could help in avoiding out-of-plane artifacts in 2D-DIC measurements.

The quality of the applied random speckle pattern is of crucial importance. It exerts some influence on the correlation results [10]. A low quality speckle pattern results in no correlation at some mesh nodes of the virtual grid. It also exerts certain influence on the accuracy at calculating correlation function. Using an optimized speckle pattern that can be printed directly on the surface of the sample is an efficient solution. Thus, the obtained speckle pattern will be independent of the operator and; hence the results would be more reproducible. To optimize the speckle pattern a factorial design to adjust the airbrush settings is proposed to form a pattern having the required average speckle element size with minimal scatter [7, 26, 27]. Paper [26] shows that it is possible to obtain a pattern with a highly controllable average and a limited scatter of speckle sizes, so as to match the ideal distribution of speckle sizes for DIC. Although the settings identified here apply only to the specific equipment being used; however this method can be adapted to any airbrush to produce a desired speckle pattern.

The lens distortion is responsible for a systematic error, which can be partially compensated through dedicated algorithms [28], or an appropriate calibration [29, 30]. A method of lens distortion correction was proposed [28] in order to improve the measurement accuracy of digital image correlation for 2D displacement measurement. The amounts of lens distortion are evaluated from displacement distributions obtained in a rigid body in-plane translation or rotation test. After detecting the lens distortion, its coefficient is determined using the method of the least squares. Then, the corrected displacement distributions are obtained. Such artefacts can be completely eliminated with telecentric lenses [19], or by exploiting the central portion of the lens angle [7]. An in-house smart solution consists in performing 3D deformation measurements with a single camera using a biprism to avoid distortion of the images [31, 32].

As shown by the test, the errors related to the variation of the lighting source may be of importance [10]. The illumination must be stable and uniform to reduce the noise and register better quality raw experimental images. Noise and its influence can be somehow reduced, but not completely eliminated, even with high-performance hardware (i.e. lenses, cameras, frame grabber, etc.) [4].

The main destination of the subset is to characterize the similarity between two patches while it depends on the speckle pattern [10]. When a subset size is too small it is likely to find wrong correlation due to the increase in the number of local minima of the correlation coefficient distribution. The increase of the subset size enhances the correlation. The increase of the subset size over a given dimension does not present a tangible enhancement of measurements quality. One should keep in mind that the increase of the subset size is synonymous to the increase of computation time. So the subset size should cover a given range and one may try several sizes for the subset to find its appropriate value that presents a good compromise between accuracy and computational time. It should be noticed that the subset size is closely linked to the dimension of the speckle pattern [10].

Filtering can also help reducing the noise in the DIC-computed strains. However, filtering should be used with extreme caution to avoid loss of information in high-gradient regions [33, 34]. A careful optimization of the entire measurement chain can reduce the errors and provide more accurate and precise outputs [7]. DIC is aimed to measuring displacements with very high accuracy and robustness. However, to obtain accurate and precise strain measurement data, it is required to take care of surface preparation, as well as finding appropriate hardware and software settings [4].

Acknowledgments
This work was supported by Russian Science Foundation under grant № 18-71-00087.

References

[1] Peters W H and Ranson W F 1982 Digital imaging techniques in experimental stress analysis *Opt. Eng.* **21** 213427
[2] Sutton M A, Wolters W J, Peters W H, Ranson W F and McNeill S R 1983 Determination of displacements using an improved digital correlation metho. *Image Vis. Comp.* **1** 133–39
[3] Wang Y, Lava P, Reu P and Debruyne D 2015 Theoretical Analysis on the Measurement Errors of Local 2D DIC: Part I Temporal and Spatial Uncertainty Quantification of Displacement Measurements *Strain* **52(2)** 110–28
[4] Palanca M, Tozzi G and Cristofolini L 2015 The use of digital image correlation in the

biomechanical area: a review. *Int. Biom.* **3(1)** 1–21

[5] Nicolella D P, Nicholls A E, Lankford J and Davy D T 2001 Machine vision photogrammetry: a technique for measurement of microstructural strain in cortical bone *J. of Biom.* **34(1)** 135–39

[6] Palanca M, Brugo T M and Cristofolini L 2015 Use of digital image correlation to investigate the biomechanics of the vertebra. *J. of mech. in med. and biol.* **15(02)** 1540004

[7] Zhang D and Arola D D 2004 Applications of digital image correlation to biological tissues *J. of Biom. Optics.* **9(4)** 691

[8] Amiot F, Bornert M, Doumalin P, Dupré J-C, Fazzini M, Orteu J-J, Poilâne C, Robert L, Rotinat R, Toussaint E, Wattrisse B and Wienin J S 2013 Assessment of Digital Image Correlation Measurement Accuracy in the Ultimate Error Regime: Main Results of a Collaborative Benchmark *Strain* **49(6)** 483–496

[9] Barranger Y, Doumalin P, Dupré J C and Germaneau A 2010 Digital Image Correlation accuracy: influence of kind of speckle and recording setup *EPJ Web of Conf.* **6** 31002

[10] Haddadi H and Belhabib S 2008 Use of rigid-body motion for the investigation and estimation of the measurement errors related to digital image correlation technique. *Opt. and Las. in Eng.* **46(2)** 185–96

[11] Lecompte D, Smits A, Bossuyt S, Sol H, Vantomme J, Van Hemelrijck D and Habraken A M 2006 Quality assessment of speckle patterns for digital image correlation *Opt. and Las. in Eng.* **44(11)** 1132–45

[12] Carriero A, Abela L, Pitsillides A A and Shefelbine S J 2014 Ex vivo determination of bone tissue strains for an in vivo mouse tibial loading model *J. of Biom.* **47(10)** 2490–97

[13] Lecompte D, Bossuyt S, Cooreman S, Sol H and Vantomme J 2007 Study and generation of optimal speckle patternsfor DIC *Pr. of the Ann. Conf. and Exp. on Exp. and Appl. Mech.* **3** 1643–49

[14] Crammond G, Boyd S W and Dulieu-Barton J M 2013 Speckle pattern quality assessment for digital image correlation *Opt. and Las. in Eng.* **51(12)** 1368–78

[15] Myers K M, Coudrillier B, Boyce B L and Nguyen T D 2010 The inflation response of the posterior bovine sclera *Acta Biom.* **6(11)** 4327–35.

[16] Wang Y, Lava P, Coppieters S, De Strycker M, Van Houtte P and Debruyne D 2012 Investigation of the uncertainty of dic under heterogeneous strain states with numerical tests *Strain* **48(6)** 453–62

[17] Freddi A, Olmi G and Cristofolini L 2015 Experimental stress analysis for materials and structures: stress analysis models for developing design methodologies Springer International Publishing Switzerland p 498

[18] Sutton M A, Yan J H, Tiwari V, Schreier H W and Orteu J-J 2008 The effect of out-of-plane motion on 2D and 3D digital image correlation measurements *Opt. and Las. in Eng.* **46(10)** 746–57

[19] Pan B, Yu L and Wu D 2013 High-Accuracy 2D Digital Image Correlation Measurements with Bilateral Telecentric Lenses: Error Analysis and Experimental Verification *Exp. Mech.* **53(9)** 1719–33

[20] Baldoni J, Lionello G, Zama F and Cristofolini L 2015 Comparison of different strategies to reduce noise in strain measurements with digital image correlation *J. of Strain Anal. for Eng. Design* **51(6)** 416–30

[21] Lava P, Cooreman S, Coppieters S, De Strycker M and Debruyne D 2009 Assessment of measuring errors in DIC using deformation fields generated by plastic FEA *Opt. and Las. in Eng.* **47(7-8)** 747–53

[22] Lava P, Cooreman S and Debruyne D 2010 Study of systematic errors in strain fields obtained via DIC using heterogeneous deformation generated by plastic FEA *Opt. and Las. in Eng.* **48(4)** 457–68

[23] Lava P, Coppieters S, Wang Y, Van Houtte P and Debruyne D 2011 Error estimation in measuring strain fields with DIC on planar sheet metal specimens with a non-perpendicular camera

alignment *Opt. and Las. in Eng.* **49(1)** 57–65

[24] Sutton M A, Ke X, Lessner S M, Goldbach M, Yost M, Zhao F and Schreier H W 2008 Strain field measurements on mouse carotid arteries using microscopic three-dimensional digital image correlation *J. of Biom. Mat. Res. Part A* **86A(2)** 569–569

[25] Gilchrist S, Guy P and Cripton P A 2013 Development of an Inertia-Driven Model of Sideways Fall for Detailed Study of Femur Fracture Mechanics *J of Biom. Eng.* **135(12)** 121001

[26] Lionello G and Cristofolini L 2014 A practical approach to optimizing the preparation of speckle patterns for digital-image correlation *Meas. Science and Techn.* **25(10)** 107001

[27] Lionello G, Sirieix C and Baleani M 2014 An effective procedure to create a speckle pattern on biological soft tissue for digital image correlation measurements *J. of the Mech. Beh. of Biom. Mat.* **39** 1–8

[28] Yoneyama S 2006 Lens distortion correction for digital image correlation by measuring rigid body displacement *Opt. Eng.* **45(2)** 023602

[29] Patterson E A, Hack E, Brailly P, Burguete R L, Saleem Q, Siebert T, Tomlinson R A, Whelan M P 2007 Calibration and evaluation of optical systems for full-field strain measurement *Opt. and Las. in Eng.* **45(5)** 550–564

[30] Sebastian C and Patterson E A 2012 Calibration of a Digital Image Correlation System *Exp. Techn.* **39(1)** 21–29

[31] Genovese K, Casaletto L, Rayas J A, Flores V and Martinez A 2013 Stereo-Digital Image Correlation (DIC) measurements with a single camera using a biprism *Opt. and Las. in Eng.* **51(3)** 278–285

[32] Genovese K, Lee Y-U, Lee A Y and Humphrey J D 2013 An improved panoramic digital image correlation method for vascular strain analysis and material characterization J of the Mech. Beh. of Biom.Mat. **27** 132–142

[33] Panin S V, Titkov V V, Lyubutin P S 2013 Efficiency of vector field filtration algorithms in estimating material strain by the method of digital image correlation *Opt., Instr. and Data Proc. 49(2)* 155–63

[34] Panin S V , Lyubutin P S, Eremin A V, Titkov V V, Chemezov V O 2017 Application of lucas-kanade algorithm with weight coefficient bilateral filtration for the digital image correlation method *IOP Conf. Series: Mat. Scien. and Eng.* 012039

Mechanical and tribological characteristics of a-C:H:SiO_x films formed by PACVD on titanium alloy VT1-0

A S Grenadyorov[1], A A Solovyev[1, 2], K V Oskomov[1] and O A Yaroslavtseva[2]

[1]Institute of High Current Electronics SB RAS, 2/3 Akademichesky Avenue, Tomsk, 634055, Russia
[2]National Research Tomsk Polytechnic University, 30 Lenina Avenue, Tomsk, 634050, Russia

E-mail: 1711Sasha@mail.ru

Abstract. This paper is devoted to the study of the mechanical and tribological properties of a-C:H:SiO_x films deposited on a titanium alloy VT1-0 by a plasma chemical deposition method using pulsed bipolar bias voltage. It was shown that after deposition of 2 μm-thick a-C:H:SiO_x film on a titanium alloy VT1-0 sample, the root-mean-square surface roughness R_q measured using atomic force microscopy decreased from 74 to 50 nm compared to the original substrate. The surface hardness H measured using nanoindentation increased from 3.3 to 12.4 GPa with an almost unchanged elasticity modulus E. As a result, the plasticity index (H/E) of titanium samples increased from 0.03 to 0.11, and the plastic deformation resistance (H^3/E^2) increased from 3 to 156 MPa. Deposition of a-C:H:SiO_x film on the titanium alloy VT1-0 surface makes possible to reduce the friction coefficient from 0.3-0.6 to 0.1 and the wear rate from $6 \cdot 10^{-4}$ to $7 \cdot 10^{-6}$ mm^3/Nm.

1. Introduction

Presently, a lot of attention is given to ensuring the durability of metal parts of machines and mechanisms by creating thin-film coatings on their surfaces. The most common coatings are nitride (TiAlN, TiAlCrYN, TiCN, TiN) and carbon-based coatings (DLC) that can reduce friction and wear rates. The addition of SiO_x to DLC allow to reduce internal stresses in DLC films that provides excellent adhesion to many types of substrates and makes it possible to form tens of micrometers thick films [1,2]. On top of that, SiO_x containing DLC (a-C:H:SiO_x) films are characterized by high hardness (10-20 GPa) and elasticity modulus (30-150 GPa) [2-4], low friction coefficient of about 0.02-0.2 [1,5] and wear rate 10^{-5}-10^{-8} $mm^3N^{-1}m^{-1}$ [5,6], high transparency in the visible and near IR wavelength region (~80-85%) [7,8], excellent biocompatibility with human body [1]. Due to the unique properties of a-C:H:SiO_x films, they can be used as wear-resistant anti-friction coatings on parts of internal combustion engines, in MEMS technologies, medicine, industry, lithography [9] and other fields.

The most common method for such films forming is the radio-frequency plasma-assisted chemical vapor deposition (RF PACVD) [5,10,11]. Koshigan *et al.* [1] studied influence of hydrogen and oxygen pressure on tribological characteristics of a-C:H:SiO_x films deposited on silicon wafers. Friction experiments were performed with a linear reciprocating pin-on-flat tribometer, using 52100 steel pins. The authors showed that there are optimal values of gas pressure (H_2 and O_2) in the process of testing at which minimum values of the friction coefficient (0.02±0.01) are observed. Neerinck *et al.* [12] studied tribological properties of a-C:H:SiO_x films also deposited on silicon substrates. Unlubricated sliding ball-on-disk experiments were performed at room temperature using AISI L3 steel bearing ball. It was shown that in air at a relative humidity (RH) 50% the friction coefficient was 0.04–0.08, while at a relative humidity of 90% and under water, it was lower than 0.15. The wear rate calculated from the profilometry measurement of the wear track area was $2 \cdot 10^{-7}$ $mm^3N^{-1}m^{-1}$ in air of 50% RH and only $3.5 \cdot 10^{-8}$ $mm^3N^{-1}m^{-1}$ under water. Venkatraman *et al.* [13] studied effect of temperature annealing on the tribological properties of a-C:H:SiO_x films deposited on silicon. The wear rate of the annealed films was measured in air at room temperature with a relative humidity of ~35% using a ball-on-disk tribotester with WC ball. The authors showed that with an increase in the annealing temperature in air, graphitization of a-C:H:SiO_x films occurs, as a result of which mechanical and tribological properties deteriorate, in particular, an increase in wear rate from $(1 \div 10) \cdot 10^{-8}$ to $(8 \div 50) \cdot 10^{-7}$ $mm^3N^{-1}m^{-1}$ is observed. Jana *et al.* [14] studied tribological properties of a-C:H:SiO_x films deposited on glass substrates at different relative humidity of the environment. The wear and friction characterization was performed using ball-on-disc tribometer with WC ball. In this paper was shown that increase in relative humidity from 35 to 80% leads to an increase in the friction coefficient from 0.005 to 0.074, while the wear rate is reduced from $9.8 \cdot 10^{-8}$ to $2.7 \cdot 10^{-8}$ $mm^3N^{-1}m^{-1}$ with increasing relative humidity from 35 to 50%. A further increase in relative humidity up to 80% leads to an increase in the wear rate. Bhowmick *et al.* [6] have studied tribological characteristics of a-C:H:SiO_x films deposited on M2 grade tool steel coupons. Pin-on-disk type sliding tests were performed with 319 Al alloy pins. At a room temperature, a-C:H:SiO_x films had friction coefficient of 0.17 and wear rate of $2.86 \cdot 10^{-5}$ $mm^3N^{-1}m^{-1}$. At temperature of 400 °C, the friction coefficient was 0.11, and wear rate was $16.23 \cdot 10^{-5}$ $mm^3N^{-1}m^{-1}$. Jedrzejczak *et al.* [5] have studied tribological characteristics of a-C:H:SiO_x films deposited on cylindrical samples of Ti6Al7Nb alloy in combination with different counterbodies. A ball-on-disc tribometer with AISI 316L and ZrO_2 balls was used. The relative humidity was 30%. In this paper, the authors have shown that a-C:H:SiO_x films have a minimal friction coefficient of 0.04-0.07 at using of both balls. In this case, the wear rate for a-C:H:SiO_x films was from $1.43 \cdot 10^{-7}$ to $3.97 \cdot 10^{-7}$ $mm^3N^{-1}m^{-1}$.

This paper is devoted to the study of the mechanical and tribological properties of a-C:H:SiO_x films deposited on a titanium alloy VT1-0 by plasma chemical deposition method using pulsed bipolar bias voltage applied to substrate. The titanium alloy VT1-0 (Ti content is ~ 99.24-99.7%) is widely used in aircraft and rocket production, medicine and other areas. However, it is known that titanium alloys have the high surface reactivity, low work hardening coefficient, low shear strength, high value of a friction coefficient and a wear rate [15].

2. Materials and Methods

Deposition of a-C:H:SiO_x films was realized on vacuum installation equipped with plasma generator with heated cathode working in mixture of argon and polyphenyl-methylsiloxane (PPMS) vapor. PACVD deposition system has been described in detail in [16]. Vacuum in the chamber to a pressure of 10^{-2} Pa was realized by turbomolecular pump.

Made from titanium alloy VT1-0 samples (size 2×2 cm^2 and thickness 0.4 cm) were used as the substrates. The distance between the plasma generator and the substrate was 100 mm. Before a-C:H:SiO_x films deposition, the substrates were pre-cleaned in argon plasma for 6 minutes. In this case argon pressure in chamber was 0.3 Pa, discharge voltage - 100 V, discharge current - 11 A, magnetic induction in the substrate area - 3 Gs. Bipolar bias voltage with negative pulse amplitude of 1000 V, with frequency of 100 kHz and the positive impulse duration of 4 μs was applied to the substrate during plasma cleaning. After that, a-C:H:SiO_x film was deposited at argon pressure of 0.1 Pa, PPMS flow rate of 95 μl/min, discharge voltage of 140 V, discharge current of 6 A and magnetic induction of 3 Gs. During the film deposition, the negative pulse amplitude of the bipolar bias voltage was reduced to 500 V. Film deposition rate was 3±0.3 μm/h. In [16], we showed that in this deposition regime, the hardest films with the highest sp^3-bonded carbon content are formed on Si substrates.

The deposited a-C:H:SiO_x films were investigated by different characterization techniques. Raman spectroscopy (Centaur U HR complex) with 532 nm Ar laser (beam cross-section of 50 μm) was used for recording of spectra. These spectra were obtained in a spectral range of 800–1800 cm^{-1} with a resolution better than 1.5 cm^{-1}. From the Raman spectra it is possible to obtain information about the carbon bonds in the film. For this, the spectra were divided into Gaussians and the position, width, and intensity of the *D* and *G* Raman peaks were determined. Hardness (*H*) and coefficient of elasticity (*E*) of the a-C:H:SiO_x films were determined by NanoTest 600 device (MicroMaterials, Great Britain) at 2 mN load by Oliver–Pharr method [17]. The surface morphology of the samples was studied by atomic force microscopy using AFM Solver P47 (NT-MDT, Russia) microscope. The friction coefficient was determined using the THT-S- AX0000 (CSM Instruments, Switzerland) tribometer in the "ball on disk" geometry. A ball with a radius of 3 mm made of VK-6 alloy was used as counterbody. The rotation speed of the disk was 5 cm/s, and rotation frequency was 2 Hz at normal load 5 N. Air temperature was 25 °C during the test, and the relative humidity was 50%. Wear rate was estimated by method described in [18]. The film thickness was measured by a Linnik microinterferometer MII 4 (LOMO, Russia).

3. Results

3.1. Atomic force microscopy results

The surface morphology of the samples was studied on initial substrate made from titanium alloy VT1-0, polished by an abrasive paper, and the same substrate coated by a-C:H:SiO_x film. The surface of initial sample had a granular structure with grains of approximately 100 nm in width and 50- 100 nm in height (figure 1,a). After deposition of a-C:H:SiO_x film with the thickness of 2±0.2 μm, the substrate surface became smoother. Root-mean-square surface roughness R_q decreased from 74 to 50 nm. Table 1 presents samples surface roughness values before and after film deposition. In [19] it was shown that the wear rate of DLC films depends on the substrate surface roughness, and latter is the decisive parameter for the wear resistance of the DLC to external loads.

Table 1. Surface roughness characteristics.

Sample	R_{mean} (nm)	R_{max} (nm)	R_q (nm)	R_a (nm)
Initial substrate	258	461	72	57
Coated substrate	134	412	54	43

R_{mean} – mean peaks height, R_{max} – maximum peaks height, R_q – root-mean-square surface roughness, R_a – arithmetic-mean surface roughness.

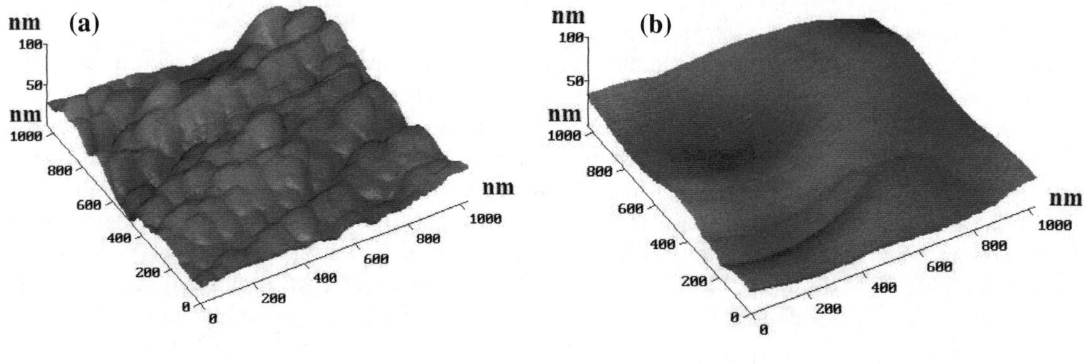

Figure 1. AFM images of initial VT1-0 sample (a) and coated with a-C:H:SiO_x film (b)

3.2. Mechanical characteristics research

The comparison was made between the hardness and the elastic modulus of the initial VT1-0 sample and a sample with a-C:H:SiO_x film (figure 2). It is noticeable that after deposition of a-C:H:SiO_x film, the surface hardness increased from 3.3 to 12.4 GPa. Herewith the elastic modulus of both samples remains almost unchanged – 112.9 and 110.5 GPa. Although hardness is important material property which defines wear resistance, it is considered that the elastic modulus also has an important influence on wear behavior. In particular, the elastic strain to failure, which is related to the ratio of hardness and elastic modulus, is a more suitable parameter for predicting wear resistance [20].

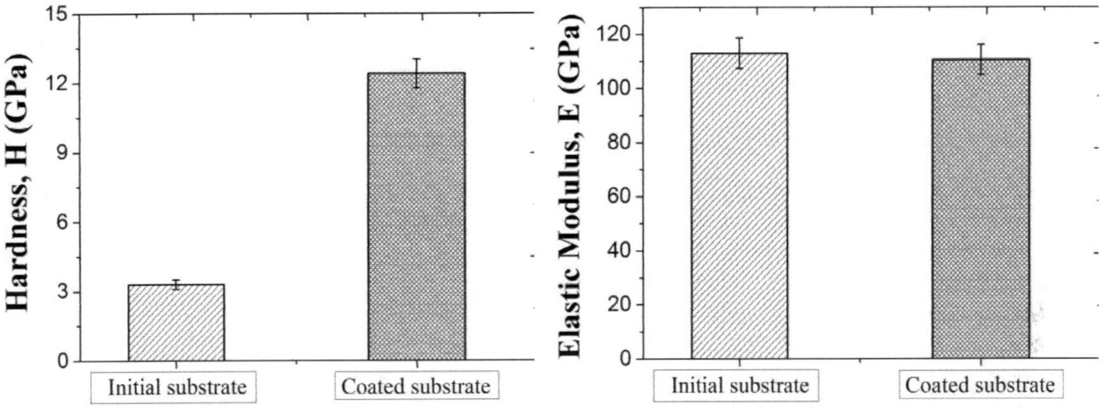

Figure 2. Mechanical characteristics of initial VT1-0 sample and coated with a-C:H:SiO_x film

Since the hardness grew almost 4 times with a practically unchanged elasticity modulus on the sample with a-C:H:SiO_x film, then the elastic-plastic characteristics of the surface (elastic index H/E and plastic resistance H^3/E^2) changed significantly. In particular, elastic index increased from 0.03 to 0.11 and plastic resistance increased from 3 to 156 MPa. The increase in these characteristics indicated an improvement in the wear resistance of the sample surface. Table 2 gives mechanical and elastic-plastic characteristics of the samples surface.

Table 2. Mechanical and elastic-plastic characteristics of the samples surface.

Sample	H (GPa)	E (GPa)	H/E	H^3/E^2 (MPa)
Initial substrate	3.3	112.9	0.03	3
Coated substrate	12.4	110.5	0.11	156

3.3. Tribological properties research

After deposition a-C:H:SiO_x film, the friction coefficient of the sample surface decreased from 0.3-0.6 to 0.1 (figure 3). The initial sample had a friction coefficient that varied in the range from 0.3 to 0.6 and was not stable with time. The reason of the friction scattering was not clear. After film deposition, the friction coefficient not only decreases, but also its oscillations become less pronounced. Except for a short running-in period at the beginning, the sliding on the films soon turns to a stable friction.

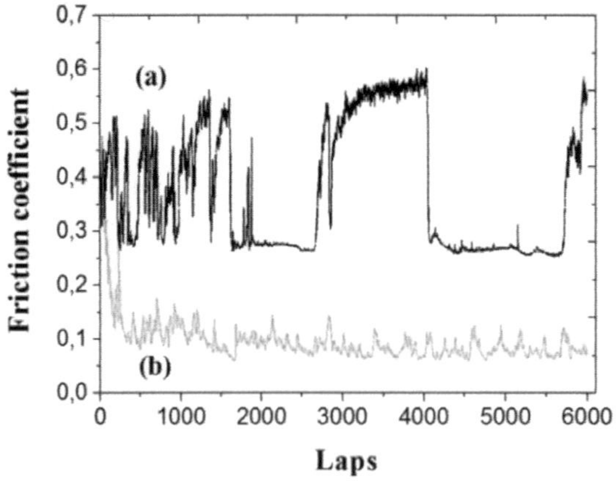

Figure 3. Friction behaviours of initial VT1-0 sample (a) and coated with a-C:H:SiO_x film (b)

Figure 4 show the optical images of wear track of both samples. Wear rate of the sample with a-C:H:SiO_x film, calculated from profile of wear track, decreased from $6 \cdot 10^{-4}$ to $7 \cdot 10^{-6}$ $mm^3/N \cdot m$ in comparison with initial sample. It is clearly seen that the wear track on the VT1-0 sample (figure 4-a) is wider and deeper, while on the sample with the film it is narrower and less deep.

We have previously shown that a-C:H:SiO_x film deposited on the GaAs substrate has higher friction coefficient (0.16) and greater wear rate ($1.3 \cdot 10^{-5}$ $mm^3/N \cdot m$) [21]. Such a difference in film characteristics may be caused by a difference in film deposition conditions, namely another negative pulse amplitude of the substrate bias voltage and Ar pressure during deposition.

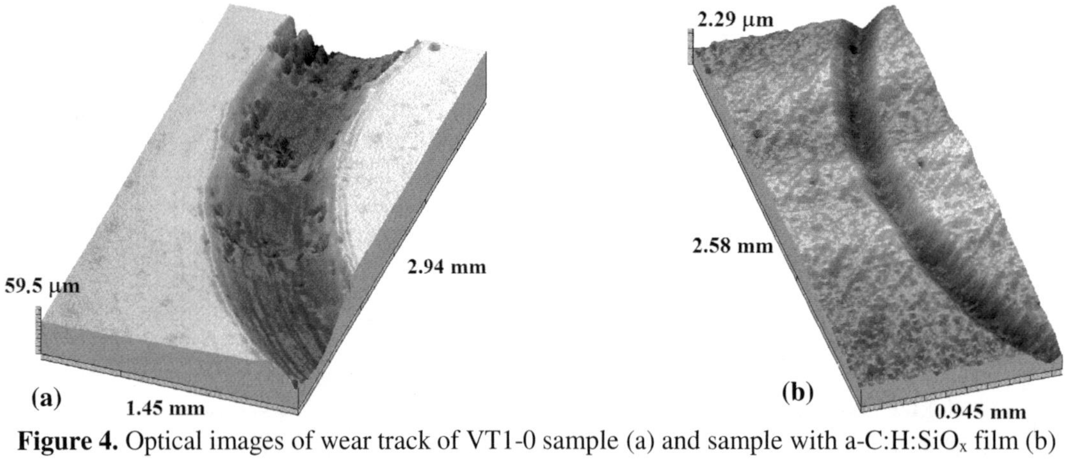

Figure 4. Optical images of wear track of VT1-0 sample (a) and sample with a-C:H:SiO_x film (b)

In paper [22], an influence of electron-ion plasma treatment on tribological properties of a titanium alloy VT1-0 sample was studied. Combined processing in a single vacuum cycle was used for the surface modification. It included nitriding in low-pressure arc plasma and subsequent TiN coating deposition. It was shown that the combined treatment provides the formation of a modified layer which decreases the friction coefficient from 0.4 to 0.3 and the wear rate from $\sim 3 \cdot 10^{-4}$ to $\sim 8 \cdot 10^{-6}$ $mm^3/N \cdot m$. It suggests that a-C:H:SiO_x films are comparable in wear rate to the nitride coating, but significantly exceed it in terms of reducing the friction coefficient.

3.4. Raman spectroscopy

Raman spectroscopy is one of the simplest methods for determining the structure of carbon-based thin films and gives information about the content of sp^3- and sp^2-hybridized carbon atoms in film. Properties of the deposited films depend on the prevalence of one or another hybridization type. Raman spectrum of a-C:H:SiO_x film has wide peak in wavenumber range 1000-1700 cm^{-1} typical for

DLC films (figure 5). This peak was decomposed on two Gaussians, corresponded to D (disordered) and G (graphite) peaks and their position, full width at half-maximum (FWHM), peak intensity ratio I_D/I_G were determined (table 3).

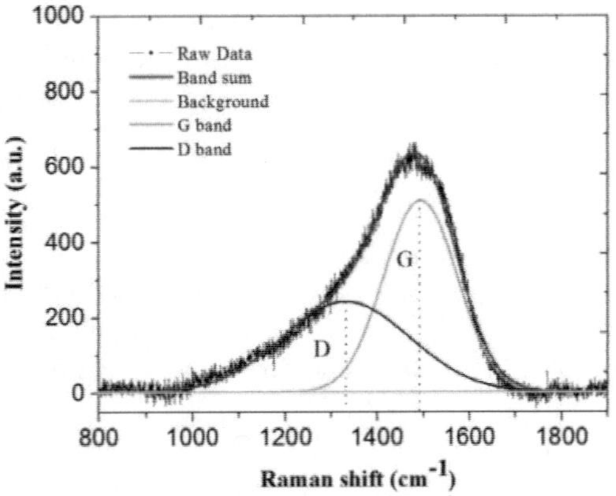

Figure 5. Raman spectrum of a-C:H:SiO_x film deposited on VT1-0 sample

Table 3. Results of analysis of Raman spectrum of a-C:H:SiO_x film.

ω_D (cm^{-1})	Γ_D (cm^{-1})	ω_G (cm^{-1})	Γ_G (cm^{-1})	I_D/I_G
1330	334	1494	189	0.83

ω – peaks position, Γ – full width at half maximum (FWHM), I_D/I_G – intensity ratio.

Conclusions

The results of this study showed that the deposition a-C:H:SiO_x films on the VT1-0 samples allows:

- to increase the surface hardness from 3.3 to 12.4 GPa with a practically unchanged elasticity modulus (110-113 GPa) and, consequently, to increase plasticity index H/E from 0.03 to 0.11 and plastic resistance H^3/E^2 from 3 to 156 MPa;
- to reduce friction coefficient and wear rate by 3.7 times and 85 times, respectively.

Consequently, obtained a-C:H:SiO_x films can be used to increase the durability of products made from titanium alloy VT1-0 in aircraft manufacturing, medicine and other industries.

Acknowledgments

This work was supported by Government task of Institute of High Current Electronics № 0366-2016-0010. The authors are thankful to Tomsk Regional Center for Collective Use of the TSC SB RAS for the provided NanoTest 600 nanoindentator and AFM Solver P47 atomic force microscope.

References

[1] Koshigan K, Mangolini F, McClimon J B et al 2015 *Carbon* **93** 851
[2] Batory D, Jedrzejczak A, Szymanski W et al 2015 *Thin Solid Films* **590** 299
[3] Randeniya L K, Bendavis A, Martin P J et al 2009 *Diamond & Related Materials* **18** 1167
[4] Mallik A, Dandapat N, Ghosh P et al 2013 *Bull. Mater. Sci.* **36** 193
[5] Jedrzejczak A, Kolodziejczyk L, Szymanski W, Piwonski I, Cichomski M, Kisielewska A, Dudek M and Batory D 2017 *Tribol. Int.* **112** 155
[6] Bhowmick S, Banerji A, Lukitsch M J and Alpas A T 2015 *Wear* **330-331** 261
[7] Santra T S, Bhattacharyya T K, Tseng F G and Barik T K 2012 *AIP Adv.* **2** 022132
[8] Jana S, Das S and Gandopadhyay U 2014 *Applied Physic A* **114** 965
[9] Meskinis S and Tamuleviciene A 2011 *Materials science* **17** 358
[10] Barve S A, Chopade S S, Kar R, Chand N, Deo M N, Biswas A, Patel N N, Rao G M, Patil D S, Sinha S 2017 *Diam. Relat. Mater.* **71** 63
[11] Nakazawa H, Kamata R, Miura S and Okuno S 2015 *Thin Solid Films* **574** 93

IOP Conf. Series: Materials Science and Engineering **511** (2019) 012019 doi:10.1088/1757-899X/511/1/012019

[12] Neerinck D, Persoone P, Goel A, Venkatraman C 1998 *Thin Solid Films* **317** 402
[13] Venkatraman C, Brodbeck C, Lei R 1999 *Surface and Coatings Technology* **115** 215
[14] Jana S, Das S, Gandopadhyay U, Mondal A, Ghosh P 2013 *Advances in Tribology* **2013** 352387
[15] Farokhzadeh K and Edrisy A 2016 *Tribology International* **94** 98
[16] Grenadyorov A S, Solovyev A A, Oskomov K V, Rabotkin S V et al 2019 *Thin Solid Films* **669** 253
[17] Oliver W C and Pharr G M 2004 *J. Mater. Res.* **19(1)** 3
[18] Maru M M and Tanaka D K 2007 *J. of the Braz. Soc. of Mech. Sci. & Eng.* **XXIX** 55
[19] Jiang J and Arnell R D 2000 *Wear* **239** 1
[20] Leyland A and Matthews A 2000 *Wear* **246** 1
[21] Grenadyorov A S, Oskomov K V, Solovyev A A and Rabotkin S V 2016 *Technical Physics* **61(5)** 690
[22] Devyatkov V N, Ivanov Yu F, Krysina O V, Koval N N, Petrikova E A, Shugurov V V 2017 *Vacuum* **143** 464

IOP Conf. Series: Materials Science and Engineering **511** (2019) 012020 doi:10.1088/1757-899X/511/1/012020

Microstructure and mechanical properties of annealed, quenched and suction cast Ti-6Al-4V

A Thoemmes, I V Ivanov and A A Ruktuev

Novosibirsk State Technical University, Pr. Karla Marksa 20, Novosibirsk 630073, Russia

alexander.thoemmes@outlook.de

Abstract. The effect of heat treatment or processing method on microstructure, mechanical properties and phase formation in Ti-6Al-4V alloys has been researched. The alloys in as-delivered, annealed, quenched and suction cast conditions were investigated using scanning electron microscopy (SEM), energy dispersive X-ray spectroscopy (EDX) and synchrotron X-ray diffraction (SXRD) analysis. The different processing- or heat treatment routes lead to significant changes in the phase composition. The quenched and suction cast alloys exhibited a mixture of martensitic α` / α`` and BCC β phase. In annealed and as-delivered alloys α` + β phase mixtures was detected. The properties of alloys strongly depend on heat treatment and / or processing route.

1. Introduction

Ti-based alloys are currently one of the most attractive materials for application in medicine, since they are characterized by perfect corrosion resistance, high strength to weight ration and high biocompatibility [1, 2]. Since the properties of the alloys are depending on their phase composition, Ti-based alloys can be classified in three groups: α - alloys, α + β alloys and β alloys. The nowadays mostly used titanium alloys, Ti-6Al-4V is a typical example of a dual phase alloy. The martensitic Ti-6Al-4V alloy, which is suitable to both plastic working and heat treatment whereby its macrostructure and mechanical properties can widely be varied [3, 4]. The mechanical properties can be optimized by controlled adaptation of microstructure by thermo- or thermomechanical processing paths. For example the mechanical behavior of α + β and / or β Ti-based alloys can be significantly improved by a uniform distribution of fine α / α' precipitations [5–10]. The transformation and precipitation pathways occurring during annealing decide the size, arrangement and morphology of the precipitation products.

In depth qualitative and quantitative studies of phase transformations and their influence on mechanical properties of the Ti6Al4V alloy are still lacking in literature, although better understanding of phase transformation kinetics during cooling from either α +β or β range seems essential. Therefore the main objective of the present work was to determine the effect of furnace cooling from 900°C (α +β range) on the mechanical properties and lattice parameters of Ti-6Al-4V.

2. Materials and Methods

The Ti-6Al-4V raw material used in this study was delivered in a form Ø 50 x 200 mm rolled bar. Small 20 x 20 x 10 mm^3 specimens were cut from the bar normal to rotation axis. The Ti-6Al-4V_SC specimen was received by remelting and suction casting one of the above-mentioned specified specimen using a

IOP Conf. Series: Materials Science and Engineering **511** (2019) 012020 doi:10.1088/1757-899X/511/1/012020

BUHLER ArcMelter. For this, the specimen was placed in a triply flushed melting chamber, during melting an argon pressure of 800 mbar was maintained. For melting a water-cooled copper crucible and a non-consumable tungsten electrode were used. A titanium ingot was used as an oxygen getter and was melted prior to every melting procedure. The specimen was remelt 6 times and flipped after every melt. The received button shaped specimen was remelt and suction cast into a water-cooled cylindrical copper crucible with a diameter of 5 mm.

The Ti-6Al-4V_AN specimen was received by annealing the sample in a vacuum furnace at 900°C for 4h, followed by cooling in the furnace. The Ti-6Al-4V_Q specimen was heated to 1000°C, held for 0.5h and then cooled to room temperature by quenching in oil. For the Ti-6Al-4V_AD specimen no further heat treatment or other processing route was conducted.

The samples were mounted into phenolic resin and mechanically ground using SiC paper up to grade 1000 followed by pre-polishing with Al2O3 suspension. Final polishing was conducted using colloidal SiO2. The microstructure of the samples was revealed by etching with a Kroll`s reagent (5 vol. % HF, 10 vol. % HNO3 and 85 vol. % H2O). The microstructural analysis was performed using an optical microscope (Carl Zeiss AxioObersver Z1m). The hardness measurements, an average of 30 readings, were carried out using a WOLPERT Group 402 MVD Vickers hardness tester under a load of 50 g and a dwell time of 10 s. Synchrotron X-ray diffraction (SXRD) was conducted at the P07 beamline of the "Deutsches Elektronen Synchrotron" in Hamburg, Germany. Diffraction patterns were recorded by a 2-dimensional (2D) image plate detector with 2048 x 2048 pixels centered on the beam. An X-ray wavelength of $\lambda = 0.124$ Å and a beam cross section of 0.5 x 0.5 mm² were used. The sample-to-detector distance and diffraction center were calibrated at RT using a LaB6 powder. Azimuthal integration was carried out over the entire 360° to minimize potential effects of texture and grainsizes. The peaks in the diffraction patterns were fitted using Pearson-VII function and the peak parameters were calculated using least square method.

3. Results and Discussion

The SXRD patterns of quenched and annealed samples are plotted in figure 1. In both samples a mixture of α / α' + β was detected. According to the higher intensities detected in the annealed sample it can be concluded that the volume ration of the β phase is higher than in the quenched sample. Lower cooling rates typical for furnace cooling give rise to the transformation of β→α or β→α` from the high temperature β phase. Considering that the initial Ti-6Al-4V samples consist at room temperature mostly of α phase (Al is a very strong α stabilizer) the weighted average lattice expansion of the two-phase mixture should be closer to that of α phase. When the β→α takes place the ratio of β to α increase and the expansion of the β phase dominate. The lattice parameters of the samples are also shown in the figure. It can be seen that after annealing the lattice parameters of both phases increase, whereas the hexagonality of the α / α' is nearly equal (1.5973 for quenched sample and 1.5974 for annealed sample).

Figure 2a shows the microstructure of the Ti-6Al-4V sample in as-delivered condition. A typical dual phase microstructure is visible. So called grain boundary α has formed along the boundaries. The suction cast sample showed in figure 2b shows α' being composed of long rectangular to each other oriented martensitic plates exhibiting an acicular morphology. The volume fraction of grain boundary α in the quenched sample, showed in figure 2c, was increased. According to the results of Ahmed and Rack this is a testimony of a reduced cooling rate [11].

Among all examined samples the Ti-6Al-4V_AD sample showed the lowest microhardness (388 $HV_{0.05}$). TheTi-6Al-4V sample in as delivered condition is well known as α + β alloy. Both stable phases, α phase as well as β phase are softer compared to their transformed phases like the martensitic α' or α'' phases. The quenched and suction cast samples showed only a slightly different microhardness of 415 $HV_{0.05}$ and 424 $HV_{0.05}$. According to the SX-ray diffraction analysis showed in figure 1 a mixture of α' + β was obtained. Therefore, it can be concluded that the cooling rate in the suction casting process is as fast as during quenching. Both cooling rates are not high enough to force the formation of metastable α'' phase in the samples. The highest microhardness is present in the annealed sample. This perceptible increase in microhardness was probably related to formation of metastable ω phase. It is assumed that the cooling rate during the furnace cooling after annealing was small enough to subject the samples to an aging treatment.

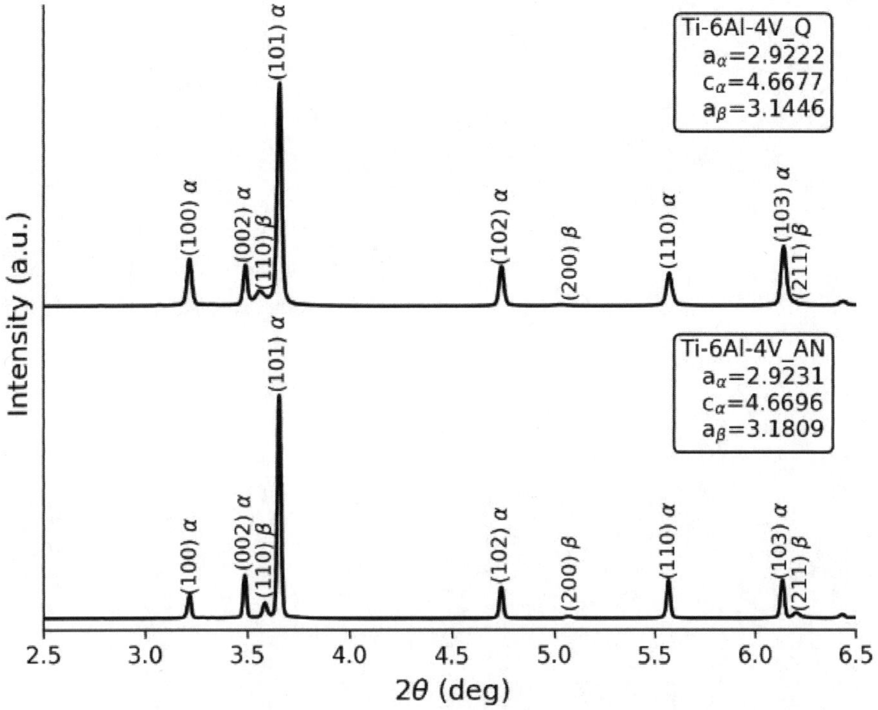

Figure 1. SXRD patterns of a) quenched and b) annealed Ti-6Al-4V samples.

4. Conclusions

Based on the results obtained from characterization of the microstructure, phase composition and resulting mechanical properties, the following conclusions can be drawn. The microstructures of dual phase Ti-6Al-4V samples is sensitive to the heat treatment respectively the processing route. The cooling rate during the heat treatment or processing have a strong effect on mechanical properties of Ti-6Al-4V samples. The volume fraction of grain boundary α increased with decreasing cooling rate and the highest volume fraction was reached in the quenched sample. Among the studied alloys, the annealed sample showed a microhardness of 424 HV0.05 which is probably related to the presence of metastable ω phase.

Acknowledgments

The work was carried out within the framework of the C-11 project of a competition among young scientists on the topic "Investigation of thermal stability and regularities in the formation of metastable phases of the Ti-Nb system for application in medicine" in 2018. Materials characterization was carried out at NSTU Materials Research Center.

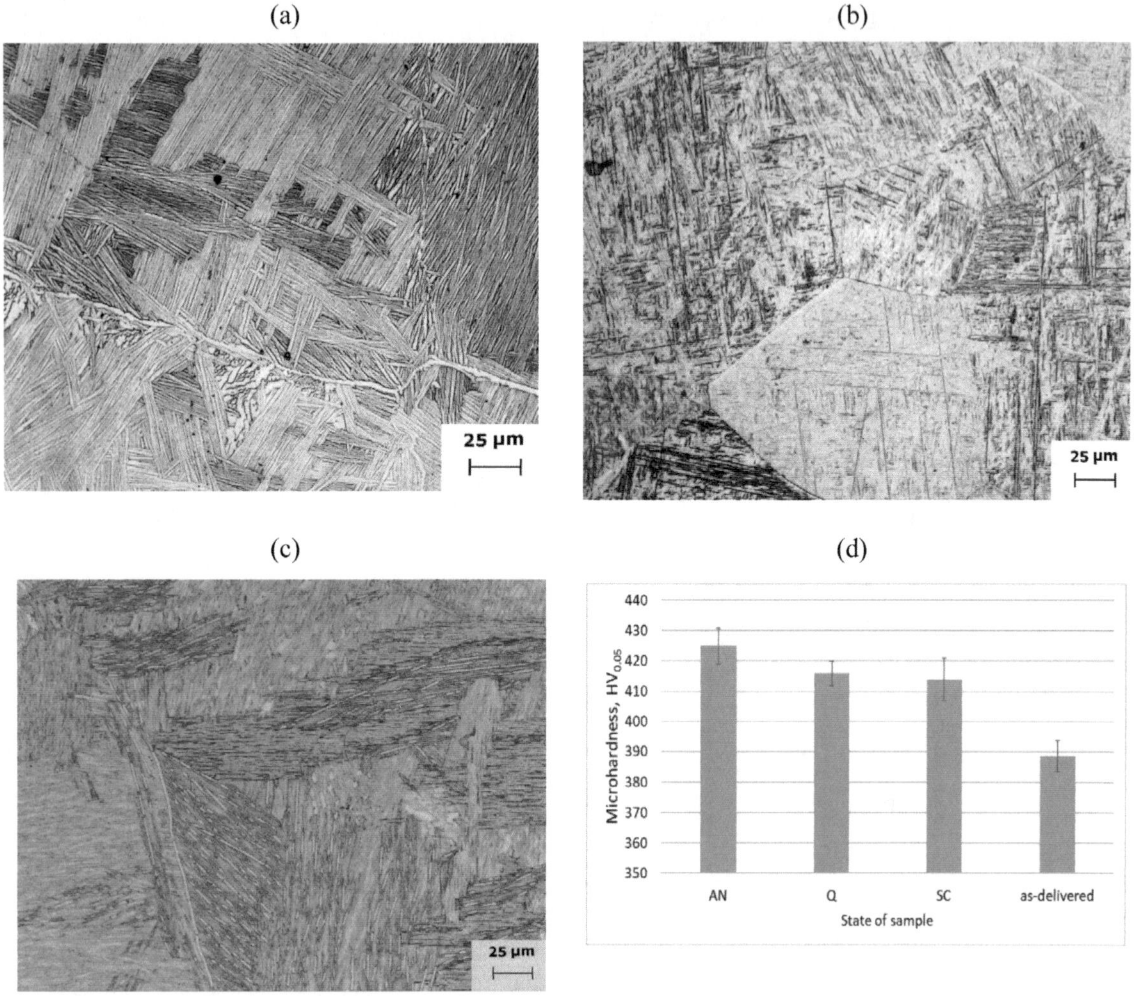

Figure 2. Microstructure of Ti-6Al4V in a) as delivered, b) Suction cast, c) quenched condition and d) microhardness of the alloys

References

[1] Lütjering G and Williams J C (eds) 2007 *Titanium (Engineering materials and processes)* 2nd edn (Berlin, New York: Springer)
[2] Donachie M J 1988 *Titanium: A technical guide* (Metals Park OH: ASM International)
[3] Hennig R G, Trinkle D R, Bouchet J, Srinivasan S G, Albers R C and Wilkins J W 2005 *Nature materials* **4** 129–33
[4] Barriobero-Vila P, Biancardi Oliveira V, Schwarz S, Buslaps T and Requena G 2017 *Acta Materialia* **135** 132–43
[5] Thoemmes A, Bataev I A, Belousova N S and Lazurenko D V 2016 26–9
[6] Thoemmes A, Ivanov I V, Lazurenko D V and Bataev I A *Metal Science and Heat Treatment* **2018** (accepted)
[7] Kovalevskaya ZG, Sharkeev YP, Korchagin MA, Khimich MA, Ibragimov EA, Saprykin AA and Bataev VA 2016 *Obrabotka metallov (tekhnologiya, oborudovanie, instrumenty) = Metal Working and Material Science* **73** 34–42
[8] Thoemmes A, Ivanov I V and Kashimbetova A A 2018 *KEM* **769** 29–34

[9] Elmer J W, Palmer T A, Babu S S and Specht E D 2005 *Materials Science and Engineering: A* **391** 104–13
[10] Biesiekierski A, Lin J, Munir K, Ozan S, Li Y and Wen C 2018 *Scientific reports* **8** 5737
[11] Ahmed T and Rack H J 1998 *Materials Science and Engineering: A* **243** 206–11

Microwave electromagnetic characteristics of polymeric composite materials containing carbonyl iron and MWCNT/Ferrites

G E Kuleshov[1,2], A V Badin[1,2], A V Sbrodov[2] and K V Bilinskiy[2]

[1]Laboratory of radiophysical and optical methods of exploring the surrounding environment, National Research Tomsk State University, Tomsk, Russia
[2]Department of Radioelectronics, Faculty of Radiophysics, National Research Tomsk State University, Tomsk, Russia

E-mail: grigorij-kge@sibmail.com

Abstract. Microwave electromagnetic characteristics of radiomaterials based on carbonyl iron with the addition of multi-walled carbon nanotubes or ferrites were investigated. On the basis of the obtained measurement results, the frequency dependences of the reflection coefficients for the samples of composite materials were calculated. Comparison of calculations with the results of the experiment is presented. Adding 2 wt.% multi-walled carbon nanotubes to a composite with 40 wt.% carbonyl iron is equivalent to increasing the concentration of carbonyl iron to 60 wt.%, which significantly reduces the consumption of materials and the weight of the composite. The results obtained make it possible to characterize the materials under study as effective absorbers of electromagnetic radiation in the microwave frequency range.

1. Introduction

Every year there are more and more modern wireless devices that operate in the microwave frequency range. This leads to a number of problems. Such problems include the threat to biological objects (electromagnetic ecology), protection against interception of transmitted information (electromagnetic safety), ensuring electromagnetic compatibility, etc. To solve the above problems, it is reasonable to use shielding or absorbing coatings. They can be used to ensure the electromagnetic compatibility of devices and the matching of the receiving-transmitting path, to protect biological resources and objects from exposure to electromagnetic radiation, to protect computer information processing systems from unauthorized access and to reduce the radar visibility of military and civilian objects. To date, there is no single-component material with high absorbing and low reflectivity in a wide frequency band of the microwave range. Therefore, the use of composites as a basis for radio absorbers is becoming increasingly important. The use of such materials is of particular interest due to the fact that by mixing materials we can achieve properties that the components of the composite do not possess separately from each other. It is quite convenient to work with composite materials, since various polymers (paints, varnishes, plastics, epoxy resins, polyethylene structures, silicone, etc.) can be used as a binder (matrix) [1]. They can be easily processed and applied to a particular surface. As fillers for modern high-frequency radiocomposites various ferromagnets and ferrimagnets [2 – 4], carbon materials (carbon black, graphene, nanodiamonds, single-wall and multi-walled nanotubes, onion structures) [5

–7], ultrafine metal powders (most often carbonyl iron) [8, 9] and microwires [10] are most often used. Fillers are used both separately and together [11, 12].

Thus, the development and study of materials and coatings based on carbonyl iron with multi-walled carbon nanotubes and ferrites are of great interest and also have great practical significance.

2. Materials and Methods

The structure of the materials was determined by X-ray analysis. Images were obtained on X-Ray Diffractometer Shimadzu XRD 6000 with copper radiation (CuKα) and Wavelength Dispersive X-Ray Fluorescence Spectrometer XRF-1800. Powder samples and composites samples were studied. Tube voltage of 40 kV, anode current of 30 mA, goniometer speed when shooting is 2 deg/min, X-ray diffraction is $2\theta=(20\div60)^{o}$.

The structure of material was studied by optical microscope and by scanning electron microscopy. Electron-microscopic measurements were carried out with a scanning electron microscope with a focused ion beam QUANTA 3D.

2.1. Materials

The choice of binder depends on the purpose of the composite coatings. As the resin base may be: silicone sealants, plastics, paints, varnishes, epoxy resins, natural and synthetic rubbers. In our case an EDP-20 grade epoxy resin with a hardener of polyethylene polyamine (PEPA) was chosen.

Powders of various radio materials were used as fillers. The main component of the radio composite was an industrially manufactured carbonyl iron (CI) of mark P-100. Additionally, a ferrimagnet (Fer) and multi-walled carbon nanotubes (MWCNT) were used. The obtained images of the microstructure of ferrite and MWCNT are shown in figure 1.

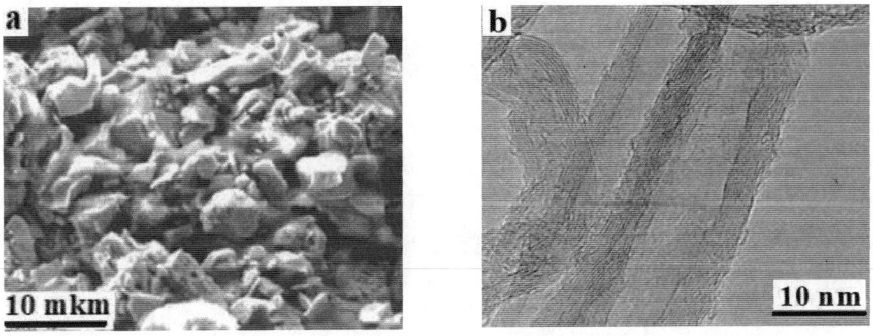

Figure 1. TEM images of the powders of ferrite (a) and MWCNT (b).

Ferrite is a magnetite spinel structure with the chemical formula Fe_3O_4. It was manufactured by the method of self-propagating high-temperature synthesis and contained up to 96 wt.% of the main phase. Later it was pressed and ground into a powder with a particle size of less than 250 microns. Used MWCNTs were synthesized by catalytic gas-phase deposition of ethylene in the presence of a FeCo catalyst at a temperature of 680 °C [13]. They have a high degree of purity and content of the main phase (Table 1). Metal impurities are the residue of the catalyst encapsulated inside the nanotubes.

Table 1. Characteristics of the MWCNT.

	Characteristic value				
Material	CNT content	Average diameter of nanotubes (diameter range)	Metal impurities	Outer specific surface	Number of layers CNT (wall thickness)
MWCNT	More than 97.5 wt.%	9.4 nm (4 – 21 nm)	< 1.7 wt.% (FeCo, Al_2O_3)	320 m^2/g	7-9 layers (4.0 nm)

2.2. Obtaining experimental samples

Manufactured samples are plane-parallel washers for measurement in a coaxial cell. Sample sizes: outer diameter d_{out} = 7 mm, internal diameter d_{in} = 3 mm, thickness h = 2.25 mm. The sample preparation scheme is shown in figure 2. For the production of the samples a thorough weighing of their component parts (filler and binder) was produced on a Shimadzu AUX-320 scales (error ~ 0.5 mg). Thereafter, the composite components were mixed in the appropriate parts (by weight). Then the mixture was blended until homogeneous state (using an ultrasonic disperser and magnetic stirrer). Ultrasonic treatment was carried out for 3 minutes at a power of 80 watts. Manufactured mixture was placed in a special form. The final composite sample is of a toroidal shape with various thickness. Polymerization of the finished product was carried out for several hours at room temperature.

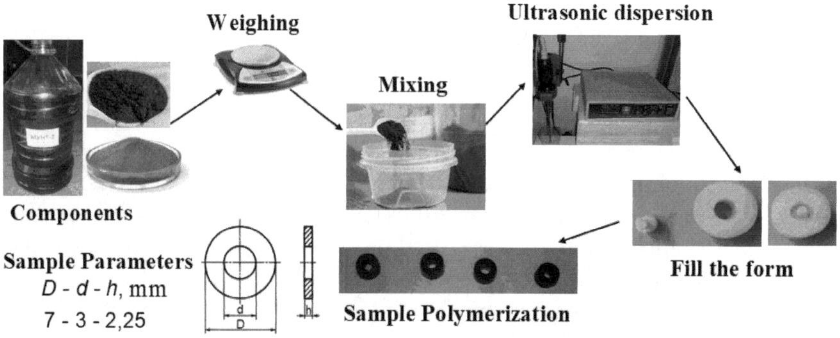

Figure 2. Production of samples of composite materials.

Thus, 9 experimental samples of composites for measurements were made. Their main characteristics are presented in table 2.

Table 2. Manufactured samples.

Sample No.	Content of epoxy resin, wt.%	Content of the filler, wt.%		
		Carbonyl iron	MWCNT	Ferrite
1	60	40	–	–
2	50	50	–	–
3	40	60	–	–
4	30	70	–	–
5	20	80	–	–
6	59	40	1	–
7	58	40	2	–
8	40	40	–	20
9	20	40	–	40

2.3. Measuring equipment

The waveguide method using a coaxial measuring cell was used to study the behavior of electromagnetic characteristics. As the instrumental basis of the experimental setup, the vector network analyzer P4M-18 produced by the company "Micran" was used. This device allows measurements in the frequency range from 0.01 to 18.00 GHz. The experimental study was carried out according to the measurement scheme "for pass" and "for reflection" (figure 3).

The "pass" scheme (figure 3a) is used to measure the scattering matrix coefficients $S11$, $S12$, $S21$ и $S22$, in other words, to measure the transmission and reflection coefficients of a wave through a sample placed in a coaxial cell, while the wave passes in one direction and in the opposite. On the

IOP Conf. Series: Materials Science and Engineering **511** (2019) 012021 doi:10.1088/1757-899X/511/1/012021

basis of the S-parameters obtained, taking into account the change in their phase, it is possible to calculate the spectra of the complex magnetic and the dielectric constant. For this purpose, the Becker-Jarvis technique [14] was used. On the basis of the obtained spectra of the complex magnetic and dielectric constant in the plane-wave approximation, the reflection coefficient from the layer of magnetodielectric located on the reflecting surface (metal surface) is calculated.

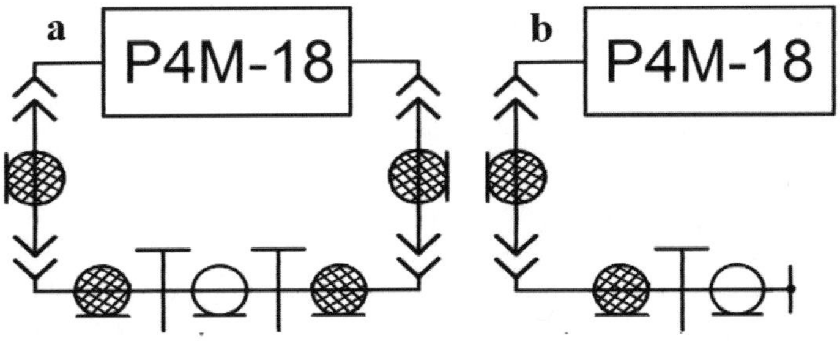

Figure 3. Schematic diagrams of measurements: a) "on pass"; b) "on reflection".

The "reflection" scheme (figure 1b) allows to directly measure the frequency dependence of the electromagnetic response on samples of composites that are placed in a coaxial cell with a short-circuit at the end. This method characterizes the absorbing properties of a sample located on a metal surface. The measurements allowed us to obtain the values of the reflection coefficient (*R*) from the layer of composites on the metal in the frequency range from 10 MHz to 18 GHz.

3. Results and discussion

Figure 4 presents the frequency dependences of the complex magnetic and dielectric constant for composite materials containing various concentrations of carbonyl iron.

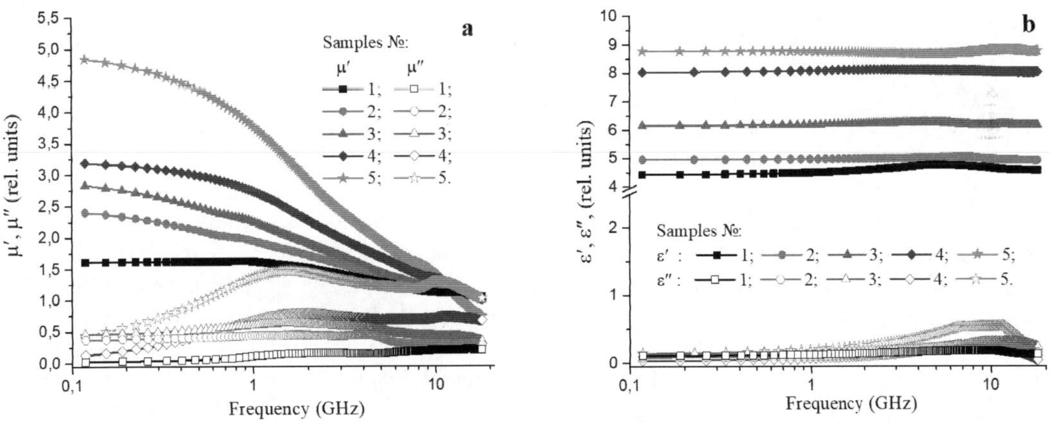

Figure 4. Frequency dependences of permeability (a) and permittivity (b) for a composites containing various concentrations of carbonyl iron.

The concentration of carbonyl iron varies from 40 wt.% to 80 wt.% in 10% increments. With an increase in the concentration of carbonyl iron, an increase in the value of the complex permeability is observed. This is especially noticeable for the real part of the permeability at low frequencies. In the range of 1-2 GHz, composites containing carbonyl iron show a sharp decline in the real part of permeability, as well as a maximum in the imaginary part of permeability, which corresponds to the phenomenon of natural ferromagnetic resonance. Spectra of complex dielectric constant are presented.

With an increase in the mass fraction of carbonyl iron, an almost linear increase in the values of the real part of permittivity from 4 to 9 relative units is observed. The imaginary part changes slightly, only at frequencies above 7 GHz and at a maximum concentration of carbonyl iron ε″ reaches 0.6 relative units. Such behavior of the complex permittivity is caused solely by the polarization characteristics of the filler.

Figure 5 shows the frequency dependences of the complex magnetic and dielectric constant for composite materials containing a fixed concentration of carbonyl iron with the addition of MWCNT and ferrite.

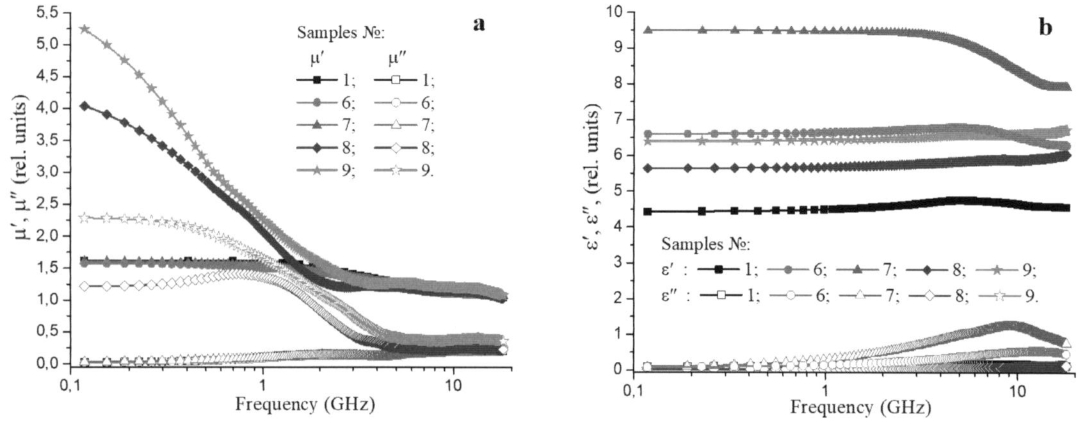

Figure 5. Frequency dependences of permeability (a) and permittivity (b) for a composite containing a fixed concentration of carbonyl iron with the addition of MWCNT and ferrite.

From the above dependences it can be seen that the addition of carbon nanotubes does not affect the value of magnetic permeability, since MWCNTs do not contain magnetic atoms (with the exception of an insignificant amount of metallic impurities) and do not possess magnetic properties. The addition of ferrite significantly increased the value of both the real and imaginary parts of permeability, but only at frequencies below 1 GHz. This indicates that the selected ferrite of the spinel structure has low-frequency magnetic properties.

Analyzing the frequency dependences of the dielectric constant, it can be seen that with the addition of multi-walled carbon nanotubes to the composite containing 40 wt.% CI, a significant increase in real part of permittivity (from 4.5 to 9.5 rel. units) and imaginary part of permittivity (from 0.2 to 1.4 rel. units) happens. However, at frequencies above 5 GHz, a slight decrease in the real part of the permittivity can be seen. This behavior is associated with both the polarization characteristics of carbon nanotubes and the significant conductive properties of this substance (decrease in dielectric constant with increasing frequency). The addition of ferrite has a less pronounced effect on the real part of permittivity P. In this case, the real part of permittivity practically does not depend on the frequency and almost linearly increases with increasing ferrite concentration (from 4.5 to 6.5 relative units). The imaginary part varies little. This behavior of the complex permittivity is caused solely by polarization characteristics of the filler.

Below are the results of measurements of the electromagnetic response from composite materials located on the metal. In addition to studying the experimentally obtained dependencies, the method of calculating the electromagnetic response from the magnetodielectric layer was used, based on the calculated spectra of complex dielectric and magnetic constant, which allowed us to obtain the frequency dependence of the reflection coefficient in the plane-wave approximation.

Figure 6 shows the frequency dependences of the reflection coefficients from experimental samples of composite materials containing carbonyl iron, multi-walled carbon nanotubes, and ferrites. With increasing concentration of carbonyl iron, the reflection coefficient decreases significantly, reaching values exceeding $R = -10$ dB, which corresponds to a decrease in reflected power by 10 times. The peak reflectance minimum is shifted to a low frequency region. The peak attenuation of –15 dB is attained at the frequency of 10.8 GHz.

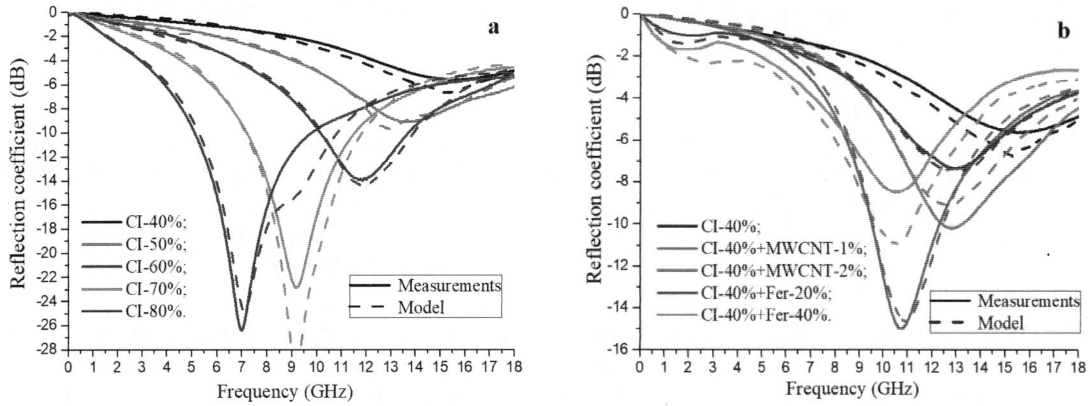

Figure 6. The frequency dependence of the reflection coefficient of the material on the metal for samples containing different concentrations of carbonyl iron (a) and samples containing a fixed concentration of carbonyl iron with the addition of MWCNT and ferrite (b).

On the frequency dependence presented in figure 6b, a weakening of the reflection coefficient is observed when MWCNT and ferrite are added to the carbonyl iron. Adding a few percent (1% and 2%) of MWCNT is almost equivalent to adding a few tens of percent (20% and 40%) of ferrite. The addition of 2 wt.% MWCNTs led to a decrease in the reflection coefficient by about 8 times, while the absorption peak also shifts to lower frequencies. The addition of ferrite allows to decrease the reflection coefficient from -0.4 dB to –2.5 dB at frequencies from 1 to 4 GHz.

The dependences obtained make it possible to characterize a composite material based on epoxy resin with the addition of carbonyl iron, multi-walled carbon nanotubes and ferrites as an effective absorber of electromagnetic radiation.

Comparing the results obtained using modeling, it can be noted that this model agrees quite well with the experimentally obtained dependencies. And this model can be used to calculate the reflection coefficient of an electromagnetic wave from an absorbing coating located on a reflecting surface, if the spectra of complex permittivity and permeability are known.

4. Conclusions

A comparison of the measured and calculated values of the reflection coefficient from the composite layer on the metal is done. They showed a good match. So using the composite with 80 wt.% CI will reduce the reflection coefficient by more than 10 times in the frequency range from 5.5 to 9.5 GHz, the composite from 70 wt.% CI - in the range from 7.5 to 11.5 GHz, and the composite from 40 wt.% CI and 2 wt.% MWCNTs range from 9.5 to 12 GHz. Adding 2 wt.% MWCNTs to 40 wt.% CI is equivalent to increasing the concentration of carbonyl iron from 40 to 60 wt.%. Which significantly reduces the consumption of materials and the mass of the composite.

Acknowledgment

The study was sponsored by RFBR in the framework of a research project №18-32-00810.

References

[1] Magisetty R P, Shukla A and Kandasubramanian B 2019 Terpolymer (ABS) cermet (Ni-

IOP Conf. Series: Materials Science and Engineering **511** (2019) 012021 doi:10.1088/1757-899X/511/1/012021

$NiFe_2O_4$) hybrid nanocomposite engineered 3D-carbon fabric mat as a X-band electromagnetic interference shielding material *Mater. Lett.* **238** 214–217

[2] Kumar S, Meena R, Chatterjee R 2016 Microwave absorption studies of Cr-doped Co–U type hexaferrites over 2–18 GHz frequency range *J. Magn. Magn. Mat.* **418** 194–199

[3] Zhuravlev V, Naiden E, Minin R 2015 Radiation-thermal synthesis of W-type hexaferrites *IOP Conf. Series: Materials Science and Engineering* **81** 012003

[4] Dunaevskii G E, Suslyaev V I, Zhuravlev V A, Badin AV, Dorozhkin K V 2016 Ferromagnetic resonance in hexagonal ferrite $Ba_3Co_2Fe_{24}O_{41}$ at the THz frequency range *41th International Conference on Infrared, Millimeter, and Terahertz Waves* 1-2

[5] Peng Q, Dearden A, Crean J et al. 2014 New materials graphyne, graphdiyne, graphone, and graphane: review of properties, synthesis, and application in nanotechnology *Nanotechnology, Sci. and Appl.* **7** 1–29

[6] Ebrahimi I and Gashti M P 2018 Polypyrrole-MWCNT-Ag composites for electromagnetic shielding: Comparison between chemical deposition and UV-reduction approaches *J. Phys. Chem. Solids* **118** 80–87

[7] Sun J, Wang Y, Wang W, Wang K and Lu J 2018 Application of featured microwave-metal discharge for the fabrication of well-graphitized carbon-encapsulated Fe nanoparticles for enhancing microwave absorption e ffi ciency *Fuel* **233** 669–676

[8] Ren X, Fan H, Cheng Y 2016 Microwave absorption properties of double-layer absorber based on carbonyl iron/barium hexaferrite composites *App. Phys. A: Mat. Sci. and Proc.* **122** 506

[9] Dai L, Xie S, Yu M and Ci L 2018 Fabrication and electromagnetic properties of carbon-based iron nitride composite *J. Magn. Magn. Mater.* **466** 22–27

[10] Faxiang Q, Hua-Xin P 2013 Ferromagnetic microwires enabled multifunctional composite materials *Prog. in Mat. Science.* **58** 183–259

[11] RaviPrakash M, Anuj S and Balasubramanian K 2018 Magnetodielectric Microwave Radiation Absorbent Materials and Their Polymer Composites *Journal of Electronic Materials.* **47** 6335–6365

[12] Huang L, Liu X and Yu R 2018 Enhanced microwave absorption properties of rod-shaped Fe2O3/Fe3O4/MWCNTs composites *Prog. Nat. Sci. Mater. Int.* **28** 288–95

[13] Kuleshov G, Zhuravlyova Y, Dotsenko O 2015 Electromagnetic response from composite radiomaterials based on multiwall carbon nanotubes at microwave frequencies *SIBCON 2015 Proc.* 7147115

[14] Chalapat K, Sarvala K, Li J, Sorin G 2009 Wideband Reference-Plane Invariant Method for Measuring Electromagnetic Parameters of Materials *IEEE Transactions on Microwave Theory and Techniques* **57** 2257–2267

Nitriding of ferrochromium-aluminum during combustion and evaluation of the photocatalytic activity of obtained composites

K A Bolgaru[1], A A Reger[1], L N Skvortsova[2]

[1]Tomsk Scientific Center SB RAS, 10/4 Akademicheskii Pr., 634055, Tomsk, Russia
[2]National Research Tomsk State University, 36 Lenin Ave., 634050, Tomsk, Russia

E-mail: antonandmakar@gmail.com

Abstract. The paper represents the possibility of nitriding of ferroalloy (ferrochromium-aluminum) by the method of self-propagating high-temperature synthesis. The main ferroalloy nitriding characteristics, such as the nitriding degree and the combustion rate are determined versus the sample diameter, nitrogen pressure and dispersity of the initial powder. Combustion of the alloy is found to take place in a layer-by-layer mode. The photocatalytic activity of the obtained ferrum-containing nitride composites during the degradation of soluble organic matters (dyes, phenol, diclofenac sodium) is evaluated. High efficiency of the obtained materials is observed by combining heterogeneous and homogeneous catalysis.

Introduction

Combustion of different metal powders under the nitrogen atmosphere was presented for the first time in the works conducted by A.G. Merzhanov [1]. The method is based on exothermic reactions proceeding in the combustion mode with the formation of valuable compounds and materials after passing the front of combustion wave. The combustion wave propagates due to the heat release of the chemical reaction in the initial layer and the transfer of heat to initiate an exothermic chemical reaction in the subsequent layer of the initial mixture [2]. The method of self-propagating high-temperature synthesis (SHS) is promising due to its significant advantages: low energy consumption, environmental friendliness, small-sized equipment, a large number of available raw materials, high-speed production of products, etc. [3]. As a rule, expensive powders are used as initial materials for SHS, which significantly increase the cost of final products [4]. The cost of the products obtained is significantly reduced when raw materials that are not subjected to deep cleaning are used as initial reagents, in particular, ferroalloys. The works [5, 6] demonstrate the use of ferroalloys for SHS and show that the properties of the combustion product are comparable with the properties of the products obtained from pure precursors. In our opinion, the most promising materials for SHS conducted in the mode of filtration combustion under the nitrogen atmosphere are ferroalloys, since these alloys can contain several nitride-forming elements and are interesting both from a theoretical and practical view. Ferrum contained in alloys catalyzes the nitriding of nitride-forming elements. The decomposition of ferroalloys allows nitrides of elements to be obtained in one stage. In this regard, ferroalloys are promising both in scientific and applied aspects, but at the same time they are complicated reagents for the synthesis of nitride-containing composites and nitrides. To control synthesis and solve the main task of SHS - obtaining the product with a required phase composition and a microstructure determining its properties, it is necessary to determine the characteristics of the nitriding process using ferroalloys as reagents. Earlier it was found that ferrum-containing metal-ceramic composites based on boron nitride, sialon, silicon and titanium [7] exhibit a high catalytic activity during oxidative degradation of soluble organic matters in the presence of activating reagents (H_2O_2, $H_2C_2O_4$, EDTA) and UV radiation, forming photoactive systems, such as Photo-Fenton, ferrioxalate and ferricomplexonate systems (Fe^{2+}/H_2O_2/UV, $Fe(C_2O_4)^{3-}_3$/UV, Fe/EDTA/UV) [8, 9]. Catalytic processes occur due to the high oxidative ability of OH-radicals [10] which are formed in the presence of composites and activating reagents. The photocatalytic activity of ferrum-containing metal-ceramic composites is related not only to the participation of ferrum in the formation of OH-radicals, but also to the presence of semiconductor phases in the ceramic matrix, which provide optical properties.

This paper presents studying the characteristics of combustion of ferrochromium-aluminum in nitrogen and evaluation of the catalytic activity of the product obtained.

Materials and methods

A ferroalloy (ferrochromium-aluminum) with a content of 8.5% aluminum, 30% ferrum, 61.2% chromium and 0.3% impurities was chosen to study the characteristics of combustion. The phase composition was determined using a Shimadzu XRD6000 diffractometer (Japan). According to X-ray phase analysis, ferrochromium-aluminum contains the phases as follows: AlFe, Cr, $AlFeCr_2$. Before synthesis, ferrochromium-aluminum was dried in a CHVS-40/3.5 vacuum oven (Russia) for 3 hours at a temperature of 150°C to remove water and volatile impurity compounds.

Ferrochromium-aluminum was placed on a cylindrical metal mesh with a diameter from 5 to 60 mm in a constant pressure unit on a non-conductive stand and burned at a pressure from 1 to 8 MPa. The combustion reaction was initiated by electric pulse supplied from the transformer through a tungsten coil. After the initiation of chemical reaction in initial ferrochrome with nitrogen, the movement of the combustion wave front was observed. After complete passage of the combustion wave, the sample was kept in the unit under the nitrogen atmosphere for 30 minutes until complete cooling, and then pressure was released and the obtained composite was studied.

The total content of nitrogen in the products was determined using a LEKO-ONH836 (USA) device.

The photocatalytic activity of nitrided composites was investigated by the following method: a sample of catalyst in the amount of 100 mg was placed in a quartz glass, and an organic pollutant solution (phenol, methylene blue, diclofenac sodium) in the amount of 10 ml and a few drops of activating reagent (H_2O_2, $H_2C_2O_4$) were added. The quartz glass with the reaction solution was covered and placed on a magnetic stirrer in front of the radiation source for a certain time. Analytical control of the concentration of pollutants was performed by the spectrophotometric method: methylene blue (MB) at $\lambda \sim$ 660 nm, eosin at $\lambda \sim$ 515 nm, diclofenac sodium at $\lambda \sim$ 275 nm, phenol with 4-aminoantipyrine at $\lambda \sim$510 nm. The degradation degree of the pollutant (R) was calculated by the formula:

$$R,\% = \frac{(c_{init} - c_{fin})}{c_{init}} 100,$$

where C_{init} is the initial concentration of SOM, and C_{fin} is the final concentration of SOM.

Results and discussion

In contrast to ferrochromium-aluminum, nitriding of ferrochrome by the SHS method is possible if preheating of an alloy to 300 C is conducted. The content of chromium in the initial mixture should exceed 75%. Studies have shown that the combustion of ferrochromium-aluminum occurs in the layer-by-layer mode. In this case, the initial mixture does not require preheating, and the content of chromium is 61%. According to X-ray phase analysis, the product obtained by nitriding of

IOP Conf. Series: Materials Science and Engineering **511** (2019) 012022 doi:10.1088/1757-899X/511/1/012022

ferrochromium-aluminum is multiphase and contains the phases as follows: CrN, α-Fe, $(Cr,Fe)_2N_{1-x}$, Cr_2N. Aluminum nitride, as a separate phase, is not detected, however, according to X-ray microanalysis, this phase is contained in chromium nitride. This is confirmed by the broadening of CrN reflexes presented in the X-ray pattern (figure 1).

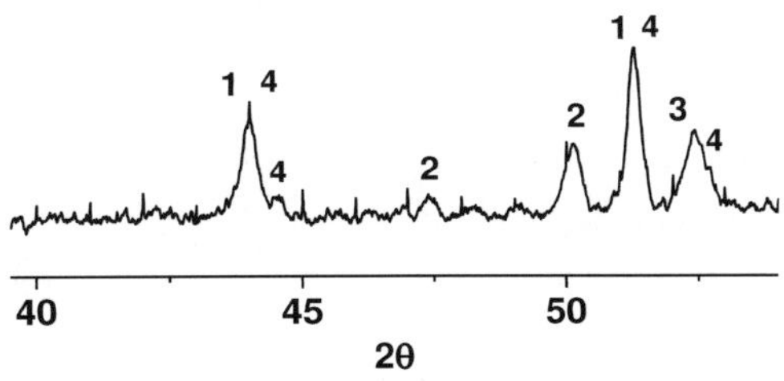

Figure 1. X-ray pattern of nitrided ferrochromium-aluminum (1 – CrN, 2 –$(Cr,Fe)_2N_{1-x}$, 3 – α-Fe, 4- AlN).

As a rule, the reaction of chromium nitriding consists of two stages:

The first stage is the Cr_2N reaction:

$$4Cr + N2 = 2Cr_2N - 25.2 \text{ kcal/mol.}$$

The second stage is the CrN reaction:

$$2Cr_2N + N2 = 4\ CrN - 28.2 \text{ kcal/mol.}$$

The CrN system is characterized by a low calculated adiabatic combustion temperature (2063 °C). The presence of $(Cr,Fe)_2N_{1-x}$ phase (Cr_2N phase, where the chromium atoms are partially replaced by ferrum in the products indicates that the nitriding process is incomplete, since not all Cr_2N turned into CrN.

For filtration combustion, one of the important factors affecting this process is nitrogen pressure. The value of initial pressure has a significant effect on the filtration characteristics and the nitriding degree of synthesized products. Usually the maximum pressure used in SHS processes is 10 MPa. Experiments have shown that the increase in pressure leads to a monotonic increase in the amount of nitrogen absorbed and in the combustion rate (figure 2). Then these values reach a maximum that corresponds to the nitriding degree of chromium. The study of the nitriding degree, combustion rate and phase composition as a function of nitrogen pressure showed that the combustion of the samples began at a minimum pressure of 1.5 MPa.

The phase composition of the products of ferrochromium combustion in nitrogen depends on the pressure of nitrogen. When the nitrogen pressure is above 8 MPa the main phase is CrN, at a pressure of 2-8 MPa a mixture of CrN and $(Cr,Fe)_2N_{1-x}$ phases is present, and at $P < 2$ MPa the predominant phase is $(Cr,Fe)_2N_{1-x}$.

The study of the nitriding degree, combustion rate and phase composition of the obtained composites versus the diameter of samples showed that the combustion reaction was initiated in a sample with a minimum diameter of 15 mm. The increase in the sample diameter leads to the slight decrease in the amount of nitrogen absorbed, and a relatively constant combustion rate is observed. These results are associated with two parallel factors: difficulties with filtration of gas towards the reaction zone and a decrease in heat loss in the reactor volume. In this case, the phase composition remains almost unchanged.

The study of the of nitriding degree, combustion rate and phase composition versus the dispersity of the initial powders showed that the samples with a particle size of more than 100 μm and up to 100 μm (40-63, 63-100) in the absence of a fine fraction did not burned. The effect of dispersity on the amount of nitrogen absorbed and the rate of combustion is presented in Figure 4. A decrease in the particle size of the initial powder leads to a significant increase in the amount of nitrogen absorbed and the rate of combustion. This is due to the fact that the heat release during the combustion of coarse powders is much lower than for fine powders.

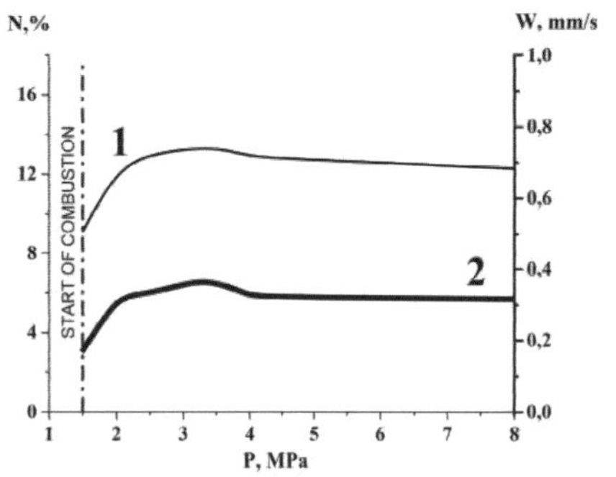

Figure 2. Content of nitrogen (1) and combustion rate (2) as a function of nitrogen pressure.

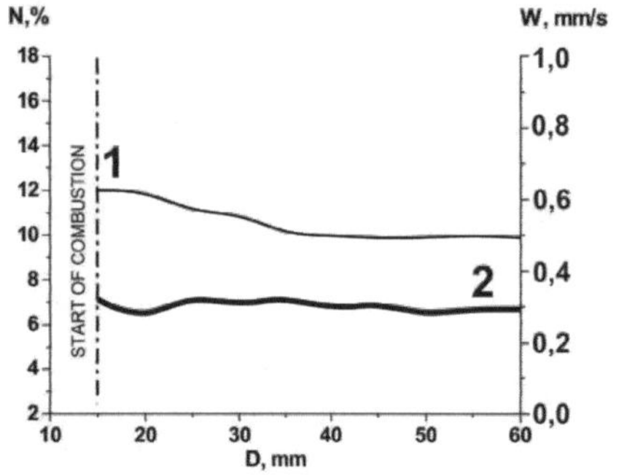

Figure 3. Content of nitrogen (1) and combustion rate (2) as a function of the sample diameter.

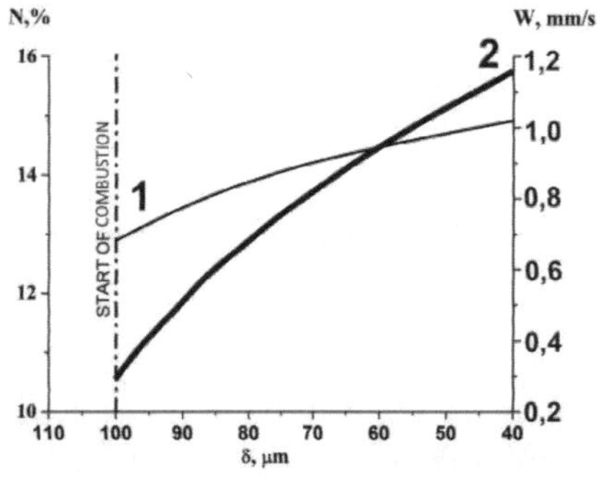

Figure 4. Content of nitrogen (1) and combustion rate (2) as a function of the particle size.

The obtained metal-ceramic composite contains a phase of CrN chromium nitride that is a narrow-bandgap semiconductor [11] and can participate in heterogeneous photocatalysis. At the same time, the composite contains the phase of metallic ferrum (α-Fe) the partial dissolving of which in the presence of H_2O_2 or $H_2C_2O_4$ forms the homogeneous photocatalytic Photo-Fenton and ferrioxalate systems generating a superoxidant - hydroxyl radicals. The catalytic activity of the synthesized sample during oxidative destruction of some organic pollutants (phenol, methylene blue, eosin, diclofenac sodium) was evaluated for different photocatalytic systems.

To identify the role of ceramic matrix in the photocatalytic process, the destruction of organic pollutants exposed to UV radiation was investigated in the absence and presence of a composite but without the addition of activating reagents. Figure 5 shows the degradation degree of diclofenac sodium as a function of time.

IOP Conf. Series: Materials Science and Engineering **511** (2019) 012022 doi:10.1088/1757-899X/511/1/012022

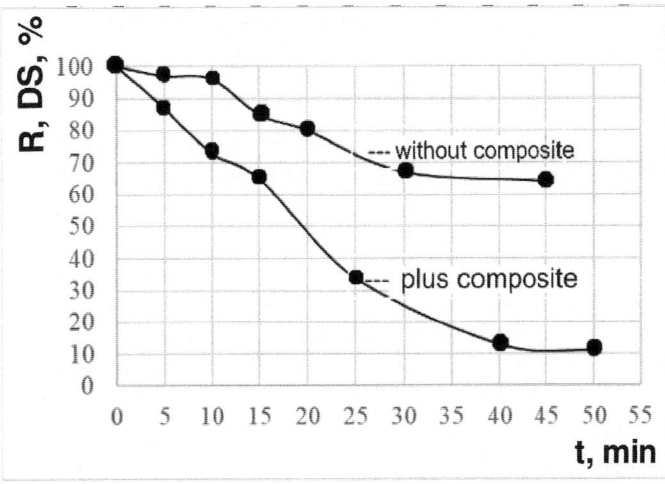

Figure 5. Effect of UV-radiation on the degradation degree of diclofenac sodium (DS) in the absence and presence of an iron-containing composite (m_{kt} = 100 mg, c_{DS} = 25 mg/l, v_{sol} = 10 ml).

It can be seen that the main destruction of organic pollutant occurs within 40 minutes. In the presence of a composite, the degree of degradation is almost 50% greater than the degradation degree of diclofenac sodium in the absence of a catalyst. This indicates the activity of ceramic matrix during the oxidative destruction of SOM.

Table 1 shows the catalytic activity of the synthesized composite in the absence and presence of H_2O_2 and $H_2C_2O_4$ additives during the degradation of the SOM studied. In the absence of activating reagents and presence of UV radiation, degradation of DOM can be conducted with the participation of ceramic matrix, since CrN is a semiconductor capable of absorbing UV energy and facilitating electron transfer processes. In the presence of UV radiation and the addition of activating reagents to the solution, the destruction degree of SOM increases significantly, which may be related to the participation of OH radicals in the catalytic process, which are generated by the photocatalytic ferrioxalate and Photo-Fenton systems formed in solution. Moreover, the efficiency of Photo-Fenton process is higher.

Table 1. Degradation (%) of SOM for different photocatalytic systems (c_{MB} = 5 mg/l, c_{phenol} = 0.5 mg/l, c_{DS} = 25 mg/l, c_{H2O2} = c_{H2C2O4} = 0,002 M, m_{kt} = 100 mg, V_{sol} = 10 ml, $\tau_{(UV)}$ = 15 min)

SOM	UV	UV/ $H_2C_2O_4$	UV/ H_2O_2
Methylene blue	21	64	97
Eosine	31	69	93
Diclofenac sodium	37	51	100
Phenol	28	48	87

Thus, the obtained composite exhibits a high activity during the degradation of different SOMs and is promising for the purification of wastewater from organic pollutants. The photocatalytic activity of Fe-containing compositions in the presence of activating reagents is related to the combination of homogeneous and heterogeneous catalysis.

The characteristics of the combustion of ferroalloy (ferrochromium-aluminum) were determined for different diameters, nitrogen pressure and dispersity of the initial powder. The main conditions were found for the initiation of reaction. The photocatalytic activity of the samples obtained during the oxidative destruction of SOM was evaluated.

Acknowledgments
The work was carried out as a part of the state task for TSC SB RAS (No. 0365-2018-0002) and partially supported by RFBR (project No. 18-33- 00387 mol_a).

References
[1] Merganov A G, Borovinskaya I P, Shkiro V M 1967 *The phenomenon of wave localization auto-inhibited solid-phase reactions (*USSR: Chernogolovka)
[2] Maximov U V, Chuxlomina L N, Braverman B S 2014 *Self-propagation high-temperature synthesis of nitrogen-containing alloys for metallurgy* (Novosibirsk: Nayka) p 232
[3] Maximov U V, Chuxlomina L N, Vereshagin V I 2012 *Self-combustion high-temperature synthesis of nitrogen-containing ceramic* (Novosibirsk: Nayka) p 260
[4] Kheirandish A R, Nekouee Kh A, Khosroshahi R A, Ehsani N 2016 *Self-propagating high temperature synthesis of SiAlON Int Journal of Refractory Metals and Hard Materials* **55** pp 68-79
[5] Ziatdinov M X, Shatoxin I M, Leontev L I 2018 *SHS technology of composition ferroalloys part 1. Metallurgical SHS process synthesis of ferrovanadium and ferrochromium nitrides* Izvestiya Visshikh uchebnykh zavedenii. Chernaya metalurgiya **61** pp 339-347
[6] Ting-an Z, Zhihe D, Hanbo Z, Zhiqi Z, Liping N, Jicheng H 2012 *Preparation of titanium based ferrous alloys with low oxygen content by reduction-SHS 3rd internation Symposium on High-Temperature Metallurgical Processing TMS (The minerals, metals & materials society)* ed T Jiang, J-Y Hwang, P Masset, O Yucel, R Padilla, G Zhoy pp 237-242
[7] Bolgary K A, Reger A A, Skvortsova L N 2018 *Combustion synthesis of metal-ceramic silicon nitride-based composites and their photocatalytic activity IOP. Series: Journal of Physics: Conf. Series* 1115(2018)042040 p 5
[8] Ebrahiem E E, Al-Maghrabi M N and Mobarki A R 2017 Removal of organic pollutants from industrial wastewater by applying photo-Fenton oxidation technology *Arabian Journal of Chemistry* **10** pp 1674–1679
[9] Sokolov E M Sheynkman L E Dergynov D V 2013 *The study of decrease of the phenolic compounds concentration in the aquatic environment using mathematical modeling Vestnik yuzhnogo naychnogo tsentra RAN* **9** pp 23-31
[10] Kozlova E A Parmon V N 2017 *Heterogeneous semiconductor photocatalysts for hydrogen production from aqueous solutions of electron donors Uspekhi khimii* **86** pp 870–906.
[11] Bhobe P A, Chainani A, Takeuchi T 2010 Evidence for a Correlated Insulator to Antiferromagnetic Metal Transition in CrN *Physical Review Letters* **104** pp 236404.

IOP Conf. Series: Materials Science and Engineering **511** (2019) 012023 doi:10.1088/1757-899X/511/1/012023

Numerical Simulation of Damage Accumulation in a Composite Flange with a Delamination Defect

P V Pisarev, A N Anoshkin and D A Ermakov

Perm National Research Polytechnic University, 614990, Komsomolskii ave., 29, Russia, Perm

E-mail: pisarev85@live.ru

Abstract. In this work, a numerical model for predicting the residual life of structures made of polymer composite materials using a structural-phenomenological model is proposed. To estimate the residual life of PCM structures under cyclic loading, it is proposed to use the approach of explicit describing the process of adhesive layers' destruction. The model is implemented in the APDL language in ANSYS Mechanical engineering complex. The model and the program testing was carried out on a flange type construction with a defect. A comparison of numerical experiments was made for flanges with and without the defect.

1. Introduction

One of the main issues arising in the process of developing promising composite products for aviation or rocket and space industries is to ensure construction reliability over the long term of operation. This issue is particularly significant when using composites in civil aviation and engine building. Unfortunately, the experience and scope of research on the use of composites in structures subjected to prolonged intense temperature and force impacts is much less than for most steels and alloys [1].

The design of power parts and assemblies made of composite materials should be based on new mathematical models of composite materials mechanics. Mathematical models should take into account the parameters of the structure, including changes in their design due to changes in the geometry or technological features of manufacturing, kinetic processes of damage accumulation during the operation under static and cyclic loads. Mathematical models should allow predicting the physic mechanical properties of the material under static and cyclic loads with different reinforcement options [2].

As part of this work, a model is proposed for predicting the residual life of structures made of polymer composite materials (PCM) with a defect in the form of delamination, using a structural-phenomenological model.

2. Description of the numerical model

In the manufacture of flanges various defects may appear, for example in the area of bending of the layers is the formation of wrinkles, which leads to the appearance of pores or delamination. To study the effect of a defect on the residual life of a PCM flange, two structural-phenomenological models with and without the defect were considered.

In order to build a structural-phenomenological model, a flange with a delamination defect was made according to the results of X-ray scanning. For the fabricated sample, an X-ray photographic survey was made, a reconstructed image was obtained, and a geometric model of the flange was constructed with an explicit description of both the composite material structure and the defect in it (Figure 1).

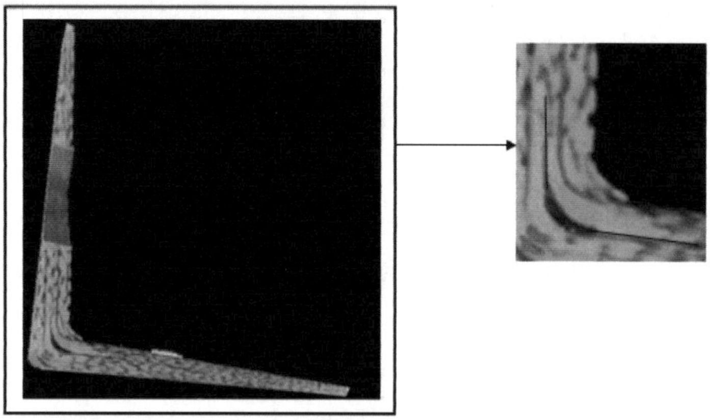

Figure 1. Shear of the reconstructed X-ray image at the location of the defect (delamination-pore)

To simulate the process of damage accumulation under cyclic loading, an approach is proposed to explicitly describe the destruction process for the adhesive layers situated in between the layers of reinforcing material. For numerical simulation of damage accumulations in the flange under a cyclic operational load, a geometric model of a flange with a delamination defect was built (Figure 2b). The model includes a system of reinforcing and adhesive layers. The number of reinforcing carbon fiber layers is 26. The adhesive layers were inserted between each reinforcing layer of the structure and are shown in purple color in Figure 2. The thickness of the adhesive layers was assumed to be 0.02 mm, which corresponds to 10% of the thickness of the structural carbon fiber layer. In the developed structural – phenomenological model with delamination, the defect was placed in the adhesive layer between 13 and 14 layers in the inflection area of the considered structure. The delamination area is shown in Figure 2b in red color.

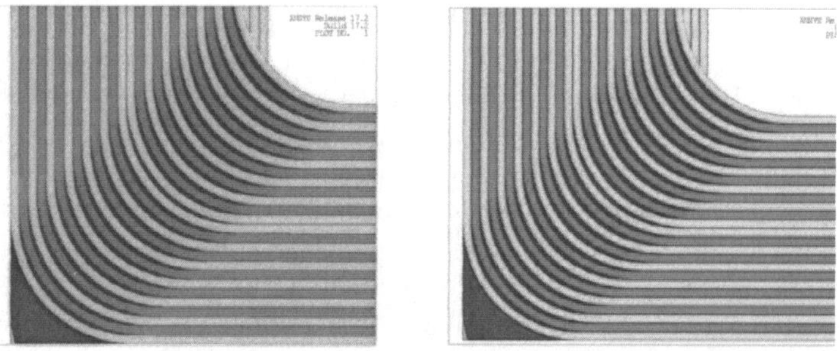

a b

Figure 2. Areas with adhesive layers in the composite flange model in the inflection area: a - without defects; b - with the defect

The calculations were carried out by the finite element method in ANSYS Mechanical software engineering complex. The finite element model of the flange was developed with an explicit account of its reinforcement scheme and the presence of adhesive layers. For better convergence of the solution and to reduce the amount of the errors of the obtained results, discretization of adhesive and reinforcement layers was carried out to eight-node PLANE182 elements. The maximum element size for the computed model of the composite flange was 0.02 mm, the minimum - 0.002 mm. The total number of finite elements was 500 thousand elements. Also, in the course of the elemental model, the carbon fiber layers' reinforcement scheme was taken into account. Layers with 0° reinforcement are painted in pale blue color, and with 45° reinforcement in dark blue (Figure 1). Technical elastic constants of the reinforcing and adhesive layers are taken from [3-4] and are presented in Table 1, where E_x, E_y, E_z are the Young's modules in the direction of the axes, υ_{xy}, υ_{yz}, υ_{xz} are the Poisson's coefficients, G_{xy}, G_{yz}, G_{xz} are the shear moduli.

Table 1. Elastic properties of materials used in the numerical calculation of the composite flange life

Material	E_x, hPa	E_y, hPa	E_z, hPa	υ_{xy}	υ_{yz}	υ_{xz}	G_{xy}, hPa	G_{yz}, hPa	G_{xz}, hPa
Reinforcement layer	63.9	20	63.9	0.3	0.3	0.04	2.7	2.7	19.5
Adhesive layer		2.9			0.356			1.07	

In the numerical experiments, the development of the "delamination" type damage was simulated, which is a typical and characteristic type of destruction for these structures, occurring both under operational load and during laboratory tests. The strength of an epoxy binder under the action of cyclic loads was taken from the experimental data described in publications. A preliminary assessment of the life of the flanges was carried out according to the weakest link criterion [5].

The calculation of the flange stress strain behavior was carried out in an axisymmetric formulation using a two-dimensional model. The following boundary conditions were used in the calculation (Figure 3): the bolted area was rigidly fixed, a load (F) was applied to the end of the right free part of the flange with a maximum value in the loading cycle of 3000 N for both models, with and without the defect. In the calculations it was assumed that the specified load is periodic, from zero with a sinusoidal loading cycle, and in the process of application of this load in the adhesive layers accumulation of damage occurs followed by the destruction.

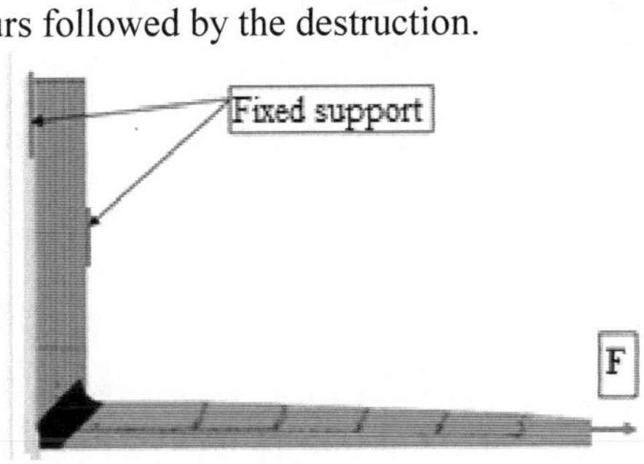

Figure3. Boundary conditions scheme

3. Description of the algorithm for predicting the residual life of a multilayer structure

In accordance with the developed algorithm, during the cyclic loading, the level of damage is fixed in the adhesive layers. When the critical level of damage is reached, the destruction of the adhesive layers is modeled. In the destroyed elements of the adhesive layers, the elastic constants are assumed to be approximately equal to zero (multiplied by 10^{-6}), which corresponds to the interlayer destruction of the laminated package. To describe the damage accumulated in an elementary volume in the area of a certain material point of the structure, a scalar time function $\psi(t)$ is used. It is assumed that the function $\psi(t)$ takes values on the interval [0, 1]. In this case, the value $\psi = 0$ corresponds to the case when there is no damage, the value $\psi = 1$ corresponds to the level of damage at which the destruction of the elementary volume occurs. It was assumed that at the initial time $t = 0$ for all points of the construction under consideration $\psi = 0$. The state $\psi = 1$ corresponds to such a time or a number of loading cycles N when a certain strength criterion is violated, depending on the stress state at a given point σ_{ij} and material constants $S_{ij}(N)$.

We write a kinetic equation of damage accumulation, which determines the value ψ (N, σ_{mn}), in accordance with the linear rule of summation of damages,

$$\psi(N,\sigma_{mn})=\sum_{k=1}^{N}\frac{1}{N_b\left(\sigma_i^{(k)}\right)}, \tag{1}$$

where N_b is the function of cycles before failure, $\sigma_i^{(k)}$ is the resultant stress in the k-th finite element.

The loading is assumed to be cyclic, symmetric with the amplitude corresponding to the maximum static load modulo. The influence of dynamic effects is neglected, the stress and strain fields in the structure for amplitude loads are considered equal to the corresponding fields under static loads. The strength of the adhesive coat under the action of cyclic loads is taken from the data reported in [6]. On the basis of these experimental data, two-link fatigue endurance strength was constructed, approximated by the equations

$$\begin{aligned}\lg\left[N_b(\sigma_i)\right]&=-0.0684\sigma_i+9.18 \qquad \sigma_i<42MPa,\\ \lg\left[N_b(\sigma_i)\right]&=-2.1029\sigma_i+94.63 \quad 42MPa<\sigma_i<45MPa.\end{aligned} \tag{2}$$

For each layer in each finite element, we obtain a nonlinear equation for N_b. After the solution, the number of cycles before the destruction of each element N_{bj}^1 is determined. The minimum value of N_{bj}^1 for all the finite elements is the number of cycles before the first act of fatigue failure in the structure $N_{b\Sigma}^1$. It is assumed thereafter that the corresponding element is destroyed and that the elastic moduli are reduced in it.

The rest elements are damaged herewith, the damages being computed by the formula corresponding to kinetic equation (1)

$$\psi_j^1=\frac{N_{b\Sigma}^1}{N_{bj}^1} \tag{3}$$

For the former amplitude of external load, a new stress-strain state of the structure with one damaged element q is calculated. Next, we determine the number of cycles before each element is destroyed N_{bj}^2 at operating stresses σ_{ij}. Destruction in this case will occur in the element where the damage, in view of the value (ψ_j^Σ) accumulated at the previous step will be equal to 1,

$$\psi_j^\Sigma+\psi_j^2=1. \tag{4}$$

The damage at the current (second) step is determined by the formula

$$\psi_j^2=\frac{N_{b\Sigma}^{2add}}{N_{bj}^2}, \tag{5}$$

where N_j^{2add} is the additional number of operating cycles in the second step. Substituting (3) into (4), we calculate N_j^{2add} for each element, the minimum value of which is the additional development of the structure up to the second act of destruction in the corresponding element. The total number of cycles of operating time is calculated by the formula

$$N_\Sigma^2=\min{}_j\left(N_j^{2add}\right)+N_\Sigma^1. \tag{6}$$

The elastic modulus tensor for the destroyed element is reduced and, for the remaining unresolved elements, the accumulated damage is calculated as

$$\psi_j^\Sigma=\psi_j^1+\frac{\min{}_j(N_j^{2add})}{N_{bj}^2}. \tag{7}$$

The next step starts with calculating the design stress-strain state with two damaged items. According to this algorithm, the fatigue life of the structure can be calculated before it exhausts the load-bearing capacity to the amplitude load determined by the static tensile strength, or by a critical reduction in the rigidity of the structure, or by the formation of a zone of damage of a given critical value. The final value N_Σ is the refined value of the fatigue life of the structure in comparison with the estimate obtained by the criterion of the weakest link.

The problem of calculating the stress strain state of the flange for the maximum (amplitude) load in the cycle, taking into account the accumulation of damage and destruction, is solved iteratively in steps. For all the elements of adhesive layers, the current value of damage is determined, which allows us to determine the location and the number of operating cycles prior to the next failure. In the considered variant, the mechanical properties of the adhesive layer material are independent on the level of damage until the critical value 1 is reached. After this value is reached in some finite element, the material was considered failed and the mechanical properties in this element have were reduced to negligible values. The structure with the changed properties in this destroyed final element is calculated thereafter. Thus, at each step of the algorithm, the fatigue damage of the structure, the deformation and stress fields are simulated. The calculations are made in the ANSYS software package with an integrated software module.

4. Analysis of the results

According to the results of the numerical calculation, a description was obtained of the processes of damage accumulation and destruction in the structure of the researched model under cyclic loading; the fields of stress intensity distribution, damage and durability were constructed and analyzed. The number of cycles to failure was estimated for all final elements of the adhesive layers. The minimum values of the operating time before failure were obtained, which for the flange without the defect is $N=2.819 \cdot 10^8$ loading cycles, and for the flange with the defect - $N=1.108 \cdot 10^8$ loading cycles. These values are a preliminary estimate of the flanges durability by the criterion of the weakest link, which refers to the first act of destruction in the structure. The maximum values of stresses in the adhesive layers for the defect-free flange are observed in the area of fastening, for the defect flange - on the border of the defect and the adhesive layer.

The estimate of the durability for this flange, obtained at the first simulation step, showed that the destruction begins for the defect-free flange on the $2.819 \cdot 10^8$ loading cycle in the first adhesive layer in the area of fastening; for the defect flange on the $1.108 \cdot 10^8$ loading cycle in the central, vertical adhesive layer in the area of the defect. Thus, the preliminary durability of the defect-free flange, evaluated by the weakest link criterion, is approximately 2.54 times more than the durability of the defect flange. The relative area of destruction was determined as the ratio of the area of the destroyed elements to the total area of the elements in the adhesive layers.

Analysis of the destruction kinetics revealed that the destruction of the defect-free flange begins in the area of fixation. At the time of $5.34 \cdot 10^8$ loading cycles, a through-delamination between the first and second layer in the area of fixation appears in the defect-free flange. The total area of destruction in this case is 1%. Calculations have shown that from the moment of the first fracture occurrence until the occurrence of the through-delamination, the flange can withstand $2.53 \cdot 10^8$ additional loading cycles.

The destruction of the defect flange begins in the area of the delamination. Once reaching $3.41 \cdot 10^8$ loading cycles, a through-delamination in the central adhesive layer appears at the defect flange. The total area of destruction was 1.673%. It has been revealed that from the moment of the occurrence of the first destruction until the occurrence of through-delamination, the defect flange withstands additional $2.31 \cdot 10^8$ cycles.

A comparative analysis of the calculation results of damage accumulation kinetics for both flanges revealed that the appearance of through-delamination in the defect flange occurs 1.5 times faster than in the defect-free one. The difference in additional cycles after the occurrence of the first area of destruction was $0.22 \cdot 10^8$ cycles.

With the emergence of through-delamination in the flanges, a significant decrease in stiffness is observed. A critical drop in stiffness (approximately more than 15%) for a defect-free flange is observed at the 200th loading step, the value of stiffness falling was about 17.65%, with a total number of operating cycles equaling $534.649 \cdot 10^6$. For a defect-free flange, a stiffness reduction of 16.32% is observed after the 250th loading step, with a total number of operating cycles equaling $341.798 \cdot 10^6$. Tables 2, 3 present the values of the maximum displacements and the stiffness reduction of the flanges at different loading steps.

Table 2. Maximum displacements and stiffness drops for a defect-free flange model

Load step	Number of loading cycles, $10^6 N$	Maximum displacement, mm	Stiffness drop, %
1	281.989	0.204	0
100	531.701	0.218	6.86
200	534.649	0.240	17.65
300	534.667	0.271	31.86
400	534.691	0.373	70.1

Table 3. Maximum displacements and stiffness drops for a model of a flange with a defect

Load step	Number of loading cycles, $10^6 N$	Maximum displacement, mm	Stiffness drop, %
1	110.849	0.230	0
100	272.844	0.249	8.26
200	341.230	0.257	11.75
300	342.298	0.339	47.39
400	343.211	0.436	89.57

Thus, the developed model and program make it possible to determine the allowable value of the stiffness reduction for PCM structures in the simulations of damage accumulation and destruction under cyclic loading processes.

5. Conclusion

Thus, in this work, a model and a program for predicting the residual life of PCM structures using a composite material structural-phenomenological model was developed and tested. A preliminary assessment of the fatigue strength of flanges by the weakest link criteria showed that the durability of the defect-free flange is approximately 2.54 times more than the durability of the defect flange. A comparative analysis of the results obtained for both flanges revealed that the occurrence of through-delamination at the defect flange occurs 1.5 times faster than at the defect-free flange. The difference in additional cycles after the occurrence of the first are of destruction was $0.22 \cdot 10^8$ cycles. In addition, it was found that with the appearance of through-delamination, a significant decrease in stiffness is observed for the flanges. It can be concluded that the developed model and program make it possible to determine the acceptable value of the stiffness reduction for PCM structures.

Acknowledgments

The study was carried out at the Perm National Research Polytechnic University with the financial support of the Russian Foundation for Basic Research (project No. 17-48-590129).

References

[1] Volegov P S, Gribov D S and Trusov PV 2015 *Physical mesomechanics* **18(3)** 11-24
[2] Volkov I A and Korotkikh Yu G 2014 *Izvestia RAN. Solid mechanics* **3** 66-78
[3] Anoshkin A N, Fedorovtsev D I, Pisarev P V and Osokin V M 2015 *PNSTU Bulletin. Aerospace engineering* **4** 43
[4] Lukina N F, Dementieva L A and Kutsevich K E 2014 *VIAM works* **6**
[5] Anoshkin A N and Tashkinov AA 1998 *Prediction of the bearing capacity of composite flanges of aircraft engine body parts* (Perm. state tech. un-t Perm) p 101

IOP Conf. Series: Materials Science and Engineering **511** (2019) 012024 doi:10.1088/1757-899X/511/1/012024

Peculiarities of chemical element redistribution near free surfaces in CoCrFeMnNi high-entropy alloys

A V Korchuganov and I S Lucenko

Institute of Strength Physics and Material Science of the Siberian Branch of the Russian Academy of Sciences, 2/4 Akademicheskii ave., Tomsk 634055, Russia

E-mail: avkor@ispms.ru

Abstract. Based on molecular dynamics and Monte Carlo simulation, the distribution of chemical elements near free surfaces of various crystallographic orientations in high-entropy CoCrFeMnNi alloys was investigated. It is shown that for studied stoichiometric compositions of the alloy it has a different character and differently affects the mechanical characteristics and the defect structure of alloys subjected to uniaxial tension. This is due to the different influence of each element on the elastic modulus of the alloy. Segregation has a greater effect on the Young's modulus and elastic limit for $Co_{30}Cr_{30}Fe_{10}Mn_{10}Ni_{20}$, while the resulting defect structures have more differences for $Co_{10}Cr_{10}Fe_{30}Mn_{30}Ni_{20}$.

1. Introduction

The study of the structure and behavior of high-entropy alloys (HEAs) during thermal and mechanical influences is one of the most relevant topics of modern materials science [1], due to their outstanding physical and mechanical properties compared to traditional alloys [2]. HEAs with equiatomic composition are most often studied, since they are considered to be more thermodynamically stable due to the maximum configurational entropy [3]. At the same time, HEAs with a non-equatomic composition also remain stable and often have better properties. Therefore, the search for new stoichiometric compositions of HEAs is an urgent material science problem [4]. In view of the complexity of a HEA, it is very important to preserve the stable alloy structure and to study the characteristics of the segregation of chemical elements and the formation of new phases that can reduce the performance of the material. Today, CoCrFeMnNi HEAs are characterized by high physical and mechanical properties and have great potential for their improvement [5,6]. The atomic mechanisms of the redistribution of chemical elements in it, as well as the effect of the stoichiometric composition on them, remain largely unclear. To solve this problem, the most effective approach is the computer simulation based on the molecular dynamics method, which explicitly takes into account the atomic structure of the material and allows simulating the behavior of the material under extreme loading conditions [7-10]. The purpose of this work is to elucidate the effect of chemical element redistribution on the characteristics of the plastic deformation nucleation in the CoCrFeMnNi HEA of different stoichiometric composition within the framework of molecular dynamics simulation.

2. Simulation details

The temperature of the samples was 300 K. The crystallites had dimensions of 10×10×10 nm. Periodic boundary conditions were used along two directions, and a free surface (001), (110) or (111) was specified along the third direction. The stretching was carried out along the [110] direction with a rate of 10^9 s^{-1}. Stress component along the other direction with periodic boundary conditions was kept zero by changing the corresponding sample size. The simulations were performed in the LAMMPS package [11]. To describe the interatomic interaction in CoCrFeMnNi, many-body potentials [12,13] were used, which were constructed in the framework of the modified embedded atom method [14]. As in [15] molecular dynamics and Monte Carlo (the Metropolis algorithm) methods were used together in the simulations for the relaxation of samples. In this approach, randomly selected atoms of different types are swapped with the probability determined by the change in the energy of the system and temperature, and the relaxation of the structure of the material is carried out within the framework of the molecular dynamics method [16]. Samples with a random distribution of chemical elements and samples after relaxation by molecular dynamics and Monte Carlo methods (after segregation of elements) were considered to elucidate the influence of segregation on mechanical behavior of HEAs. Two stoichiometric compositions of the alloy, $Co_{10}Cr_{10}Fe_{30}Mn_{30}Ni_{20}$ and $Co_{30}Cr_{30}Fe_{10}Mn_{10}Ni_{20}$, with significantly different concentrations of chemical elements were chosen. The identification of the local type of the crystal lattice was carried out on the basis of an algorithm that determines the symmetry of the nearest environment of each atom of the sample – Common Neighbor Analysis (CNA) [17]. According to the CNA analysis, the twin boundary and the intrinsic stacking fault in the fcc lattice are determined as one and two layers of atoms with the hcp symmetry of the nearest environment, respectively. The structure of the loaded samples was visualized in the OVITO [18] package.

3. Results and discussion

It was found that redistribution of chemical elements occurs near the sample surface. For both compositions of the samples Mn goes to the free surface. At the same time, the Fe content decreases by an amount approximately comparable to the increase in the fraction of Mn in the surface layer, figure 1(a). Cr concentration slightly increases in the sub-surface region and decreases significantly on the surface of the crystallites, however, the average value does not change in the region of 7Å thickness. Ni content for the (001) and (110) surfaces decreases, while for (111) surface it increases, figure 1(b). The fraction of Co changes oppositely to Ni fraction, so that the total content of Ni and Co remains approximately constant, figure 1(c).

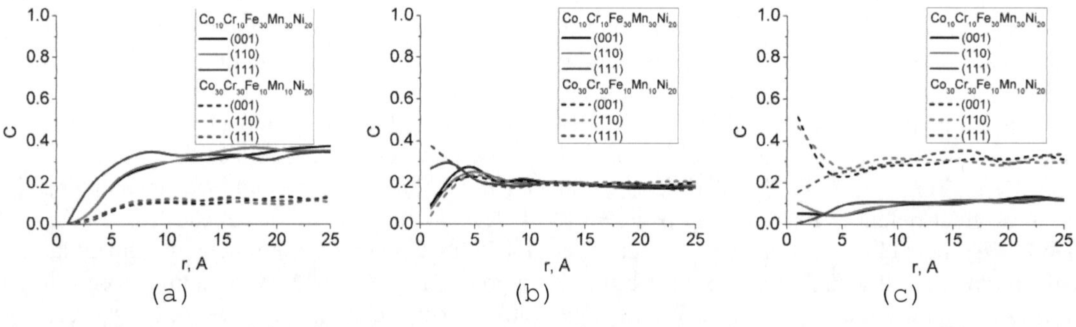

Figure 1. The fraction of chemical elements versus the distance to the free surface: (a) Fe, (b) Ni, (c) Co.

After segregation of chemical elements, the elastic limit and Young's modulus decrease for $Co_{30}Cr_{30}Fe_{10}Mn_{10}Ni_{20}$ alloys, and the elastic limit increases for $Co_{10}Cr_{10}Fe_{30}Mn_{30}Ni_{20}$ alloys with (001) and (110) surfaces, figure 2. This is due to the fact that different elements affect the Young's modulus of the alloy differently and are differently distributed near the surface. In this case, structural defects nucleate on the surface of the samples. As shown by previous studies, Cr and Fe reduce the Young's modulus of CoCrFeMnNi HEAs, and Co and Ni increase the modulus [15]. Since the total fraction of Ni and Co near the surface of samples of both compositions does not change after segregation of elements, and the Fe content decreases more significantly for $Co_{10}Cr_{10}Fe_{30}Mn_{30}Ni_{20}$, the effective Young modulus on its surface increases, which leads to an increase in the elastic limit. The decrease in the elastic limit of $Co_{30}Cr_{30}Fe_{10}Mn_{10}Ni_{20}$ samples may be due to the influence of Mn escaped to the surface together with other elements: Ni/Co or Cr. So, for example, the bulk modulus of the CoMn compound is 114 GPa, which is much less than the modulus of the equiatomic CoCrFeMnNi alloy – 183 GPa.

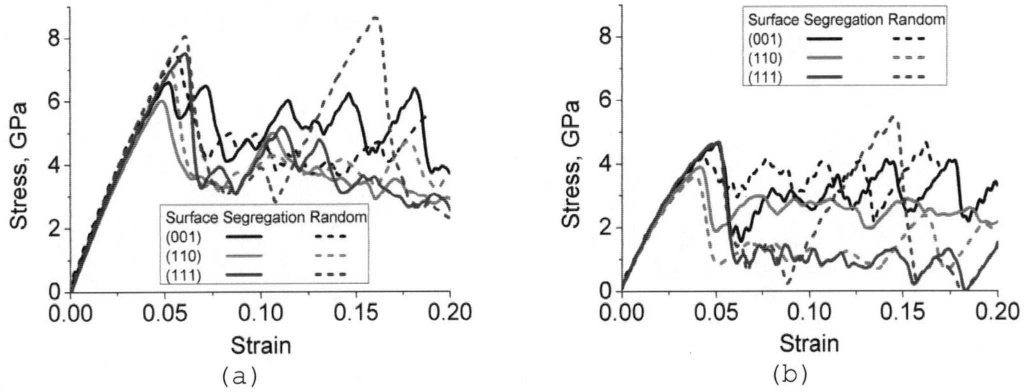

Figure 2. Stress versus strain for samples of different stoichiometric composition with random distribution and segregation of chemical elements: (a) $Co_{30}Cr_{30}Fe_{10}Mn_{10}Ni_{20}$; (b) $Co_{10}Cr_{10}Fe_{30}Mn_{30}Ni_{20}$.

Even for different initial distributions of chemical elements loaded samples with the given free surface orientation and stoichiometric composition are characterized by a certain type of nucleated structure defects. However, different initial distributions of chemical elements result in different number, size and position of defects. Segregation of chemical elements on the free surface affects these quantities. Moreover, this effect is different for considered stoichiometric compositions. Figure 3 show the fraction of atoms with hcp symmetry of the nearest environment in the samples with the (001) surface stretched by 20%. This value is calculated for all atoms corresponding to intrinsic stacking faults and bands with hcp lattice; atoms lying on twin boundaries were excluded from the calculation. As can be seen from figure 3(a), the fraction of hcp atoms is higher for samples $Co_{10}Cr_{10}Fe_{30}Mn_{30}Ni_{20}$ after segregation, which is expressed in greater thickness and number of hcp bands in these samples, figures 4 (a,b). On average, the fraction of hcp atoms does not change before and after segregation for $Co_{30}Cr_{30}Fe_{10}Mn_{10}Ni_{20}$ samples, however, there is a smaller scatter for samples with segregation, figure 3(b). The weak effect of segregation on the fraction of hcp atoms in the $Co_{30}Cr_{30}Fe_{10}Mn_{10}Ni_{20}$ samples is due to the fact that twinning is the main mechanism of deformation of these alloys, figures 4 (c,d).

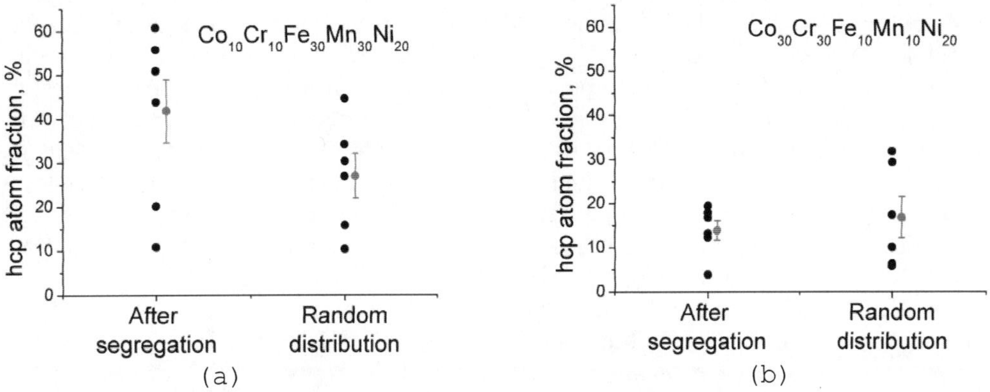

Figure 3. Fractions of hcp atoms corresponding to intrinsic stacking faults and bands with hcp lattice for samples of different stoichiometric composition with random distribution and segregation of

chemical elements on the (001) surface: a) $Co_{10}Cr_{10}Fe_{30}Mn_{30}Ni_{20}$, b) $Co_{30}Cr_{30}Fe_{10}Mn_{10}Ni_{20}$. Red points correspond to average values with the standard error.

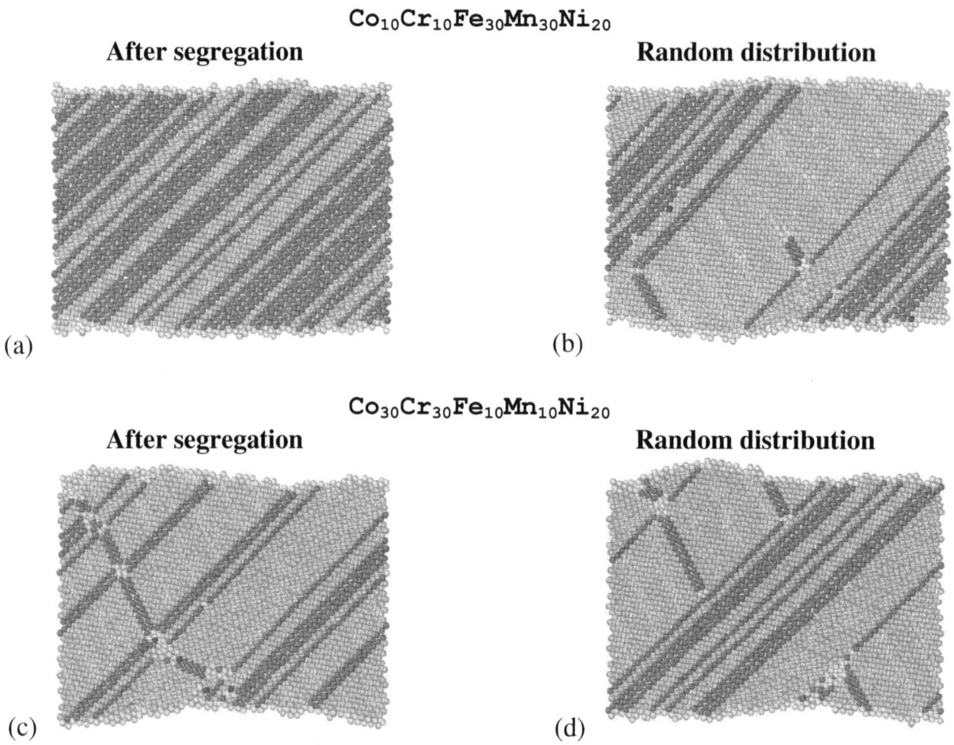

Figure 4. Defect structure at 0.2 strain for samples of different stoichiometric composition with random distribution and segregation of chemical elements on the (001) surface: (a,b) $Co_{10}Cr_{10}Fe_{30}Mn_{30}Ni_{20}$, (c,d) $Co_{30}Cr_{30}Fe_{10}Mn_{10}Ni_{20}$. Projection on the (110) plane. Green, red, blue, and grey colors of the atoms denote fcc, hcp, bcc, and undefined symmetry of their nearest environment, respectively.

4. Conclusions

Within the framework of molecular dynamics and the Metropolis Monte Carlo methods, the effect of segregation of chemical elements near the free surface on the features of nucleation and development of plastic deformation under uniaxial tension of high-entropy CoCrFeMnNi alloys of different stoichiometric composition is investigated. It was found that a redistribution of almost all chemical elements of the alloy occurs near the sample surface. For the considered alloy compositions, the Mn fraction on the surface increases, and the Fe content decreases by about the same value. The average Cr content does not change. Co and Ni fractions change opposite to each other. The redistribution of elements leads to a change in the elastic limit and the Young's modulus compared to samples with a random distribution of the elements. Moreover, it has a different effect on these values for different stoichiometric compositions. After segregation of chemical elements, the elastic limit and Young's modulus decrease for $Co_{30}Cr_{30}Fe_{10}Mn_{10}Ni_{20}$ alloys, and the elastic limit increases for $Co_{10}Cr_{10}Fe_{30}Mn_{30}Ni_{20}$ alloys with (001) and (110) surfaces. The effect of chemical element segregation on the defect structure of $Co_{30}Cr_{30}Fe_{10}Mn_{10}Ni_{20}$ alloys with the (001) surface is insignificant: the main mechanism of plasticity under tension is twinning, both before and after segregation. In $Co_{10}Cr_{10}Fe_{30}Mn_{30}Ni_{20}$ alloys with the (001) surface and random distribution of chemical elements, twins and bands with a hcp structure are formed under tension. After segregation the majority of defects is bands with hcp structure.

Acknowledgments
The work was carried out with the financial support of Russian Science Foundation (project No. 17-79-10108).

References

[1] Miracle D B and Senkov O N 2017 *Acta Mater.* **122** 448
[2] Lim X Z 2016 *Nature* **533** 306
[3] Guo S 2015 *Mater. Sci. Technol.* **31(10)** 1223
[4] Miracle D, Majumdar B, Wertz K and Gorsse S 2017 *Scripta Mater.* **127** 195
[5] Gludovatz B, Hohenwarter A, Catoor D, Chang E H, George E P and Ritchie R O 2014 *Science* **345** 1153
[6] Li Z, Tasan C C, Springer H, Gault B and Raabe D 2017 *Sci. Rep.* **7** 40704
[7] Golovnev I F and Golovneva E I 2017 *Phys. Mesomech.* **20** 353
[8] Nikonov A Y, Zharmukhambetova A M, Ponomareva A V and Dmitriev A I 2018 *Phys. Mesomech.* **21** 43
[9] Kryzhevich D S, Zolnikov K P and Korchuganov A V 2018 *Comput. Mater. Sci.* **153** 445
[10] Zolnikov K P, Korchuganov A V and Kryzhevich D S 2018 *Comput. Mater. Sci.* **155** 312
[11] Plimpton S 1995 *J. Comput. Phys.* **117(1)** 1
[12] Choi W M, Kim Y, Seol D and Lee B J 2017 *Comput. Mater. Sci.* **130** 121
[13] Wu C, Lee B J and Su X 2017 *Calphad* **57** 98
[14] Lee B J, Baskes M I, Kim H, and Koo Cho Y 2001 *Phys. Rev. B* **64** 184102
[15] Korchuganov A V 2018 *Mathematica Montisnigri* **42** 65
[16] Sadigh B, Erhart P, Stukowski A, Caro A, Martinez E and Zepeda-Ruiz L 2012 *Phys. Rev. B.* **85(18)** 184203
[17] Honeycutt J D and Andersen H C 1987 *J. Phys. Chem.* **91** 4950
[18] Stukowski A 2010 *Modell. Simul. Mater. Sci. Eng.* **18(1)** 015012

Physicomechanical properties of Ti-Ta-based surface alloys synthesized on the NiTi shape memory alloy

F A D'yachenko[1], S N Meisner[1] and E V Yakovlev[2]

[1]Institute of Strength Physics and Materials Science SB RAS, Tomsk, Russia
[2]Institute of High-Current Electronics SB RAS, Tomsk, Russia

E-mail: frozennonetroll@mail.ru

Abstract. This paper presents a research data on investigation of physicomechanical properties of Ti-Ta-based surface alloys, synthesized on the NiTi-substrate. The Ti-Ta-based surface alloys were formed in a single vacuum cycle, that comprises preliminary low-energy high-current pulsed electron beam (LEHCPEB), Ti-Ta film deposition by magnetron sputtering, and pulsed electron beam melting of the system "Ti-Ta-based surface alloy/NiTi-substrate". Experimental investigation and analysis of physicomechanical properties of Ti-Ta-based surface alloys on the NiTi-substrate were performed by method of nanoindentation. Our data suggest that the electron beam synthesis of Ti-Ta-based surface alloys, different in thickness and composition on the NiTi-substrate, leads to a gradient change of physicomechanical parameters. Modification of the surface of NiTi alloy changed the functional (pseudoelastic) properties by no more than 10%.

1. Introduction

Nowadays in the field of physics of hardening the surface of metallic materials different ways of combined processing are being developed. It can be electron beam and ion-plasma technologies [1]. The surface treatment of miniature devices (*actuators*), that are applicable in the field of microelectronics [2], by magnetron sputtering of alloying components and subsequent pulsed melting by a low-energy high-current pulsed electron beam (LEHCPEB) makes it possible to form thin layers on the surface of miniature devices. They are termed surface alloys (SA) and provide high adhesion to metallic substrate [3].

However, it is known that modification of the surface of miniature devices based on nickel titanium alloy (NiTi) does not always have a positive influence on their integral physicomechanical properties. This is important to consider during the processing of the functional NiTi alloy, which has the unique shape memory effect (SME) and superelasticity (SE). Therefore, it is necessary to understand the impact of treatments on functional (pseudoelastic) SME-SE properties during the modification of physicomechanical characteristic of the surface of miniature devices.

At the moment, a precision method of investigation and analyzing physicomechanical properties is a nanoindentation. Especially, this method is useful to evaluate the physicomechanical characteristics of modified layers. Such testing can identify differences in elastic and plastic behavior of materials with thin films and coatings, including those with a multilayer gradient structure [4, 5].

The purpose of this work is to investigate physicomechanical properties of Ti-Ta-based surface alloys synthesized on the NiTi shape memory alloy.

2. Experimental

2.1. Material

The material investigated here is commercial NiTi alloy produced as rolled sheets by vacuum induction melting (MATEK-SMA, Russia). The chemical composition of the alloy was: Ti (balance); 55.08 Ni; 0.051 C; 0.03 O; 0.002 N (wt %). The substrate was spark cut to dimensions of 10x10x1 mm by electrical discharge machining (EDM), chemically cleaned, electrolytically polished and washed in an ultrasonic bath with distilled water.

2.2. Formation of Ti-Ta-based surface alloys

The Ti-Ta-based surface alloys were formed on modified automatic setup RITM-SP (Microsplav, Russia), described in detail in Ref. [3, 6]. To avoid crater formation and local separation of the Ti-Ta film during pulsed melting the NiTi-substrate was preliminary irradiated with LEHCPEB at a pulse duration $\tau = 2\text{-}2.5$ μs, beam energy density $E_s = 3.4 \pm 0.7$ J/cm^2, and number of pulses $n = 32$. Thereafter, the NiTi-substrate was positioned by a manipulator along the magnetron sputter axis for deposition of Ti-Ta film of ~ 50 nm thickness, and along the electron beam axis for pulsed melting of the film-substrate system at regime: $E_s = 2 \pm 0.2$ J/cm^2, and $n = 5$.

The Ti-Ta-based surface alloys were created on the basis of films of two compositions: $Ti_{70}Ta_{30}$ and $Ti_{60}Ta_{40}$ (at %). The number of deposition-melting cycles were $N = 20$ and 30, respectively, so that the melted films were in total no thicker than ~ 1 and ~ 1.5 μm. Further in the text, these modified surface layers denoted as $Ti_{70}Ta_{30}$-based SA, and $Ti_{60}Ta_{40}$-based SA.

2.3. Nanoindentation

Investigation of strength and elastoplastic parameters of Ti-Ta-based surface alloys on the NiTi-substrate were examined by nanoindentation on a Nano Hardness Tester (CSM, Switzerland). As a result of the experiment, a series *P-h* diagrams (loading/unloading) were obtained. At a figure 1 *P-h* diagram demonstrates the displacement of the indenter as function of applied load.

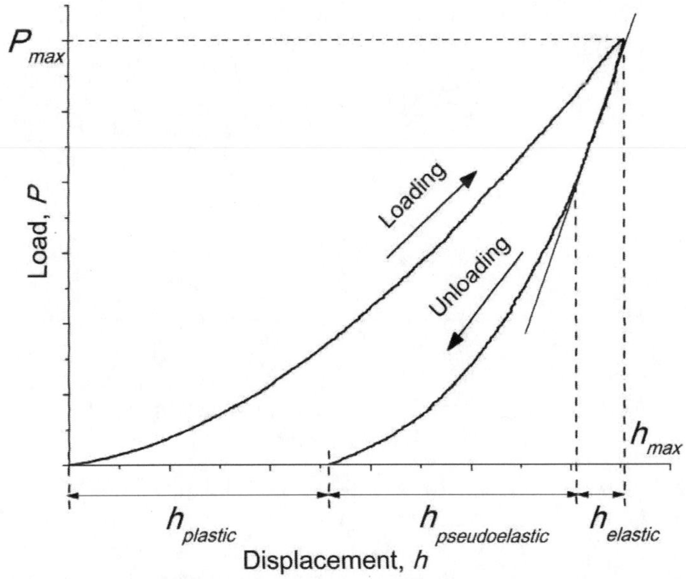

Figure 1. Example of a load-displacement curve for a single loading cycle of indentation ($h_{plastic}$ – residual depth after unloading; $h_{pseudoelastic}$ – pseudoelastic depth recovery; $h_{elastic}$ – elastic depth recovery; h_{max} – maximum indentation depth).

According to the obtained *P-h* diagrams, using the Oliver-Pharr method [7], the dynamic microhardness H_{OP}, and Young's modulus E were determined.

The plasticity characteristic δ_H of surface layer shows the propensity of the material to deform under the action of the load. This parameter was estimated using the methodology [8]:

$$\delta_H = 1 - 14{,}3\,(1 - \nu - 2\,\nu^2)\frac{HV}{E} \qquad (1)$$

where ν – Poisson's ratio; HV – Vickers microhardness; E – Young's modulus of examined material.

Pseudoelastic properties of surface layer were estimated using the parameter η. It characterizes the degree of pseudoelastic recovery of the indentation area [9]:

$$\eta = \frac{h_{max} - h_{plastic}}{h_{max}} \times 100 \qquad (2)$$

where h_{max} – maximum indentation depth; $h_{plastic}$ – residual depth after unloading (figure 1).

3. Experimental results and discussion

3.1. Physicomechanical properties of NiTi-substrate

The physicomechanical properties of surface layers of initial NiTi can be addressed from figure 2 showing Young's modulus *E (1)*, dynamic microhardness H_{OP} *(2)*, and plasticity characteristic δ_H *(3)* as a function of maximum indentation depth h_{max} under increasing load *P* from 5 up to 300 mN.

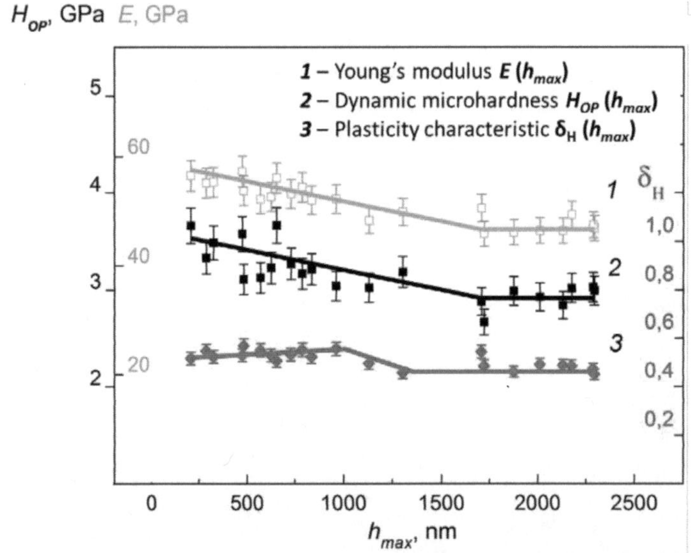

Figure 2. Dependences of *E (1)*, H_{OP} *(2)*, and δ_H *(3)* on the maximum indentation depth h_{max} for initial NiTi-substrate.

It is seen that near the surface of the initial sample the physicomechanical parameters are: $E \approx 55$ GPa, $H_{OP} \approx 3.5$ GPa, $\delta_H \approx 0.5$. Next, these parameters linearly change at a depth h more than ~ 1.5 μm to the values: $E \approx 45$ GPa, $H_{OP} \approx 3$ GPa, $\delta_H \approx 0.45$. An increase of microhardness and Young's modulus in the surface layers being indicative of surface hardening. This is due to the manufacturing of NiTi alloy that was fabricated as rolled sheets by vacuum induction melting.

3.2. Physicomechanical properties of the systems "Ti-Ta-based surface alloy/NiTi-substrate"

Figure 3*a* illustrates the physicomechanical properties of the $Ti_{70}Ta_{30}$-based SA (~ 1 μm). It is seen that the modification of the NiTi alloy leads to gradient change of physicomechanical characteristics.

In the surface layer of ~ 700 nm thickness, the gradient of their variation is greater than in the sublayer, which is being located at a depth of more than ~ 1 μm (i.e. at a depth more than calculated thickness of the $Ti_{70}Ta_{30}$-based SA). Comparing the research data on investigation of the structure (TEM/EDS) [10] with the results of this work, it follows that the structure of the $Ti_{70}Ta_{30}$-based SA (~ 1 μm) provides a more significant effect on the physicomechanical properties than the sublayer structure, which is being located at depth h from ~ 1 to ~ 2 μm. It was established [10] that this sublayer corresponds to a transition (diffusion) zone between the Ti-Ta-based surface alloy and NiTi-substrate. This means that the presence of an extended intermediate zone at a depth h from ~ 1 to ~ 2 μm provides a gradual change in the physicomechanical characteristics and high adhesion of the Ti-Ta-based surface alloy to the NiTi-substrate. Further, according to [10], the layer of NiTi-substrate located a depth of ~ 2 to ~ 2.5 μm. In this layer the structure was modified during formation of SA by multiple pulsed heating in the pre-melting mode (without actual melting).

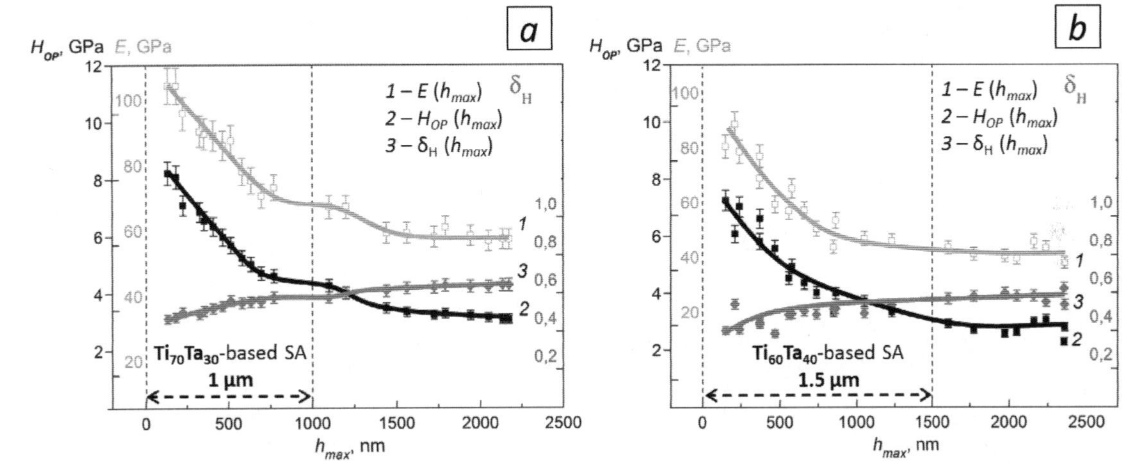

Figure 3. Dependences of E *(1)*, H_{OP} *(2)*, and δ_H *(3)* on the maximum indentation depth h_{max} for $Ti_{70}Ta_{30}$-based SA/NiTi-substrate *(a)*, and $Ti_{60}Ta_{40}$-based SA/NiTi-substrate *(b)*.

Figure 3*b* demonstrates the physicomechanical properties of the $Ti_{60}Ta_{40}$-based SA (~ 1.5 μm). It is seen that gradient change in the properties is more gradual and continuous (more than ~ 1 μm), unlike $Ti_{70}Ta_{30}$-based SA (figure 3*a*). When the depth of the calculated thickness of $Ti_{60}Ta_{40}$-based SA is reached, the physicomechanical properties in the sublayer (at depth ~ 1.5 to 2.5 μm) stabilize and do not change their values. This character of change of the physicomechanical parameters of the $Ti_{60}Ta_{40}$-based SA is obviously due to the peculiarity of the influence of modification to the structure, which requires independent investigations.

A comparison of the physicomechanical characteristics of surface alloys, different in thickness and composition, showed (figure 3*a*, *b*) that near the surface (at a depth of up to ~ 500 nm) the physicomechanical parameters differ: $E \approx (95\pm5)$ GPa, $H_{OP} \approx (7\pm1)$ GPa, and $\delta_H \approx 0.5$ for $Ti_{70}Ta_{30}$-based SA; $E \approx (75\pm15)$ GPa, $H_{OP} \approx (6\pm1)$ GPa, and $\delta_H \approx 0.45$ for $Ti_{60}Ta_{40}$-based SA. These parameters become close to each other in magnitude at a depth of more ~ 1.5 μm.

3.3. Functional (pseudoelastic) properties of the systems "Ti-Ta-based surface alloy/NiTi-substrate"

As is known [11], modification of the surface of nickel titanium alloy, including the pulsed electron beam action, leads to hardening of the surface layers of the material. This causes the suppression of the important functional properties of the alloy, such as SME-SE. An efficient way to investigate functional properties in surface modified layers is to use the method of estimating the degree of pseudoelastic shape recovery η of the indentation area, which confirmed itself in [4, 9].

Figure 4 shows the results of the dependence of shape recovery ratio η on the maximum indentation depth h_{max} for initial NiTi-substrate (figure 4, *curve 1*), and for NiTi samples after formation of Ti-Ta-based surface alloys (figure 4, *curve 2, 3*).

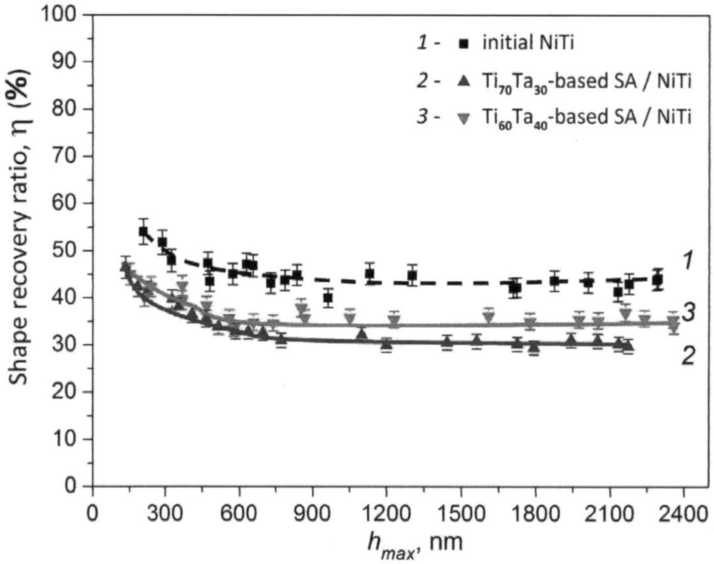

Figure 4. Dependences of shape recovery ratio η on the maximum indentation depth h_{max} for: initial NiTi-substrate *(1)*, $Ti_{70}Ta_{30}$-based SA/NiTi-substrate *(2)*, and $Ti_{60}Ta_{40}$-based SA/NiTi-substrate *(3)*.

The parameter η varies within the layer of ~ 2 μm thickness in both Ti-Ta-based surface alloys. That is the presence of Ti-Ta-based surface alloys changed the functional (pseudoelastic) properties in modified layers (at depth of up to ~ 1 μm) by no more than ~ 10%. In the underlying sublayers (at depth of more ~ 1 μm), shape recovery ratio η remains constant and less than in the surface layer by no more than ~ 5%.

4. Conclusion

The main conclusions obtained in the work:

1. Electron beam synthesis of Ti-Ta-based surface alloys, different in thickness and composition on the NiTi alloy, leads to a gradient change of physicomechanical parameters.
2. The strength and elastic-plastic parameters of Ti-Ta-based surface alloys are different from each other, but become close in magnitude at a depth of more than ~ 1.5 μm.
3. Modification of the surface of NiTi alloy changed the functional (pseudoelastic) properties by no more than ~ 10%.

Acknowledgments

The authors would like to thank D. Sc., Professor Meisner L.L. for assistance in setting tasks, organizing researches and helpful discussions of the results of the work.

Electron-ion-plasma treatment and investigation of the physicomechanical properties of NiTi alloyed with Ti-Ta were supported by grant of the Russian Science Foundation No. 18-19-00198 (26.04.2018).

References

[1] *Surface Modification and Alloying: by Laser, Ion, and Electron Beams* ed Poate J M, Foti G, Jacobson D C (New York: Plenum Press) 1983
[2] Mohd Jani J, Leary M, Subic A, Gibson A M 2014 *Mater. Des.* **56** 1078–1113

[3] Markov A B, Mikov A V, Ozur G E, Padei A G 2011 *Instrum. Exp. Tech.* **54(6)** 862–866
[4] Pan G, Cao Z, Shi J, Wei M, Xu L, Meng X 2014 *Sens. Actuators,* A **217** 75–80
[5] Ding J, Meng Y, Wen S 2000 *This Solid Films* **371** 178–182
[6] Meisner L L *et al.* 2018 *J. Alloys Compd.* **730** 376–385
[7] Oliver W C, Pharr G M 2004 *J. Mater. Res.* **19(1)** 3–20
[8] Milman Yu V 2008 *J. Phys.* D*: Appl. Phys.* **41** 074013
[9] Crone W C, Shaw G A, Stone D S, Johnson A D, Ellis A B 2003 *SEM Annual Conference Proceedings* 1–6
[10] Meisner S N, Yakovlev E V, Semin V O, Meisner L L, Rotshtein V P, Neiman A A, D'yachenko F 2018 *Appl. Surf. Sci.* **437** 217–226
[11] Raharjo P, Uemura K, Murayama H, Souba R 2010 *15th SHCE Conference Proceedings* 492–496

Polarization properties in EHF range of carbon-containing compositional structures obtained by 3D printing method

D S Bodazhkov, R A Kremzer and A V Badin

Faculty of Radiophysics, Tomsk State University, 36, Lenina av., Tomsk, 634050, Russia

E-mail: bdstsu@mail.ru

Abstract. The polarization properties of carbon-containing compositional structures obtained by 3D printing method in extremely high frequency (EHF) range were investigated. Values of the complex dielectric constant of the acrylonitrile butadiene styrene (ABS) plastic and carbon-containing ABS plastic at frequency range of 115-258 GHz were obtained. The angular dependences of the transmittance of composite structures based on the dielectric matrix and ordered carbon-containing plastic structures at frequency range of 62-128 GHz were presented.

1. Introduction

With the development of technologies in various fields of activity, the demand for high-quality materials at low prices is becoming ever stronger. The advantage of creating parts by the method of layer-by-layer addition of material is the ability to produce 3D-objects modeled using an automated design system. In our days there are many materials with different electrophysical properties for additive technology allowing you to create objects with various geometrical, elecrophysical and mechanical properties. The creation of modern elements of extremely high frequencies (EHF) technique with the using of additive technologies are actuality now. The using of carbon-based additives with ABS plastic allows to produce filaments for 3D printing with significant conductivity [1-7]. Carbon composite materials contain carbon reinforcement in the form of continuous filaments, discrete fibers, fabrics with flat and bulk weaving. The advantage of such structures is low density, high heat capacity, high strength and rigidity, resistance to thermal shock, erosion and radiation, low friction coefficient, high corrosion resistance, a wide range of electrical properties (from conductors to semiconductors). It is of interest to apply the additive technology to research the polarization properties of carbon-containing composite structures in the EHF range.

2. Experimental work

For the experiment, two samples based on ABS plastic and carbon-containing ABS plastic were produced by 3D printing using fused deposition modeling (FDM) technology. The pre-measured frequency dependence of the complex dielectric constant of these materials is shown in Figures 1-2.

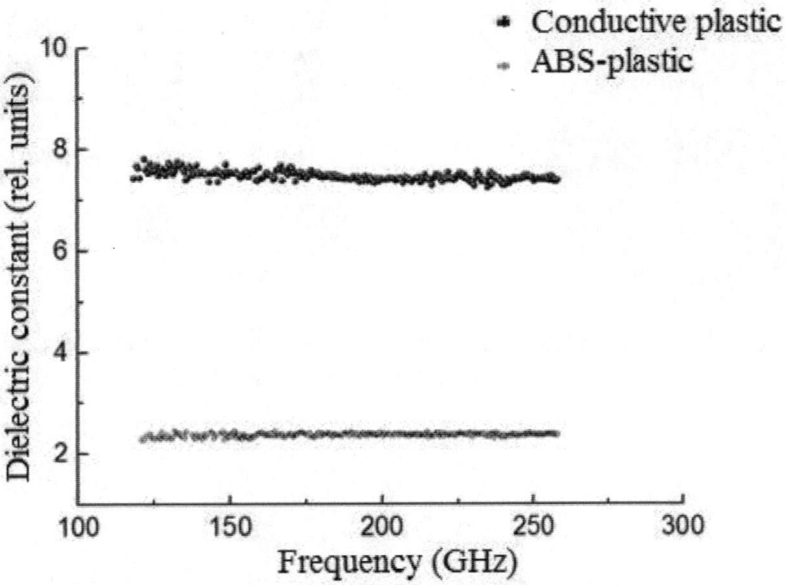

Figure 1. Dependence of the real part of the dielectric constant of ABS plastic and conductive plastic (ABS with carbon) at the EHF frequency range.

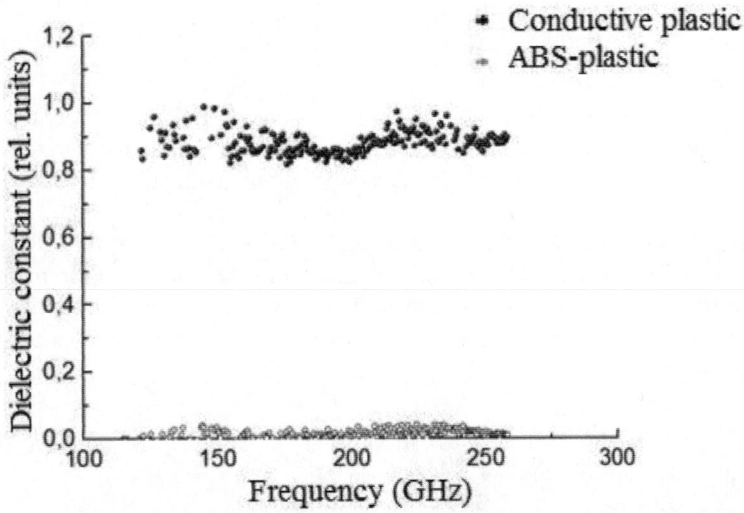

Figure 2. Dependence of the imaginary part of the dielectric constant of ABS plastic and conductive plastic (ABS with carbon) at the EHF frequency range.

Based on the graphs presented in Figure 1, the real part of the dielectric constant of the conductive plastic in the EHF range is approximately equal to 7.8. At the same time, the imaginary part is 0.9, and the specific conductivity of the plastic with a graphite content was 0.0125-0.02 μS/m with a filament diameter of 1.75 mm. In turn, the real part of the dielectric constant of the ABS plastic is 2.2, imaginary part of the dielectric constant is 0.01 - 0.03.

To test the polarization properties, two test polarizers with periods of 2 mm and 3 mm were manufactured. The samples consisted of a dielectric substrate with overlaid strips of carbon-containing plastic. The image and structure of these composites are shown in Figure 3.

IOP Conf. Series: Materials Science and Engineering **511** (2019) 012026 doi:10.1088/1757-899X/511/1/012026

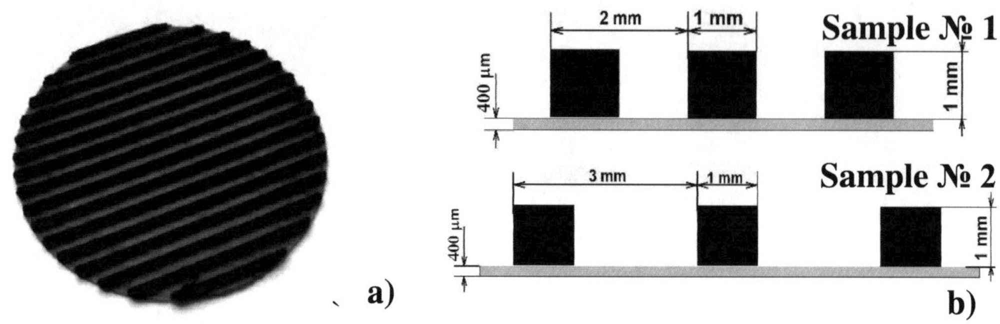

Figure 3. Image of the 3D printed material sample (a). Geometrical parameters of conductive strips (black color) in compositional material (b).

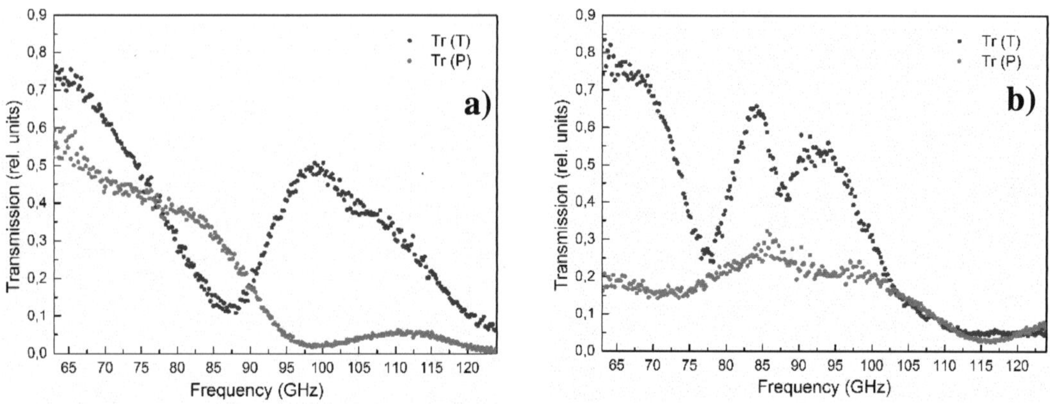

Figure 4. Measured frequency dependence of transmission coefficient of 3D printed composite materials: a) sample № 1; b) sample № 2. Label (P) and label (T) means polarization direction parallel and transverse to the direction of the conductive strips.

At Figure 4 measured frequency dependence of transmission coefficient of 3D printed composite materials (samples №1-2) at parallel and transverse polarizations is shown. Measurements of the angular dependences of the transmission coefficient of the studied material samples in the frequency range of 62-128 GHz (Figure 5) were made by using an STD-21 terahertz spectrometer modified with a precision angular positioning system.

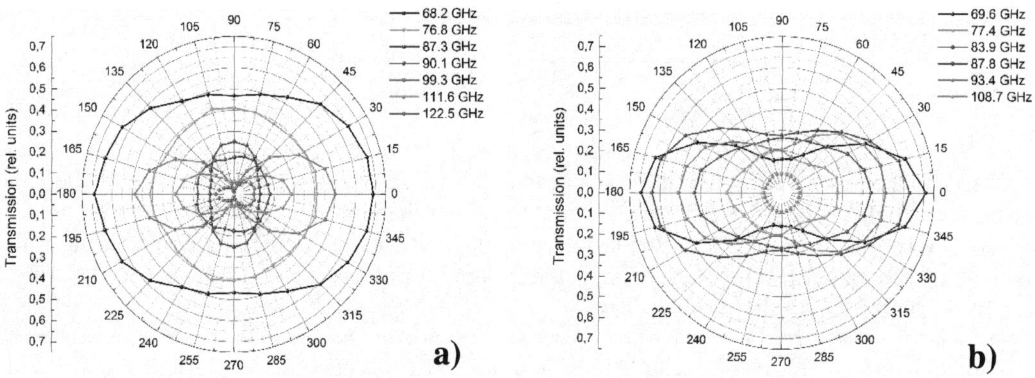

Figure 5. Angular dependence of the transmission coefficient of compositional structure: a) sample № 1; b) sample № 2.

At the frequency ranges 68-76.8 GHz and 90.1-128 GHz transmission coefficient of sample №1 at transverse polarization is greater than at parallel polarization (Figure 4a), at the frequency range of 76.8-90.1 GHz the value of the transmission coefficient with parallel polarization is higher than the transverse. Two frequencies (77 GHz and 87 GHz) of the minimum transmission coefficient of the sample №2 were found (Figure 4b).

Conclusion

One can see from the graphs at Figure 5 the maximum value of the transmission coefficient reaches at position of sample of 0 degrees when the orientation of the conducting strips is orthogonal to the vector of the electric field of the incident wave. For sample №1 there is found a change in the direction of the maximum of the angular dependence of the transmission coefficient by 90 degrees at frequency 87.3 GHz (Figure 5a). This property is due to the fact that at this frequency for a periodic structure the resonance condition is fulfilled with the absorption of an electromagnetic wave with transverse polarization when the orientation of the conductive strips is orthogonal to the vector of the electric field strength of the incident wave. These composite materials can be used to create mechanically controlled EHF attenuators, as well as in the development of frequency-selective elements.

Acknowledgment

Authors are grateful to an engineer of the Shared Use Center "Centre of radiophysical measurements, diagnosis and research of parameters of natural and artificial materials" of National Research Tomsk State University, Dorozhkin K.V. for assistance in carrying out the experiments.
The reported study was funded by RFBR according to the research project № 18-32-00810.

References

[1] Gnanasekaran K, Heijmansa T, Bennekom S, Woldhuis H, Wijnia S, With de G and Friedrich H 2017 3D printing of CNT- and graphene- based conductive polymer nanocomposites by fused deposition modelling *Applied materials today* **9** 21–8

[2] García-Tuñon E, Barg S, Franco J, Bell R, Eslava S, D'Elia E, Maher R C, Guitian F, Saiz E 2015 Printing in three dimensions with graphene *Advanced materials* **27** (10) 1688–93

[3] Badin A V, Kuleshov G E, Dorozhkin K V, Dunaevskii G E, Suslyaev V I and Zhuravlev V A 2018 Anisotropy of electrical properties of 3D-printing MWCNT composites at the THz frequency range *proc. in 43th International Conference on Infrared, Millimeter, and Terahertz Waves (IRMMW-THz 2018)* 1-2

[4] Osa de la G, Pérez-Coll D, Miranzo P, Isabel Osendi M and Belmonte M 2016 Printing of graphene nanoplatelets into highly electrically conductive three-dimensional porous macrostructures *Chemistry of Materials* **28** (17) 6321–28

[5] Gardner J, Sauti G, Kim J, Cano R, Wincheski R, Stelter C, Grimsley B, Working D and Siochi E 2016 3-D printing of multifunctional carbon nanotube yarn reinforced components *Additive Manufacturing* **12** 38-44

[6] Leigh S, Bradley R, Purssell C, Billson D and Hutchins D 2012 A simple low-cost conductive composite material for 3D printing of electronic sensors *PloS one.* **7** (11) 1-6

[7] Busch S F, Weidenbach M, Fey M, Schäfer F, Probst T, Koch M 2014 Optical properties of 3D printable plastics in the THz regime and their application for 3D printed THz optics *Journal of Infrared, Millimeter, and Terahertz Waves* **12** (35) 993–7

Polymer composites based on oligosulfones

R Ch Bazheva[1], Z I Inarkieva[2], Z Kh Sultygova[2] and A M Kharaev[1]

[1]Kabardino-Balkarian State University named after Kh.M. Berbekov, Nalchik, Chernyshevsky street, 173, Russia
[2]Ingush state University, Nazran, I.B. Zyazikov, 7, Russia

E-mail: r.bazheva@mail.ru

Abstract. Halogen-containing oligosulfones based on 1,1-dichloro-2,2-di (4-hydroxyphenyl) ethylene and 4,4'-dichlorodiphenylsulfone on various degrees of condensation were synthesized in the solution by high-temperature polycondensation. Using the obtained oligosulfones, we conducted a physical modification of industrial bisphenol A polycarbonate in a wide range of concentrations. Physical and mechanical properties of composites are investigated. The compatibility of the obtained oligosulfones with polycarbonate was studied using viscometry, differential scanning calorimetry, and probe microscopy. It was shown that the introduction of oligosulfones into the polycarbonate matrix promotes an increase in the glass transition temperature, found by differential scanning calorimetry, at 3 to 200 ° C, depending on the composition, the composites have good dielectric and technological properties. The temperature dependences of the dielectric properties of PCs and PC-based composites with different oligosulfone content are characterized by the presence of a single.

1. Introduction

Recently, a noticeable expansion of the applications of rigid chains of glassy polymers characterized by low resistance to cracking is largely due to the development of composites based on them.
Bisphenol A (PC) polycarbonate, along with a complex of valuable properties, has a number of disadvantages, which significantly limit its fields of application. In particular, high internal (residual) stresses, which lead to cracking of products during operation, slow-going relaxation processes, low adhesion, low resistance to alkaline environment, high melt viscosity, and hence the processing difficulty associated with this [1-5].

Despite numerous scientific studies on modification of polycarbonate, its assortment is insignificant. Numerous attempts to modify polycarbonate with low-molecular compounds did not give positive results. Over 50 compounds of various classes have been tested. However, to date, modification of polycarbonate in order to improve a number of performance characteristics is relevant. In this direction, the most promising way is the physical modification, i.e. development of compositions based on polycarbonate and various modifiers. In addition, it is known that the introduction of halogen atoms into a polymer contributes to an increase in fire resistance [6–11].

In order to create polymer composites based on polycarbonate, oligosulfones based on 1,1-dichloro-2,2-di (4-hydroxyphenyl) ethylene and 4,4'-dichlorobiphenyl sulfone were synthesized and some properties of the composites were studied.

2. Experimental

In a three-necked flask equipped with a mixer, a reflux condenser with a Dean-Stark trap, a gas supply bubbler and a thermometer, are entered 5.62281 g (0.02 mol) 1,1-dichloro-2,2-di (4-hydroxyphenyl) ethylene, 40 ml of dimethyl sulfoxide (DMSO) and 30 ml of toluene. When stirring, nitrogen is passed in and the temperature is raised to 70 °C. After complete dissolution of 1,1-dichloro-2,2-di (4-hydroxyphenyl) ethylene, 3.98406 ml of 10.04 normal solution of sodium hydroxide is added. The temperature is raised to 140-145 °C and the azeotropic toluenel-water mixture is distilled off until the water is completely removed. The reaction mass is cooled to 40-50 °C and 2.87294 g (0.01 mol) of 4.4'-dichlorodiphenylsulfone is added. The reaction is carried out at 140-145 °C for 2 hours. The resulting mass is diluted with 10 ml of DMSO and precipitated into acidified distilled water. The precipitate formed is filtered off and washed with distilled water until the filtrate is negatively reacted with chlorine ion. The resulting oligoether sulfone is dried at 100 ° C under vacuum for 24 hours.

Similarly, by changing the ratio of 1,1-dichloro-2,2-di (4-hydroxyphenyl) ethylene and 4,4'-dichloro-diphenylsulfone, oligosulfones of various degrees of condensation were synthesized. In all cases, the amount of 1,1-dichloro-2,2-di (4-hydroxyphenyl) ethylene is 1 mol more than the amount of 4,4'-dichlorobiphenylsulfone.

Ubbelohde viscometer was used to determine the intrinsic viscosity.

The concentration of the solutions was determined by the formula:

$$C = \frac{aV_1 \cdot 100}{V(V_1 + V_2)},$$

where *a* is the polymer sample weight, g;

V, V_1, V_2 - respectively, the volumes of the solvent consumed to prepare the solution, the initial solution placed in the viscometer, and the solvent added to the viscometer upon dilution, ml.

Study of the surface of the samples on a scanning probe microscope. The sample is mounted on a polycore substrate 20x25x10 mm in size and then attached to the scanner in the horizontal position. A measuring head with removable probes is installed above the sample. Rapid supply of the probe is carried out by a stepper motor in about 1-3 minutes. The measurement time depends on the speed and field of the scan and is approximately 0.5-5 minutes.

Melt indices of polymers were measured by a capillary viscometer with a constant piston at a temperature of 220-280 °C. Every 5 minutes, the extruded melt was cut out of the capillary with a knife and weighed.

The test result was taken as the arithmetic average of two determinations on three pieces of material, the difference in weight, between which did not exceed 5%. The melt flow rate (melt index) was calculated by the formula:

$$i = \frac{10 \cdot Q}{t}$$

where Q is the weight of the polymer, g;

t - time extrusion, min.

Results and discussions

Oligosulfones based on 1,1-dichloro-2,2-di (4-hydroxyphenyl) ethylene and 4,4'-dichlorodiphenylsulfone were synthesized by nucleophilic substitution in an aprotic dipolar solvent dimethyl sulfoxide at a temperature of 140-145 ° C for 2 hours. Toluene was used to form an azeotropic mixture. The resulting halogen-containing oligosulfone had the formula:

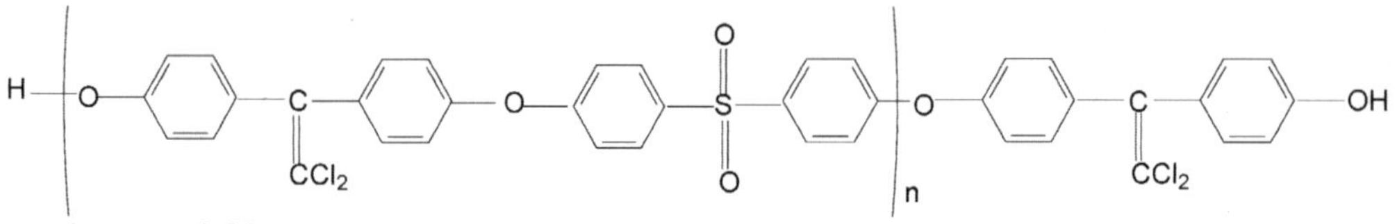

where n = 1-20.

The main properties of these oligosulfones are described in detail in [12].

Unambiguous methods of studying the compatibility in polymer-plasticizer systems, polymer-polyn find difficult. In the block state, compatible polymers form transparent films and fibers, which in a pha: contrast microscope with high magnifications or in an electron microscope do not show a heterogeneo structure under any methods of contrasting. In addition, blends of compatible polymers should have t same glass transition temperature, regardless of research methods. These criteria are, in princip unambiguous criteria for compatibility, but in practice there may be some difficulties in using them. particular, incompatible polymers form transparent films if the refractive indices of both polymers are t same or if the refractive indices, although different, but the polymers can form a two-layer film tl appears transparent when it is obtained by evaporating the solvent from the polymer solution. Such ca: are not particularly difficult for analysis, since transparent films from blends of incompatible polym(are characterized by two glass transition temperatures corresponding to the glass transition temperatuı of the components, if the latter differ sufficiently and can be determined using the method of investigati used.

The above makes it critical to refer to any measurements on polymer blends, since it is very difficult determine if the mixture is in an equilibrium state. Usually, without evidence, it is assumed that a fi from a mixture of polymers obtained from a solution is in a more equilibrium state than samples obtain by mixing polymers in a block state. Everything said predetermined the need to determine t compatibility of polycarbonate with the obtained oligosulfones.

To determine the compatibility, as well as to determine the nature of the distribution of the synthesiz oligosulfones in the PC matrix, we used the method of scanning probe microscopy.

Researches have shown, that properties of a surface depends on its structure, as illustrated in figure However, in samples with a modifier content of up to 10% by weight. there is good compatibility betwe the components. Oligosulfones are evenly distributed in the polycarbonate matrix. The particle size of t modifier is from 0.31 to 3.13 microns and depends on the composition.Later, oligosulfone with a degı of polycondensation n = 5 (OC-5C-2) was used to modify PC.

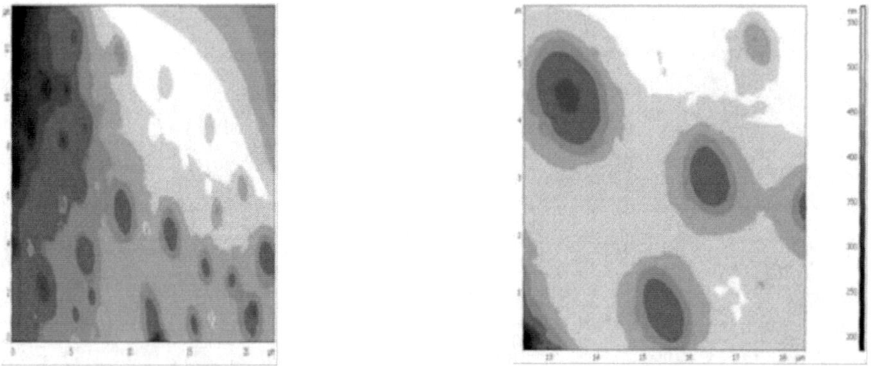

a b

Figure1. SPM photographs of a PC composition with OC-5C-2, containing 5 (a) and 10 (b)% of the weight.

X-ray phase analysis showed that all compositions of composites are amorphous.

IOP Conf. Series: Materials Science and Engineering **511** (2019) 012027 doi:10.1088/1757-899X/511/1/012027

To determine the compatibility of the oligosulfones obtained, we also used the viscometric method. The studies were carried out in the Ubbelohde viscometer at a temperature of 25 ° C. At this temperature, polymer-solvent systems containing from 1 to 10 weight % of oligosulfone were investigated. Methylene chloride was used as a solvent.

It turned out that for these systems the maximum compatibility is observed at a ratio of 95% by weight. PC and 5% of the weight . OS-5C-2. The data are presented in figure 2.

The glass transition temperatures of composites found by differential scanning calorimetry are shown in figure 3. As can be seen from the figure, with an increase in the content of oligosulfone in the composition, Tg increases. An increase in the glass transition temperature probably indicates the interaction of the filler with the matrix.

For the compositions obtained, the temperature dependences of the dielectric constant (ε ') and the tangent of the dielectric loss angle were studied at a frequency of 104 Hz. The values of the dielectric constant of all the investigated samples of composites are ~ 3–3.6 and are stable in the temperature range from 20 to 200°C. Curves of dielectric properties as a function of temperature, for PCs and PC-based composites with different oligosulfone content, have the same dielectric loss tangent. The values of ε ' for PC and all composites do not depend on the composition and, within the limits of error, coincide and correspond to values of 2.4-2.5.

Comparison of melt indices of the original PC and composites based on them (figure 4) showed that the introduction of small amounts of OC-5C-2 in PC significantly affects the melt flow rate (MFR) of these composites. Thus, the introduction of only 1-3% of the weight OS increases the PTR PC 1.5-2 times. Moreover, the introduction of small amounts is more effective - up to 5% by weight OS.

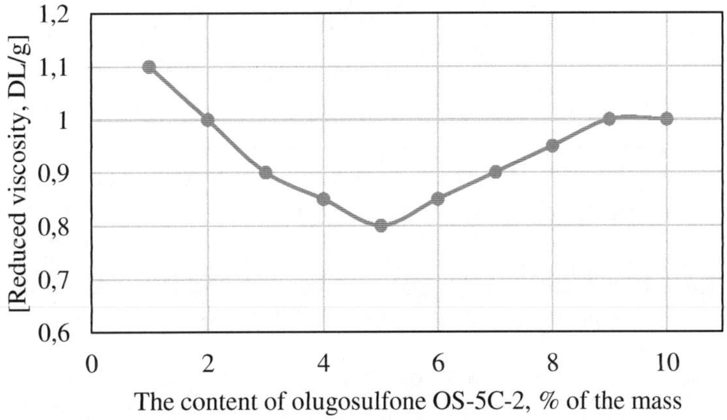

Figure 2. The dependence of the reduced viscosity of the solutions of composites from the content of OC-5C-2.

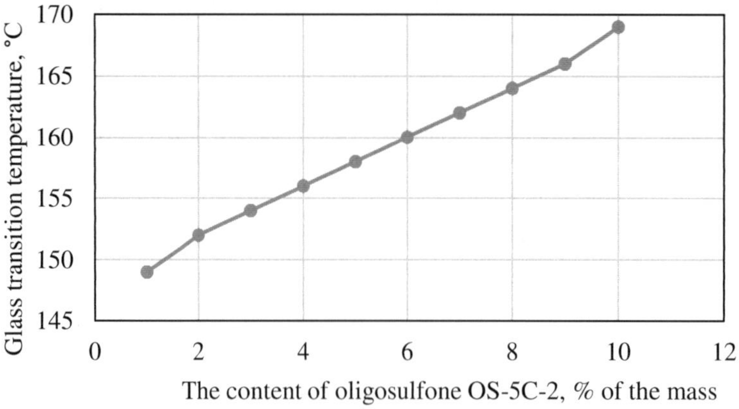

Figure 3. The dependence of the glass transition temperature of composites on the content of OS-5C-2.

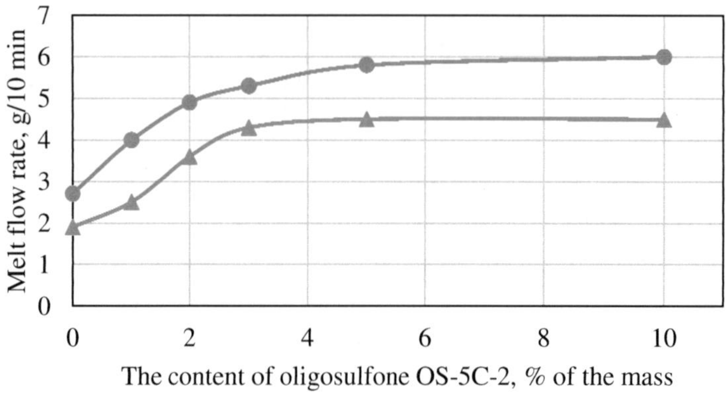

Figure 4. The dependence of the melt flow index on the content of OC-5C-2 at a processing temperature of 260° C (▲) and 280° C (•).

All the composites obtained are dissolved in inexpensive and easily volatile solvents, for example, in chloroform, methylene chloride, dioxane, dimethylformamide, etc. This will allow their processing into film products from solutions, for example, in available and volatile methylene chloride.

Conclusion

Composite polymeric materials based on aromatic polycarbonate bisphenol A synthesized on the basis of 1,1-dichloro-2,2-di (4-hydroxyphenyl) ethylene and 4,4'-dichloro diphenyl sulfone oligosulfones of various molecular weights were developed. The study of some properties of PC compositions with oligosulfone showed that these OS can be used to improve some of the performance properties of polycarbonate.

The compatibility of obtained OS with industrial poly-carbonate has been studied by various methods. It was shown that the introduction of oligosulfones into the polycarbonate matrix promotes an increase in the glass transition temperature by 3 to 200 ° C, depending on the composition. The resulting composites have good dielectric properties.

Comparison of the flow rates of PC melts and composites under various processing conditions allows us to conclude that these composites can be processed by extrusion and injection molding under milder conditions than bisphenol A polycarbonate.

References

[1] Amerik V V, Radzinsky S A, Zolkina I Yu, Andreeva T I, Simonov-Emelyanov I D, Fedotova T I and Levchuk A V 2013 Polycarbonate - market analysis and development prospects (Plastics vol 11) pp 10–13

[2] Ponomareva V T and Likhacheva N N 2000 State and prospects of structural thermoplastics market (Plastics vol 6)

[3] Bolotina L M and Chebotarev V P 2003 Development of research in the field of chemistry and technology of aromatic polysulfones (Plastics vol 11) pp 3–7

[4] Alperstein D, Knani D, Borchmann N, Spekowius M and Hopmann C 2014 Prediction of environmental stress cracking in polycarbonate by molecular modeling (Polymers for Advanced Technologies vol 25) I 12

[5] Jinhua Wang and Gregory B. McKenna 2015 Viscoelastic properties of ultrathin polycarbonate films by liquid dewetting (Journal of Polymer Science Part B: Polymer Physics vol 53) I 22

[6] Bazheva R Ch, Kharayev A M and Mikitayev A K 2006 Temperature transitions in polycarbonate-polytetramethylenoxide block copolymer resins Polymers, Polymer Blends, Polymer Composites and Filled Polymers. Syntesis, Properties, Application (Nova Science Publishers, Inc. New York) pp. 115–120

[7] Kharaev A M, Bazheva R Ch., Chaika A A and Barokova E B 2006 Chemical modification of polycarbonate (review) (Plastics vol 9) pp 25–30

[8] Bazheva R Ch, Bashorov M T, Kharaev A M and Mikitaev A K 2007 Study of temperature transitions in polycarbonate and polycarbonate-polytetramethylene oxide block copolymers (Chemical industry today vol 8) pp 35–40

[9] Bazheva R Ch, Kharaev A M, Bazhev A Z, Inarkieva Z I and Beslaneeva Z L 2016 Fire-resistant copolycarbonates (Plastics vol 5) pp 26–30

[10] Bazheva R Ch, Kharaev A M Inarkieva Z I and Beslaneeva Z L 2017 Copolycarbonates containing dichloroethylene groups in the main chain (Plastics vol 3) pp. 32–35

[11] Kharaev A M Bazheva R Ch and Chaika A A 2006 Composite materials based on polycarbonate (Plastics vol 8) pp 26–31

[12] Barokova E B, Haraev A M, Bazheva RC and Haraeva RA 2010 Oligosulfones on the basis of 1,1-dichlor-2,2- di(4-oxyphenyl)ethylene and 4,4'-dichlorodiphenylsulfone obtained by high temperature polycondensation *Handbook of Condensed Phase Chemistry* pp 233–236

Acknowledgments
The reported study was funded by RFBR according to the research project № 18-29-18063.

IOP Conf. Series: Materials Science and Engineering **511** (2019) 012028 doi:10.1088/1757-899X/511/1/012028

Possibilities for the polymer composites physicomechanical characteristics improvement

A V Kochneva and S A Pronin

Irkutsk National Research Technical University, Lermontova Str., 83, Irkutsk, Russian Federation

E-mail: koch.alex061@mail.ru

Abstract. A group of composite materials based on a polyvinyl chloride matrix and dispersed industrial wastes is presented. Production of these materials will significantly reduce the anthropogenic burden on the environment by reducing the primary resources consumption. Studies of the new materials physicomechanical properties dependence on the composition and quantity of the dispersed filler are presented. The industrial wastes characteristics influence on the composites structure formation is considered. The data obtained indicate the competitiveness of the developed composite materials and the feasibility of their production.

1. Introduction

The construction industry development leads to the need to expand the nomenclature of tailor-made building materials. Modern researchers have learned to create composites according to given characteristics that satisfy the specific customer needs. However, no less important is the construction industry environmental component, associated not only with impact mitigation, but also with the non-renewable natural resources preservation. That is why today the production technologies of building materials on the basis of or with the content of waste and recyclable materials are widely used.

In turn, the waste use in the production of building materials is accompanied by additional difficulties associated with the need to study the used industrial raw materials safety and its properties. In addition, among consumers, there are still stereotypes associated with the danger of building products containing waste. However, the recycled materials use in production has several undeniable advantages:

- finished products cost-cutting by 30-50%, due to the waste low cost compared to primary resources [1];
- decrease in the consumed primary resources volume by 40%;
- reducing the negative impact on the environment by reducing the accumulated waste volume, reducing the seized land area, reducing the pollutants emission into the air, water and soil, etc.;
- creating the manufacturer benign image that increases the demand for products.

Along with the cost and construction products environmental friendliness, consumers are interested in the material performance, which directly determine its purpose and application scope. Decorative coatings and building mixtures should have adhesion to the surface to be coated, and concrete should be frost resistant and have a slight shrinkage. However, it can be said with confidence that the applicability of most building materials is to some extent determined by their strength.

2. Experimental procedure

In general, strength is a material ability to resist an applied load without breaking. Nevertheless, depending on the applied load nature and direction, many different values can be obtained. This article considers the improving the strength characteristics of composite building materials on the example of determining the static compressive strength, static bending strength and static rupture strength by the breaking strength.

Studies were conducted on the example of the mineral-polymer building composite "Vinizol", developed on the basis of the Irkutsk National Research Technical University. The composition includes a polyvinyl chloride (PVC) matrix, dispersed filler in the fly ash form from thermal power plants and a modifying additives set, which are aimed at improving the mixture processability, reducing friction in the raw materials preparation and processing, as well as to modify the product final properties, in particular, strength and elasticity. The composition is processed by extrusion with obtaining molded building products for construction and finishing purposes [2, 3].

The research was based on the study of changes in the mineral-polymer composite strength characteristics, depending on the filler type and amount. As the latter, dispersed materials of different composition and origin, including man-made materials, were used. The basic composition for all materials remained unchanged: suspension PVC grade SI-64 – 100 pts. wt., modifying additives – 11 pts. wt. The materials composition is presented in table 1.

Table 1. The fillers content in the studied materials

	Filler type	Filler content, pts. wt.
Composite 1	Vermiculite	40
Composite 2	Marble dust	75
Composite 3	Marble dust	110
Composite 4	Fly ash	65
	Carbon nanoparticles	0.0016
Composite 5	Fly ash	30
	Aluminosilicate microspheres	48
Composite 6	Fly ash	50
	Aluminosilicate microspheres	23
Composite 7	Aluminosilicate microspheres	48
Composite 8	Aluminosilicate microspheres	23

All fillers as dispersed materials have different composition and origin. Marble dust is the marble mining and processing waste - fine white powder with a predominant grain size of 4-30 microns ($\approx$ 75%). The chemical composition is represented mainly by Ca and Mg carbonates. Vermiculite is a layered natural mineral with Si, Mg, Al oxides predominance. Exfoliated vermiculite obtained by mineral rapid roasting, as a result of which the grains are increased up to 15–20 times in volume, was used as a filler. Fly ash is the coal combustion products finest fraction, caught by electrostatic precipitators. Fly ash has a particle size wider distribution - more than 80 % of particles have a 6-200 microns in size. The fly ash composition is mainly represented by Si, Al, Fe, Ca, Mg oxides.

The aluminosilicate microspheres use is of particular interest [4]. Their chemical composition is similar to that of fly ash, however, the grains shape is close to a hollow ideal sphere, which leads to high particle strength, low thermal conductivity and chemical inertness. The microspheres chemical composition peculiarity is the acidic aluminum oxides high content (over 20.0 wt.% Al_2O_3) and silicon (over 62.0 wt.% SiO_2), but low iron content (less than 3.5 wt.% equivalent to Fe_2O_3). Of no less interest are carbon nanoparticles. In this work, nanoparticles (carbon nanotubes) were isolated from silicon dust. The specific surface area is 500-1500 m^2/g. The energy dispersive X-ray microanalysis method determined a significant (up to 35%) quartz content and the carbon content up to 12 %.

IOP Conf. Series: Materials Science and Engineering **511** (2019) 012028 doi:10.1088/1757-899X/511/1/012028

The conditions for the raw materials preparation (two-stage mixing) were identical for all compositions, and the extrusion processing technological modes were adjusted individually, which is associated with significant differences in fillers that affect the melt viscosity in the extruder chamber [5].

To study the fillers effect on the composite building materials physicomechanical properties, tests were conducted to determine the composites strength characteristics. For all compositions, the maximum load that leads to the sample destruction during compression, rupture and bending was determined. On the basis of this load the ultimate strength was determined. All tests were carried out in accordance with applicable standards - GOST 11262, GOST 4651 and GOST 4648. The test results are presented in table 2. The table also shows the ultimate strength values of the source material Vinizol and the object analogue - wood-polymer composite (WPC), which uses wood flour as a filler.

Table 2. The composite building materials physicomechanical characteristics study results

	Compressive strength, MPa	Tensile strength, MPa	Flexural strength, MPa
Composite 1	42.28 ± 0.46	21.81 ± 1.41	49.29 ± 3.04
Composite 2	45.51 ± 0.42	18.14 ± 1.57	48.21 ± 3.03
Composite 3	46.31 ± 0.3	19.61 ± 0.51	34.78 ± 1.27
Composite 4	42.34 ± 0.83	16.31 ± 0.35	27.13 ± 1.93
Composite 5	35.09 ± 0.72	11.54 ± 0.87	24.02 ± 2.5
Composite 6	40.1 ± 0.65	13.69 ± 0.48	35.78 ± 0.52
Composite 7	35.67 ± 0.39	12.55 ± 1.21	34.24 ± 0.68
Composite 8	30.88 ± 0.52	5.91 ± 0.55	15.62 ± 6.48
Vinizol	-	18.9	35.93
WPC	-	9.0	25.0

From the table 2 it can be seen difference between the obtained values of ultimate strength of various composition materials. At the same time, no single composition surpasses the rest at the same time in all characteristics. However, it can be noted that composites 1-3 have the best strength properties combination and are significantly superior to other compounds, including Vinizol and WPC.

The composite materials properties are largely due to their structure, which is a heterogeneous system. The composites physical and mechanical properties are largely characterized by the matrix-filler bond strength, which in turn depends on many factors: the interface area, the particle size and shape, the wettability of the filler polymer matrix, the adhesion, etc. Consider the studied materials micrographs with the aim of studying their structure and the components interaction nature (figure 1).

As can be seen from figures 1-2, the polymer matrix structure is not changing in relation to the filler composition. However, shapes, particles size and their incorporation nature into the matrix are different. Figure 1c shows a composite 1 micrograph containing 40 weight parts. expanded vermiculite. The figure shows how large the vermiculite particle is (on the right side of image), especially when compared to the filler particles in other photos. It should also be noted that the polymer is tight is tight to the filler particle.

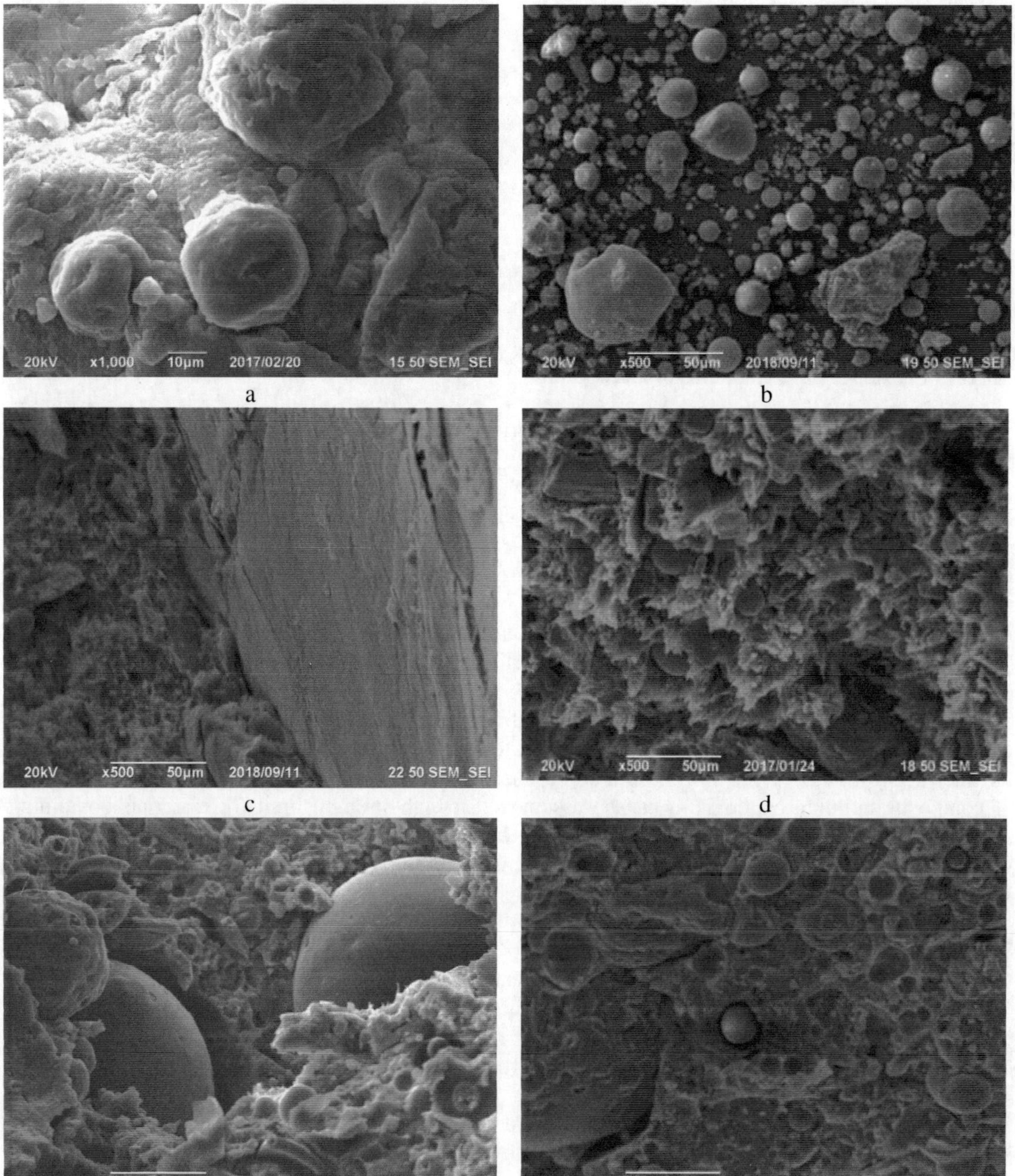

Figure 1. Micrographs of the starting components and materials with various fillers: a) PVC without filler; b) fly ash; c) PVC and expanded vermiculite; d) PVC and marble dust; e) PVC and microspheres; f) PVC and fly ash.

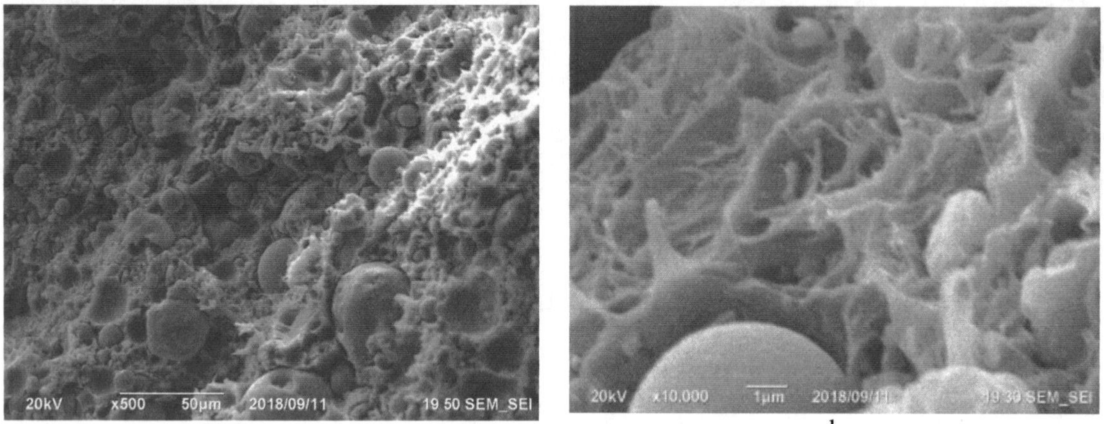

a b

Figure 2. Micrographs of the materials with fly ash and nanoparticles.

Figure 1d shows a composite 3 micrograph with a content of 11 weight parts including marble dust. The scale bar in the picture allows you to note how small the marble particles are in the composite. The particles shape is closest to that of a prism, the filler is also tightly embedded in the PVC matrix structure. Figure 1e shows the material containing microspheres. The picture shows that the filler particles shape is almost perfect, however, the density of adhesion of the polymer to the microspheres is not the same. The same applies to fly ash in the composition (figures 1f, 2). Ash particles also resemble spheres in shape, but much less regular in shape. The ash particles size varies considerably, which confirms the data on the fly ash particle size distribution.

In general, it should be noted that a filler particles significant amount is present on the studied composites breakdown, which confirms the well-known fact of the primary composites destruction at the interface [6]. Thus, taking into account the results of the study of physical and mechanical properties of composite building materials, it can be assumed that high-strength limit for materials containing expanded vermiculite and marble dust as a filler largely depends on the fillers characteristics. For vermiculite, this is a particles layered lamellar structure, and for marble dust there is a grains dispersion high degree and a prismatic shape.

3. Conclusion

1. Using the technogenic raw materials as a filler allows to improve the composite building material physical and mechanical characteristics, while reducing the primary resources consumption and the final product cost.
2. The best physical and mechanical properties combination have compositions filled with exfoliated vermiculite and marble dust.
3. Composites 1-3 high strength properties are determined by the materials structure and the fillers characteristics: the particles are tightly embedded in the polymer matrix and due to the high specific grains surface and shape form a strong particle-polymer bond.

A detailed study of the properties of the developed materials and the establishment of patterns of changes in the properties of the composition of the composition and characteristics of the filler, as well as the development of a line of highly durable highly filled composite building materials is the subject of further research by the authors.

Acknowledgments

The materials were prepared with the financial support of the Ministry of Science and Higher Education of the Russian Federation - state assignment № 5.11496.2018 / 11 and state assignment № 11.8090.2017/8.9.

IOP Conf. Series: Materials Science and Engineering **511** (2019) 012028 doi:10.1088/1757-899X/511/1/012028

References

[1] Dvorkin L I and Dvorkin O L 2007. *Building materials from industrial waste*. (Rostov-on-Don Phoenix)
[2] Burdonov A V, Barakhtenko V V, Zelinskaya E V, Suturina E O, Burdonova A V and Golovnina A V 2012. *Magazine of Civil Engineering* **9** 14
[3] Barakhtenko V V, Burdonov A V, Zelinskaya E V, Tolmacheva N A, Golovnina A V and Samorokov V E 2013. *Fundamental Research* **10** 2599
[4] Danilin L D, Drozhzhin V S, Kuvayev M D, Kulikov S A, Maksimova N V, Malinov V I, Pikulin I V, Redyushev S A and Khovrin A N 2012. *Cement and its application* **4** 100
[5] Ershova O V, Mullina E R, Chuprova L V, Mishurina O A and Bodyan L A 2014. *Fundamental Research* **12-3** 487
[6] Berlin A A and Basin V E 1974. *Basics of Polymer Adhesion* (Moscow: Chemistry)

Research of the Piezo Actuator Location Influence on the Twist Angle of a Model Helicopter Blade Made of Polymer Composite Materials

E Barkanov[1], A N Anoshkin[2], P V Pisarev[2] and V A Ashihmin[2]

[1]Institute of Materials and Structures, Riga Technical University, Riga, Latvia
[2]Perm National Research Polytechnic University, Perm, Russia

E-mail: pisarev85@live.ru

Abstract. The article formulates a mathematical model for calculating the stress-strain state (SSS) of a model helicopter blade equipped with piezo actuators. In determining the twist angle of the model blade, the principle of thermo-piezoelectric analogy was used, which allowed moving from the related boundary problem of electro elasticity to the boundary problem of thermo elasticity. A comparison was made of the deformation fields for model blades with various arrangement of piezo actuators and different materials of the spar. We analyzed the convergence of the finite element model of the blade. The limiting twist angles of the model blade were determined in the range of the control voltages operating values applied to the piezo actuators. Recommendations for the placement of piezo actuators on a model helicopter blade were formulated.

1. Introduction

Currently, among all types of aircrafts, helicopters are the ones most exposed to vibration. The helicopter rotor's characteristics worsen with increasing airspeed due to the effect of compressibility of air, the appearance of a stall on the retreating blades, an increase in the loads applied to the blades control system and an increase in vibration and noise levels [1, 2].

Many works [3–8] are devoted to the problem of numerical simulation of active control systems; most of them are based on simple two-dimensional models that apply combinations of numerical simulations for longitudinal and transverse analysis of the stress-strain state (SSS) along the blade length. These simplified models are used to estimate the distribution of stiffness and mass of the blade with variable geometry. One of the first models of a thin-walled laminated profile equipped with active controls was described in [3]. In this paper, the task was set on the basis of the classical theory of shells, the model was verified. Other models appeared later, describing the dynamic behavior of rotating solid blades with embedded piezoelectric elements. These models are taking into account the anisotropy of properties, as well as centrifugal and Coriolis inertia force [4]. However, the rotor blade of the helicopter is a very flexible structure, and therefore a blade numerical simulation model should consider non-linear deformations of flexible structures, including the contact with the control elements [5-6]. Mathematical models for calculating the mechanical characteristics and electro elastic fields in piezoelectric composite materials with different arrangement of control piezo elements are presented in [7-10].

IOP Conf. Series: Materials Science and Engineering **511** (2019) 012029 doi:10.1088/1757-899X/511/1/012029

The tasks of multiparameter optimization of the composite structure based on integral models (in unrelated formulation) were solved in the works [11–14], in these works the parameters of active (control) and power elements were determined.

Overall, it can be concluded that the approaches used today for the numerical simulation of SMART structures are largely based on simplified models and integral methods of calculation. These approaches determine a narrow range of parameters with which the performance characteristics of SMART structures can be adjusted. These models have almost exhausted their capabilities. In turn, the capabilities of modern computing, numerical methods and the corresponding development of the experimental base makes it possible to make a new qualitative leap in the design and creation of SMART structures with optimal characteristics, which in turn will allow achieving the maximum reduction of noise and vibrations of aircrafts.

To successfully solve the problems of reducing vibration in aircrafts using piezo actuators, it is necessary to determine the constructional features of the control elements, as well as their locations. It is important to develop new mathematical models for calculating the SSS of piezoelectric layered polymer composite materials (PCM), equipped with piezo actuators.

For this work, a numerical research has been carried out to determine the influence of the location of piezo-actuators on the twist angle of a model blade made of PCM with different material of the spar. Three types of arrangement of piezo actuators on the blade were considered: lower, upper, and combined arrangement.

2. Numerical model

To conduct a numerical research of the piezo actuators location influence on the twist angle, a three-dimensional geometric model of a helicopter blade was developed. The NACA 23012 profile was used as an aerodynamic profile. The researches blade profile has the following geometrical characteristics: length — 4528 mm, chord length — 310 mm. The general view and dimensions of the model blade profile are shown in Figure 1. This profile has a C-shaped spar made of aluminum, fiberglass skin (GFRP), MFC actuators embedded in the skin, foam aggregate (Rohancell 51FX) and balance weight (lead) [15].

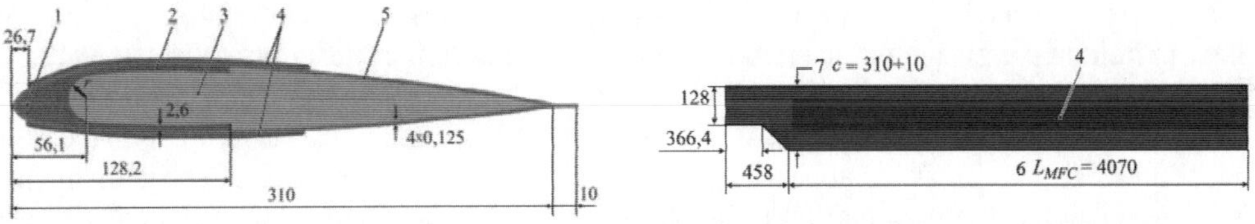

Figure 1. Cross-section and shape of a full-scale rotor blade. 1 - the balancing load; 2 –the spar; 3 –the foamed aggregate; 4 –the MFC actuator; 5 –the skin; 6 - the center of rotation; 7— the chord of the blade c = 310 + 10; 8 –the length of the MFCL MFC actuators= 4070; 9 - the radius of the rotor Rr = 4900. The dimensions are in millimeters [16]

The researched helicopter blade was made of textile fiber reinforced plastic with reinforcement scheme [0°/90°]. Technical elastic constants of fiberglass were calculated using ANSYS PC and comprised $E_x = E_y = 18.25$ GPa, $E_z = 6.05$ GPa, $\nu_{xy} = 0.148$, $\nu_{yz} = \nu_{xz} = 0.299$, $G_{xy} = 4$ GPa, $G_{yz} = G_{xz} = 3$ GPa. Technical elastic constants of the blades and actuator materials are taken from [4] MFC – $E_x = 30$ GPa, $E_y = E_z = 15.5$ GPa, $\nu_{xy} = 0.35$, $\nu_{yz} = \nu_{xz} = 0.4$, $G_{xy} = 5.7$ GPa, $G_{yz} = G_{xz} = 10.7$ GPa, $d_{31} = d_{32} = -1.98*10^{-10}$ M/B, $d_{33} = 4.18*10^{-10}$ MW; Rohancell 51FX – E = 13.79 MPa, G = 14 MPa, ν = 0.25; Pb – E = 13.79 GPa, G = 2 GPa, ν = 0.44.

For the numerical research the following boundary conditions (BC) were formulated: from the end of the structure attached to the rotor, homogeneous BCs of the second kind were specified. For piezoactive elements, the given heating temperature was set, equivalent to the control voltage $\Delta\hat{T} =$

U_{cont}. In addition, a volume force was applied to the blade – gravitation force. A complete adhesion was set in the area of the layers' and the construction structural elements' junction.

For better convergence of the solution and reduction of the errors of the results obtained, a computational grid was generated, the cells of which have a prismatic shape. Local adaptation was carried out in the areas of high stress gradients, namely in the area of the spar, the size of the adaptation element was 1 mm. The maximum element size was 5 mm, the minimum - $8*10^{-3}$ mm, for each material layer, the total number of design elements was 1.1 million. In addition, the construction and local grinding of the mesh did not allow sharp differences in the geometric dimensions of adjacent elements.

In the framework of the numerical experiments, the mechanical behavior of the model blade equipped with piezo actuators was simulated at a control voltage level of 1000V. The plan of the numerical experiments is presented in table 1.

Table 1. Plan of the numerical experiments for the model blade

Number of the numerical experiment	Reinforcement scheme	Piezo actuator arrangement	The voltage applied, V (^{0}C)
1	[0°/90°]	Upper	1000
2	[0°/90°]	Lower	1000
3	[0°/90°]	Combined	1000

To describe the inverse piezoelectric effect in the framework of the developed mathematical model, it is proposed to use the thermo-piezoelectric analogy between the piezoelectric and thermo-induced deformations. The analogy allows us to move from the need to solve a related boundary problem of electro elasticity to the solution of a much simpler unbound boundary value problem of thermo elasticity. The analogy is based on the hypothesis of the electric non-uniform field invariance in the piezoelectric element (electric field arising from the control voltage on its electrodes) to the deformation fields inside the volume of the piezoelectric element and, as a result, to the deformation of the structure, on the surface of which the piezoelectric element is fixed, piezo actuator. Therefore the hypothesis suggests the presence of the "inverse piezoelectric effect" in the piezoelectric element in the absence of the "direct piezoelectric effect". Such an assumption is admissible for the cases when the "controlling" component of the electric field in the piezoelectric element, caused by the action of the control electric voltage U_{cont} on the electrodes of the piezoelectric element, is significantly greater than the "deformation" component of the electric field caused by the "direct piezoelectric effect" from the effects of the essentially inhomogeneous deformation fields appearing and initially unknown in the piezoelectric element. Note that the heterogeneity of the deformation fields in the volume of the piezoelectric element is caused by many factors, in particular, the shape and location of the electrodes on the surfaces of the piezoelectric element, the characteristics of the control voltage on the electrodes and the attachment of the piezoelectric element to the surface of the structure [17, 18].

The tensor of reduced coefficients of linear thermal expansion is determined by the relation:

$$\alpha_{3i} = \frac{d_{3i}}{\Delta_{el}} (i = 1,2,3), \quad (1)$$

where Δ_{el} is the effective distance between the electrodes, the piezoelectric strain factor, d_{3i} characterizes the electromechanical properties of the piezoelectric material and describes the deformation of the material depending on the orientation of the electric field. The first index indicates the direction of the electric field, and the second index indicates the direction of deformation.

The values of the three coefficients of the piezoelectric deformations d_{31}, d_{32}, and d_{33} are enough to fully characterize the electromechanical properties of the piezoelectric material.

3. Results of the numerical simulation

According to the results of the numerical simulation of the model blade mechanical behavior, stress fields and axial displacements were obtained (Figure 2).

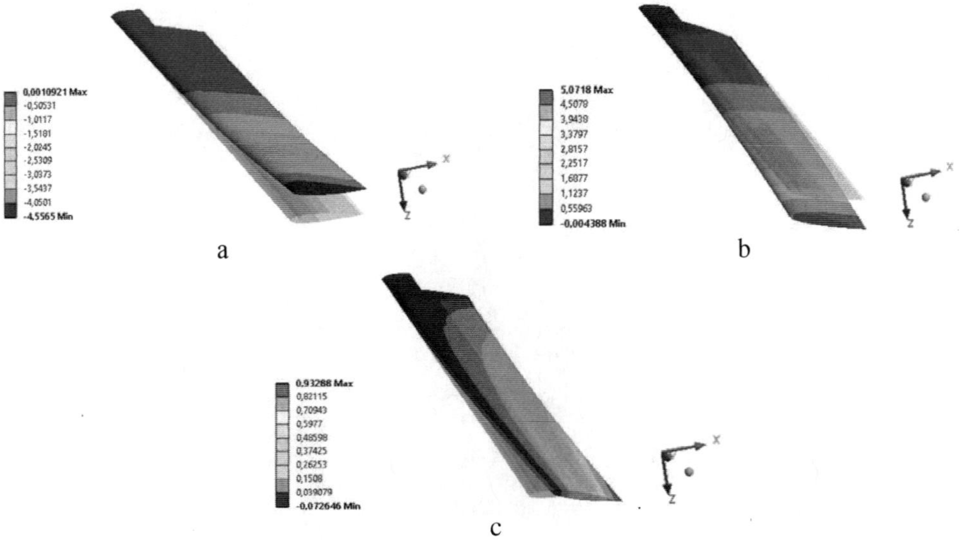

Figure 2. Fields of distribution of axial displacements U_z with the value of the control voltage 1000 V a - upper; b - lower; c - combined

Table 2 presents the results of the numerical simulation for a blade equipped with piezo actuators.

Table 2. Results of the blade numerical simulation

Reinforcement scheme	Spar material	Piezo actuator arrangement	U_x, mm	U_y, mm	U_z, mm
[0°/90°]	GFRP	Combined	1.79	0.304	1.4831

Analysis of the results of the model blade SSS numerical simulation showed that with the upper and lower placement of the torsional piezo actuator on the profile of the model blade, mutual compensation of opposing bending moments occurs, along with the predominance of torque and, as a result, twisting deformations. When designing rotor blades with variable geometry, it is recommended to place piezo actuators on both sides of the construction profile. The maximum value of the twist angle of the model blade at a control voltage level of 1000 V, with a combined (upper and lower) arrangement of the piezo actuator was φ=3.1°.

4. Verification of the numerical model

As part of the developed mathematical models' verification, a comparison was made of the results of the numerical research (performed using ANSYS software) with the results of laboratory tests of model layered PCM samples, equipped with bending piezo actuators, mechanical behavior. Figure 3 shows a general view of the samples and the experimental laboratory setup.

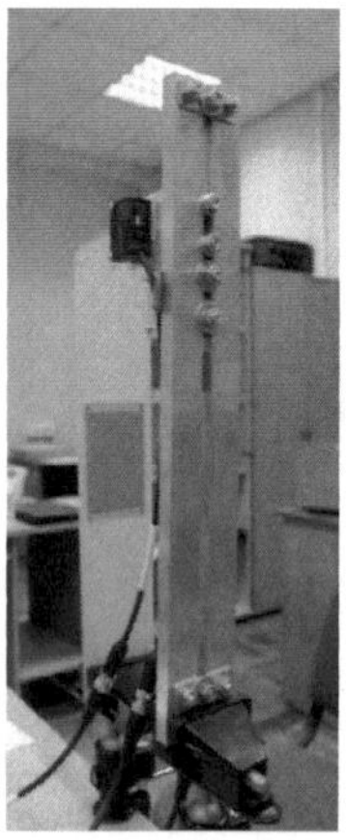

Figure 3. General view of the sample and the experimental laboratory setup

For numerical simulation, three-dimensional models of the samples were constructed, corresponding to the real ones. The samples consisted of four anisotropic layers of fiberglass on the basis of T13 glass fabric, made by the infusion method. The geometrical characteristics of the samples are 400x45x1.1, selected on the basis of the size of the M8557-F1 torsional piezo actuators used.

Comparison of the numerical research results with the results of laboratory tests was carried out at the twist angle of the free end. The twist angle was determined by the formula:

$$\sin\alpha = \frac{a}{c}, \tag{2}$$

where α is the twist angle; a - adjacent leg; c - hypotenuse.

The numerical simulation and the laboratory experiment were carried out with a control voltage of 25-125V

The following displacement fields were obtained from the results of numerical simulation of torsion tests for a fiberglass sample (Figure 4):

Figure 4. U_z displacement fields on the sample equipped with a torsional piezo actuator

Analysis of the obtained fields of displacements along the Z axis showed that the maximum twist angle occurs at the free end of the sample and equals 24', which correlates well with the experimental values.

Based on the results obtained, a graph of the sample's twist angle dependence on the control voltage applied to the piezo actuator was constructed (Figure 5).

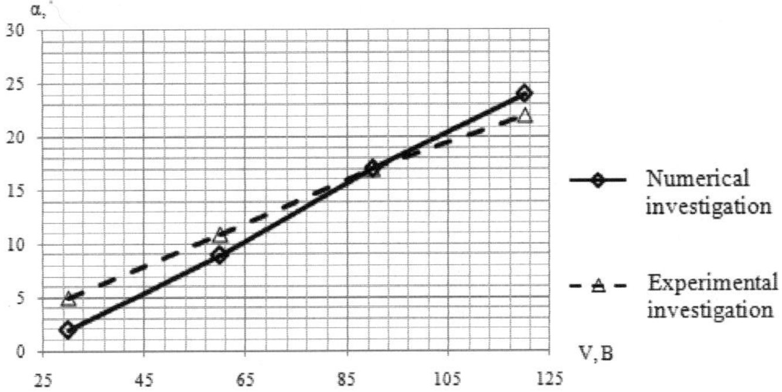

Figure 5. Graph of the dependence of the twist angle of the free end of the sample on the control voltage applied to the piezo actuator

The analysis of dependencies revealed that the results of numerical simulation are within the range of the results of torsion tests. The error in comparison with the results obtained for torsion tests does not exceed 10%.

5. Conclusion

Thus, according to the results of numerical researches, a new mathematical model of a helicopter blade, equipped with control piezo actuators, was developed. Research of the convergence of the finite element model of the blade has been carried out. The simulation of the mechanical behavior of the blade, equipped with piezo actuators, was carried out with the applied control voltage 1000V. Analysis of the axial displacement fields U_z showed that the two-way arrangement of piezo actuators in torsion is most effective. The limiting twist angles of the model blade were determined in the range of the control voltages operating values applied to the piezo actuators.

Acknowledgments

The reported study was partially supported by the Government Perm Krai, research project No. C-26/797 dated 12/21/2017.

References

[1] Animits V A, Borisov E A, Kritsky B S and Mirgazov R M 2016 *Trudy MAI* **85** 1-25
[2] Kritsky B S, Mirgazov R M and Le V Ch 2016 *Scientific Herald of MSTU GA* **19** **6** 68-76
[3] Chattopadhyay A, Liu Q and Gu H 1999 *Composite Part B: Engineering* **30** **6** 603 – 612
[4] Cesnik C E S and Shin S J 2001 *Smart Materials and Structures* **10** **1** 53 – 61
[5] Büter A and Breitbach E 2000 *Aerospace Science and Technology* **4** **5** 309–319
[6] Ghiringhelli G L, Masarati P and Mantegazza P 2001 *Composite Structures* **52** **1** 113–122
[7] Park J S and Shin S J 2007 *ICCES* **4** **4** 265–269
[8] Sekula M K, Wilbur M L and Yeager W T 2005 *Proc. of the 61st Int. Annual Forum of the American Helicopter Society* (Grapevine, USA) p 14
[9] Cesnik C E S, Mok J, Parekh A and Shin S 2004 *Proc. of the 45th AIAA/ASME/ASCE/ASC Structures* (Palm Springs, USA) p 10
[10] Cesnik C E S, Mok J, Morillo J A and Parikh A S 2004 *Proc. of the 30th European Rotorcraft*

IOP Conf. Series: Materials Science and Engineering **511** (2019) 012029 doi:10.1088/1757-899X/511/1/012029

Forum (Marseille, France) p 14
[11] Shin S, Cesnik C E S, Wilkie W K and Wilbur M L 2008 *Journal of Intelligent Material Systems and Structures* **19** **12** 1443-56
[12] Park J S, Shin S J and Kim D K 2008 *International Journal Aeronautical and Space Sciences* **9** **2** 18–33
[13] Sekula M K and Wilbur M L 2014 *NASA Technical Reports Server (NTRS)* p 13
[14] Barkanov E, Gluhih S and Kovalov A 2008 *Mechanics of Advanced Materials and Structures* **15** **3-4** 325-334
[15] Airfoil database list [Electronic resource] URL: http://airfoiltools.com/search/airfoils?m=a
[16] Kovalev A, Barkanov E, Ruchevsky S and Vseolovski M 2017 *Mechanics of composite materials* **53** **2** 1-20
[17] Makarova I S 2012 *Herald of the Samara State Technical University. Series of Physics and Mathematics* **3** **28** 191-195
[18] Pan'kov A A 2009 *Statistical mechanics of piezocomposites* (Perm: Publishing house of Perm State Technical University) p 480

Structure and optical-mechanical properties of the Zr-Y-O coatings deposited by pulse magnetron sputtering on optical glass

A V Ivanova[1], T I Dorofeeva[2], M P Kalashnikov[1,2], I A Bozhko[1,2], M V Fedorishcheva[1,2], V P Sergeev[1,2] and Y F Khristenko[3]

[1]Tomsk Polytechnic University, 30, Lenina ave., Tomsk, 634050, Russia
[2]Institute of Strength Physics and Materials Science of the Siberian Branch of the Russian Academy of Sciences, 2/4, Akademicheskiy ave., Tomsk, 634021, Russia
[3]Scientific research institute of applied mathematics and mechanics of Tomsk state university, Tomsk, 634050, Russia

E-mail: fed_mv@mail.ru

Abstract. The results of the research of the phase composition and the mechanical properties of the coatings on the basis of the Zr-Y-O system produced by pulsed magnetron sputtering on the K208 glass substrates are presented in this paper. By the X-ray diffraction method it has been discovered that there is the ZrO_2 phase in two modifications in the coatings. The deposition of the Zr-Y-O coating system allows not only to increasing the microhardness of the surface layer of the K208 glass up to ~19 GPa, but also maintain a high level of the elastic properties (We> 0.70). The laboratory tests have been carried out on the impact of the high-speed flows of iron microparticles on the Zr-Y-O protective coating produced by pulsed magnetron sputtering. When a high-speed flow of iron microparticles impact the samples with the Zr-Y-O coatings the surface density of the craters decreases ~ 7 times compared with initial state.

1. Introduction

In near-Earth space, outer elements of spacecraft are almost continuously subjected to impacts of high-speed micrometeoroids and particles of anthropogenic pollution. Optical elements [1], such as glass windows, lenses, as well as photoelectric converters of solar batteries suffer the most. As a result of such collisions, mechanical damage in the form of craters and cracks is formed on the glass surface. An increase in their density during long-term operation leads to a significant erosion of the surface, a decrease in transparency and a degradation of the mechanical properties of optical elements.

It was shown [2], that surface modification by protective Al-Si-N coatings can significantly increase the erosion resistance of quartz glasses used in spacecraft windows. However, the significant difference in the thermal expansion coefficient of the Al-Si-N coating and the K208 glass [3] complicates the task of deposition of high-quality coating on thin plates of optical glass. The Zr-Y-O system as a material for deposition of a protective coating has a closer coefficient of thermal expansion to K208 glass. It also has a unique set of properties [4], such as high strength, wear resistance and transparency. It may be a suitable option for the K208 protecting glasses of the solar cells of the spacecraft from the impact of high-speed microparticles.

The purpose of the work is to study the structural state, optical and mechanical properties of coatings based on the Zr-Y-O system, as well as the effect of these coatings on the resistance of K208 optical glasses against the impact of solid microparticles moving with a speed close to the first cosmic one.

2. Experimental procedure

The investigated coatings were deposited on substrates, which were polished discs ∅15 × 4 mm in size made from the K208 radiation-resistant glass, using pulsed magnetron sputtering on a UVN-05MD KVANT vacuum unit (Techimplant Ltd) [5] in an Ar + O_2 gas mixture using a target based on Zr-Y. The temperature of the substrates during the coating process was 573 K. The thickness of the coatings was 5–6 μm.

The structural-phase state of the samples was investigated by X-ray using the DRON-7 device (Burevestnik Innovation Center, JSC). X-ray investigation of the coatings was carried out under continuous 2θ - scanning with the Bragg - Brentano focusing at Co Kα radiation (wavelength λ= 1.78897 A). The data base JCPDS PDF-2 (The Joint Committee on Powder Diffraction Standards) was used for interpretation of the diffractograms.

Analysis of the optical properties of magnetron coatings of the Zr-Y-O system on quartz glass substrates was based on measurements of the transmittance $T = f(\lambda)$ in the wavelength range λ = 200-1100 nm using the SF-256 UVI spectrophotometer (LOMO Photonics). The principle of measuring the ratio of two light fluxes that have passed through the investigated and the standard sample respectively was taken as a basis.

The mechanical properties of the coatings and the glass substrates were measured using the NanoHardnessTester (CSM Instruments, Switzerland) under the load on the indenter 20 mN, with the depth of indentation of the coated samples not exceeding 220 nm. In order to obtain reliable results for each sample at the above loading no less than 10 prints were made and the final result was the arithmetic mean value. The microhardness, the modulus of elasticity and the elastic recovery coefficient of the coatings were calculated on the basis of the analysis of the unloading branch obtained by the Oliver - Pharr method [6].

To assess the impact-protective properties of the coatings, the experimental samples were bombarded by high-speed microparticles in the light-gas gun MPH23/8 [7]. The gun was developed in the Institute of Applied Mathematics and Mechanics affiliated to Tomsk State University. The experimental samples were bombarded by iron microparticles of the spherical form with an average size of 56 μm at the velocity of 5-8 km/s. After the shock impact test the images of the craters formed on the surface of the samples were investigated using a scanning electron microscope LEO EVO-50XVP (Karl Zeiss Group, Germany), and the surface density of the craters (ρ) was calculated.

3. Results and discussion

The X-ray diffraction analysis previously established that there were ZrO_2 phases in various structural modifications in the coating based on Zr-Y-O [8,9]. The phase ZrO_2 was represented by two modifications: tetragonal (t) and monoclinic (m). The quantity of tetragonal and monoclinic phases was 44 and 56%, respectively. Figure 1 shows the X-ray diffraction patterns of the Zr-Y-O coating with a thickness of coating about of 6 μm.

TEM shows that the grain structure of the coating based on the Zr-Y-O system is columnar. The longitudinal grain size of the ZrO_2 phase coincides with the growth axis of the coating and is determined by the thickness of the layer, and the transverse grain size does not exceed ~ 80 nm. Figure 2 shows the electron microscopic image of the columnar structure of a coating based on the ZrO_2.

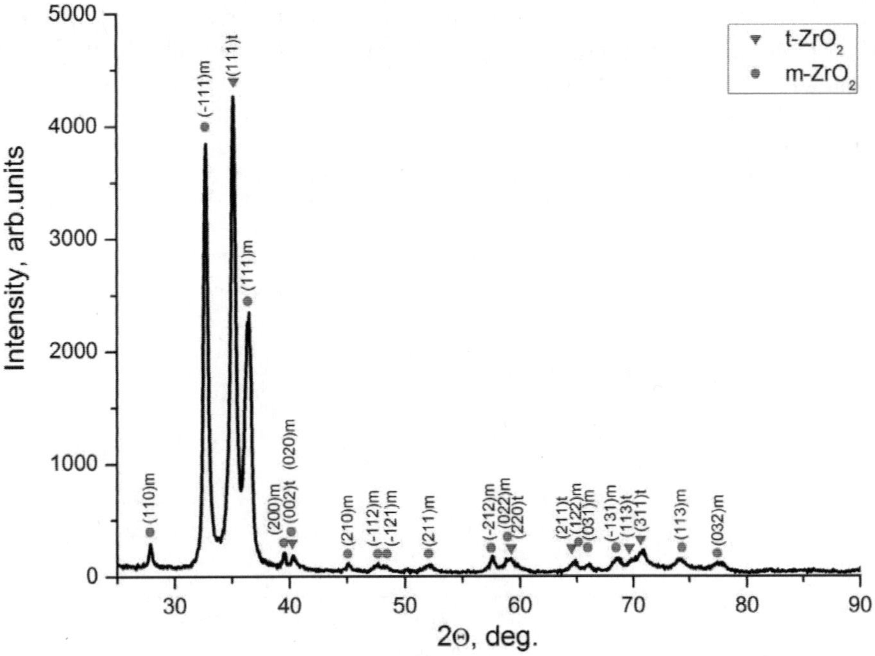

Figure 1. X-ray patterns of coatings on the basis of Zr-Y-O on the K208 glass substrate.

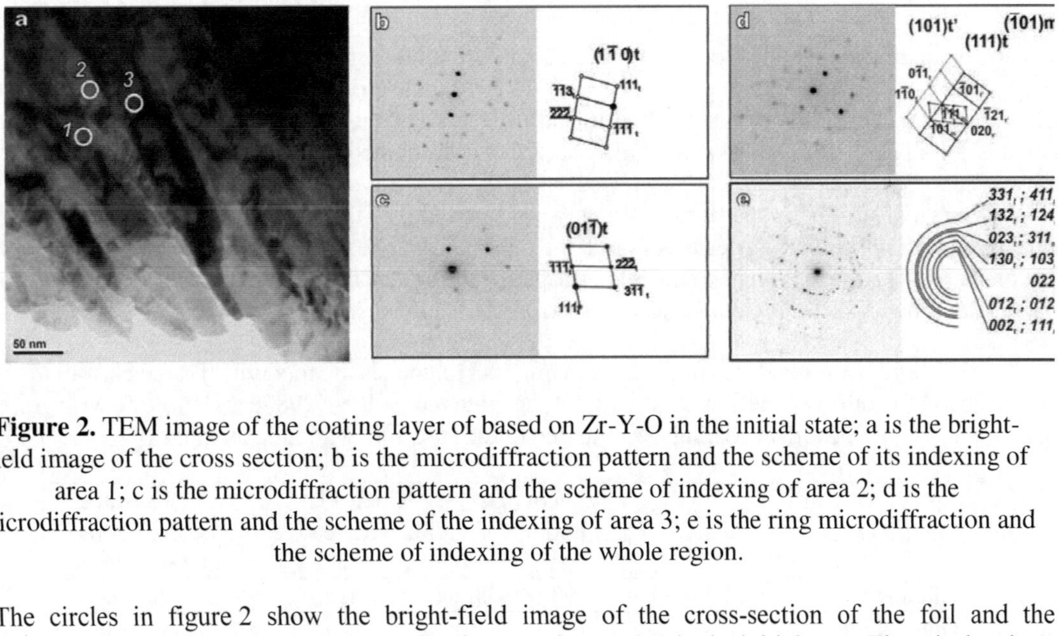

Figure 2. TEM image of the coating layer of based on Zr-Y-O in the initial state; a is the bright-field image of the cross section; b is the microdiffraction pattern and the scheme of its indexing of area 1; c is the microdiffraction pattern and the scheme of indexing of area 2; d is the microdiffraction pattern and the scheme of the indexing of area 3; e is the ring microdiffraction and the scheme of indexing of the whole region.

The circles in figure 2 show the bright-field image of the cross-section of the foil and the microdiffraction pattern of the Zr-Y-O coating from sections 1, 2, 3 in the initial state. There is the phase ZrO_2 in area 1 with the crystalline lattice parameters a = 5.12 Å and c = 5.25 Å. In area 2 there is the same phase but one can see another plane. In area 3 there are two tetragonal phases (t and t' with a smaller lattice parameter a = 3.64 Å, b = 3.64 Å, c = 5.27 Å) and a monoclinic modification of ZrO_2 with the lattice parameters a = 5.31 Å, b = 5,21 Å, c = 5.15 Å. These data are consistent with the results of the X-ray diffraction analysis.

Figure 3 shows the transmission spectra of the investigated samples in comparison with the transmission of light for a k208 glass substrate. One can see that the initial glass K208 (figure 3, curve 1) has a high transmittance of 87-90% in all investigated visible wavelength interval. The deposition of protective coatings of the Zr-Y-O system results in an insignificant decrease in the light transmittance of K208 glass samples (down to 60-80%) in the visible region of the spectrum (figure 3, curve 2). The oscillating character of the experimental transmission spectra observed for samples of K208 glass with protective coatings in the visible region of the spectrum is due to the phenomenon of interference of light in the coatings.

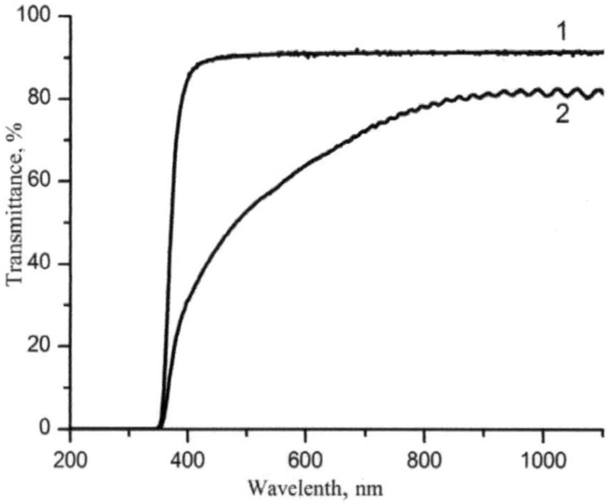

Figure 3. Spectra of transmission of K208 glass samples with and without protective coatings: 1–K208 glass; 2 – the same with Zr-Y-O coatings.

The results of the investigation of the mechanical properties of the Zr-Y-O are shown in Table 1. One can see that the microhardness for the Zr-Y-O coatings exceeds the microhardness of the initial K208 glass more than 2 times and the coefficient of elastic recovery - by 1.3 times. It is important to note that for all Zr-Y-O coatings a high microhardness is combined with a high coefficient of recovery ($W_e > 0.70$).

It is known [1, 2] that at the impact of high-speed particles at the target surface a dispersing shock wave is formed, which propagates deeply into the material. As a result, a crater is formed of the impact spot on the target surface. Figure 3 shows the images of the craters formed on the surface of the samples as a result of the exposure to high speed iron microparticles flow.

Table 1. Average values of the microhardness H_m, the elastic modulus E*, the elastic recovery ratio W_e, the surface density of the craters ρ on the K208 glass substrate with the Zr-Y-O coating as compared to the substrate without a coating.

Sample	Microhardness, H_m GPa	Young's modulus, E*, GPa	Coefficient of elastic recovery, R_e	The surface density of the craters ρ, mm^2
Glass K208	8.40 ± 0.15	93.0 ± 0,8	0.53	20.19
Glass K208 with Zr-Y-O coatings	18.81±0,69	318.6±8,8	0.72	3.04

One can see that the craters formed on the K208 glass surface have a shape characteristic of brittle materials. There are a large number of craters having various sizes (figure 4a) on the surface of the initial K208 glass. On glass substrate with the Zr-Y-O protective coating the number of the craters decreases by ~ 7 times. The quantity of the craters whose cross sectional dimension does not exceed 50 μm decreases to large degree.

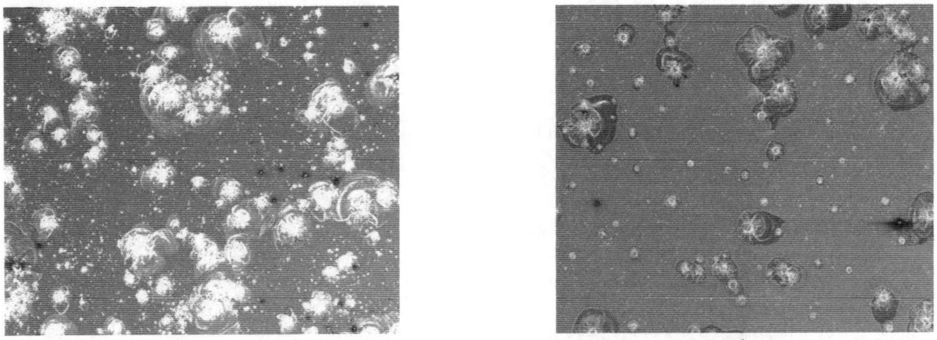

a b

Figure 4. SEM images of craters formed on the samples surface as a result of high-speed impact flow by iron microparticles: a - uncoated sample, b - Zr-Y-O coated sample (craters are circled by red circles).

4. Conclusion

The Zr-Y-O coatings deposited on K208 glass by pulsed magnetron sputtering have a crystalline multiphase structure including the ZrO2 phases in monoclinic and tetragonal modifications. The deposition of these coatings leads to significantly increase in the erosion resistance of K208 glass. The decrease in the surface density of craters on the surface of the samples after impact of high-speed iron microparticles by ~ 7 times compared to the initial ones indicates this. The microhardness of the K208 glass substrate with deposited Zr-Y-O coating was 18.81 GPa.

Acknowledgments

Work was supported in the scope of basic scientific research of the Russian Federation state academies of sciences for 2013-2020 and performed within the scope of the state task "Space Materials Science" NRTPU for 2016-2018.

References

[1] Novikov L S 2009 Exposure of solid particles of natural and artificial origin on the spacecraft (Moscow: Moscow State University Press) 104
[2] Sergeev V, Psakhie S, Chubik P, Cherniavsky A, Soloviev V, Solntsev V, Khristenko Yu, 2017 Vacuum **143** 454
[3] Babichev A P, Babushkina N A, Bratkovsky A M 1991 Physical quantities. Directory., ed. I S Grigorieva and E Z Meilikhova (Moscow: Energoatomizdat) 1232
[4] Panin V E, Sergeev V P, Moiseenko D D, Pochivalov Yu I 2011 Physical mesomechanics **14** 5
[5] Sergeev V P, Yanovsky V P, Paraev Yu N, Kozlov S A, Zhuravlyov S A, 2004 Physical Mesomechanics **7** 333.
[6] Pharr G M, Oliver W C, Brotzen F R, 1992 J. Mater. Res **7** 613
[7] Gerasimov A V, Pashkov S V, Khristenko Yu F 2011 Bulletin TSU. Mathematics and Mechanics **16** 70
[8] Petra Langjahr A, Rainer Oberacker and Michael J. Hoffmann, 2001 J. Am. Ceram. Soc., **84** 1301
[9] Sergeev V P, Fedorischeva M V, Neufeld B V, Kalashnikov M P, 2014 Adv. Mat. Res. **880**, 184

IOP Conf. Series: Materials Science and Engineering **511** (2019) 012031 doi:10.1088/1757-899X/511/1/012031

Study of mechanical and protective characteristics of composite materials based on aramid fabrics modified in a microwave electromagnetic field

I V Zlobina and N V Bekrenev

Department "Technical mechanics and machine parts", Yuri Gagarin Saratov State Technical University, 77 Politechnicheskaya street, Saratov, 410054, Russia

E-mail: irinka_7_@mail.ru

Abstract. The studies of the protective characteristics of samples of armor packages, which are compositions of aramid fabric TSVM-J and glue 88, after influenced by the microwave electromagnetic field, have been carried out. A reduction in the number of maximum punctured layers to 7, and an armor effect to 10 mm was stated. In this case, compared with the control sample, a significant change in the mechanical characteristics of the composite armor package was found. Density increased by more than 10%. At the same time, the tensile ductility decreased by (40-47)%, and the elastic deformation with indentation increased almost 5 times. The results obtained after their refinement and substantiation of the physical mechanism can be used as the basis for the development of technologies for enhancing the characteristics of armor protection elements made of composite ballistic materials and structural organoplastics for aircraft.

1. Introduction

1.1. Status of the question

The development of means of destruction and the expansion of regional conflicts, the increase in the threat of international terrorism, and the increase in the number of man-made and natural disasters are raising the demands on the means of protecting the personnel of the armed forces, the Ministry of Internal Affairs and the Emergencies Ministry of Russia. The importance of developing methods and technologies to counter these threats is evidenced by the inclusion in the priority areas of development of science and technology of the Russian Federation in the direction "Security and Counteraction to Terrorism" (Presidential Decree of July 7, 2011 No. 899). According to Russian classification, all protective equipment is divided into 6 classes. The most difficult and reliable are systems consisting of metal or ceramic plates mounted on a sublayer of ultra high molecular weight polyethylene. However, these tools protecting from bullets of automatic weapons, including SVD-type sniper rifles, have considerable weight and hamper human movement. At present, in the structures of the Ministry of Emergency Situations, as well as for equipping the flight crew of the Air Force, light flexible personal protective equipment of classes 1-2 based on aramid woven materials and their compositions with adhesive compositions, as well as ultra high molecular weight polyethylene, are widely used. Good protective properties of aramid fabrics combined with light weight and flexibility led to their widespread use in the army, aviation and navy [1]. In Russia, one of the most common ballistic materials based on aramid filaments are SCRMs and new material Rusar [2, 3]. Compared to carbon and glass fibers, aramid filaments are less fragile and suitable for processing on conventional textile manufacturing equipment [4, 5]. In this case, the elements of clothing are made of a certain number of unrelated layers of fabric (usually 30). However, protective helmets and part of the structural elements of the modern Russian "Warrior" complex (knee pads, elbow pads, etc.) are in fact the composite structures of a polymer matrix reinforced with layers of aramid fabric. Aramid fabrics also have significant drawbacks associated with weak resistance to low-speed solid projectiles [4, 5], weakening of the protective properties when wet, exposure to direct sunlight, a large enough armor effect, leading to serious injuries.

1.2. Relevance of research
Studies of increasing the strength of such fabrics and compositions based on them are aimed at improving the technologies for obtaining the initial components and the formation of fabrics by optimizing weaving [6, 7], which complicates and increases the cost of the process and also leads to the need for restructuring or re-equipment of production. Previously, we carried out the studies that showed the effectiveness of reinforcing modification of finally formed products made of composite materials reinforced with carbon fibers in a microwave electromagnetic field [8].Since, the plan of the structure and differences in the physicomechanical properties of the matrix and filler of the composition containing aramid fabrics as a reinforcing component, differ little from the materials we studied, it is of interest to consider the effect of the microwave electromagnetic field on the mechanical properties of the compositions of aramid fabrics and additionally evaluate the changes in their protective characteristics.

1.3. Statement of the problem
The purpose of the research was to determine the possibility of increasing the protective properties of light armor elements made of polymer composite materials based on aramid fabrics by modifying them in a microwave electromagnetic field.

2. Research methods and equipment
We carried out studies of the effect of microwave processing on the physicomechanical and protective properties of samples formed from 24 layers of aramid tissue TSVM-DJ article 56319A with a binder in the form of glue 88. While preparing samples, each layer was impregnated with a layer of glue, placed under a press and held for 24 hours After the composition was cured, one part of the samples was left as a control, another part was placed in a microwave electromagnetic field of average power density for 3 minutes. The treatment was carried out using the installation "Zhuk-2-02" with a horn radiating antenna manufactured by LLC AgroEkoTech (Obninsk, Kaluga region). The power of the installation magnetron is 1200 W, the frequency is 2450 MHz. The distance to the plane of the aperture of the horn was 150 mm, which corresponds to the average power level. According to our earlier studies [9], it is the average microwave power that provides the greatest efficiency in increasing the breaking strength of aramid filaments. A total of three specimens were made with dimensions of 60x60 mm (figure 1).

Tests of samples before and after processing were carried out on a computer setup equipped with strain gauge force sensors and a worm loading mechanism. Sensor signals were transmitted via an analog-to-digital converter to a computer. Processing the results of measuring the increase in the load applied to the sample according to a special program installed (LabVIEW, Orel) made it possible to obtain load (moment) graphs in dynamics from the moment of application to the maximum value. At the same time, the deformation of the sample was measured using a dial gauge with a scale value of 0.01 mm. According to the ratio of the deformation to the loading force, the compliance of the material was determined. The density of the samples was determined by calculating the volume according to the results of measurements and weighing on an electronic balance RM200 with an accuracy of 0.0001 g. The tensile deformability was determined on a tensile testing machine IR-5082-100.

a *b*

Figure 1. Samples of armored packages of 24 layers of fabric TSVM-DJ and glue 88: control (a), after modification in the microwave electromagnetic field (b).

The protective properties of the samples were determined by firing an armored package from a PM pistol from a distance of 3 meters in the dash of the Military Department of the Saratov State Technical University named after Gagarin Y.A. The bullet velocity was previously determined using a chronograph (figure 2).

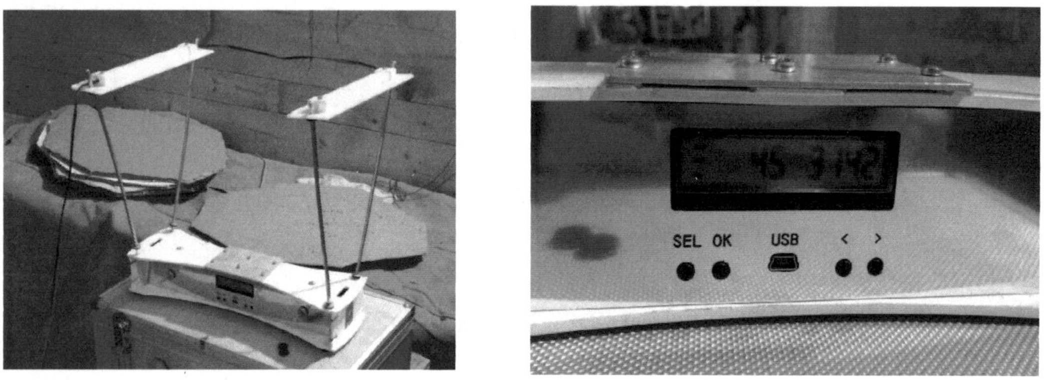

a *b*

Figure 2. Chronograph: general view (a); indicator with fixed bullet speed (b).

3. Results and discussion

As a result of the research, the following main data was obtained.

The thickness of the formed samples was 12 mm. In this case, the sample subjected to microwave exposure, has a more uniform thickness. The dispersion of the thickness of the control samples is 0.48 mm^2, processed - 0.4 mm^2. The density of samples after microwave processing increased from 0.55 g/mm^3 to 0.602 g/mm^3, i.e. more than 9%. This result can be associated with the dehydration of the adhesive matrix and the stretching of aramid fibers, which leads to their convergence and, as a result, to an increase in the density of the structure.

The ductility of the control samples was (0.5-0.4) mm/N, treated - (0.3-0.34) mm/N. Thus, the ductility of composite materials modified in the microwave electromagnetic field, reinforced with layers of aramid fabric, decreased by (40-47) %, which is quite consistent with an increase in the density of the sample.

At the same time, the elastic deformation of stretching of the treated samples increased from 0.04 mm to 0.22 mm, i.e. in 5 times. This may be due to the increase in the number of bonds between the filaments and fibers in the filaments through fibrils, which are stimulated by the activation of large molecules when exposed to microwave radiation, which prevents their rupture and helps the sample to

IOP Conf. Series: Materials Science and Engineering **511** (2019) 012031 doi:10.1088/1757-899X/511/1/012031

return to its original state after the penetration load ceases. For the control sample, predominantly plastic deformation is typical due to the rupture of a small number of interfibril bonds and stretching of the plastic adhesive binder. When determining the protective parameters of armored packages, the samples were fixed on the target in the dash of the Military Department of the Yuri Gagarin State Technical University (figure 3a).

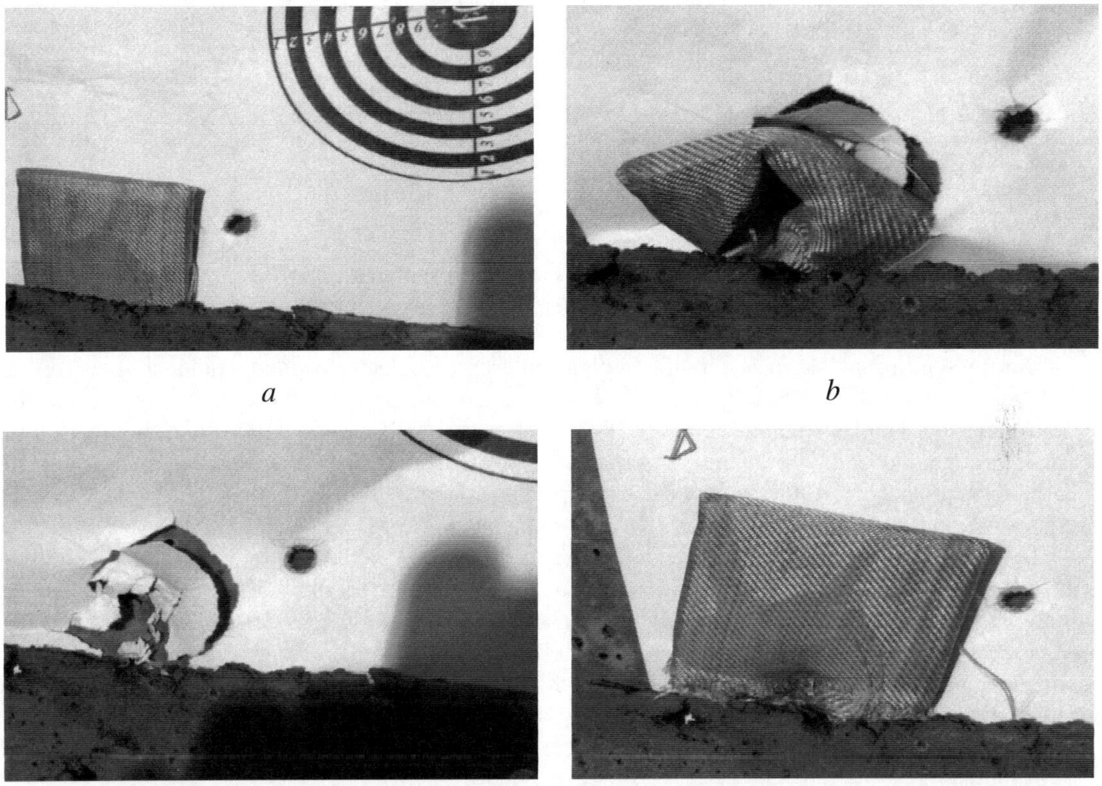

a *b* *c* *d*

Figure 3. The sample on the target (a). control sample after a bullet hit (b), target after shelling a control sample (c), treated sample after a bullet hit (d).

The speed of a bullet fired from a PM pistol, determined by the chronograph, was 314.2 m/s. The actual bullet caliber (figure 4b), determined by measuring its diameter, was 9.2 mm. As a result of the shelling, when the bullet hit the control samples, they were severely deformed (figure 3b) and shot through, and one sample was broken off and separated from the target (figure 3c). This may be due to the above-noted low ductility and low elastic deformability of the control armor packages.

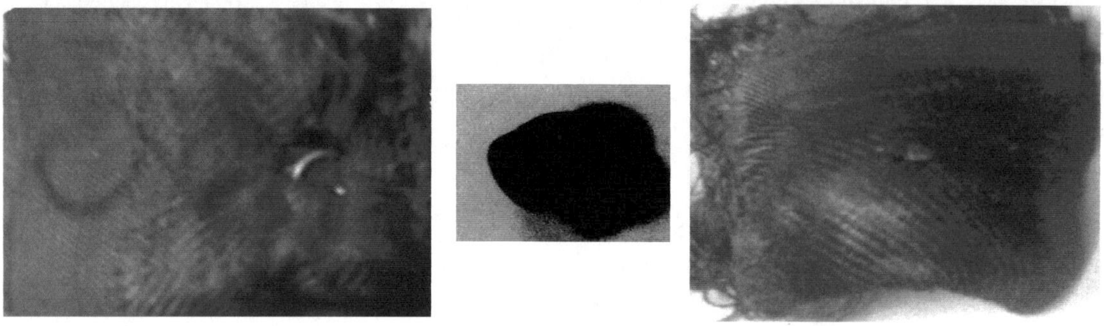

a *b* *c*

Figure 4. Results of shelling of an armor package modified in a microwave electromagnetic field: a bullet stuck in the structure of layers (a), a bullet extracted from an armored package (b), armor effect (c)

During shelling of samples modified in the microwave electromagnetic field, a small residual deformation was established (figure 3d). In this case, the bullets were stuck in the structure without destroying the samples (figure 4a). When disclosing an armored package after removing the bullet, a break through of 7 layers of 24 and a slight damage of the next 2-3 layers were recorded. Armor effect in the form of buckling of the back of the armor package did not exceed 10 mm (figure 4c). Allowed by domestic standards the armor effect is 20 mm with the number of layers equal to 30.

The obtained results can be associated with the 2-fold increase in the tensile strength of aramid fibers, as well as a significant (up to 50%) increase in the strength of polymeric materials after their modification in a microwave electromagnetic field [9-11]. The cumulative effect of these factors leads to a change in the elastic-strength parameters of the armored package, increasing its density simultaneously with elastic deformability, which prevents destruction when exposed to a bullet and reduces the value of the armor effect.

4. Conclusion

The results obtained after the refinement and substantiation of the physical mechanism can be used as the basis for developing technologies of improving the reliability of light armor protection elements made of composite ballistic materials based on aramid fibers.

Acknowledgments

The studies were carried out with the support of the RFBR grant No. 17-03-00720 "Methodology of optimization micro-design of composite materials for objects of complex shape with increased dynamic strength, which are layer-by-layer formed by electrotechnological methods".

References

[1] Bhatnagar A 2006 Lightweight ballistic composites *Woodhead Publishing Limited (Cambridge England)* 416 p

[2] Chistyakov E 2013 Russian body armor and new armor materials (*Technique and weapons* vol 8) pp 2–6

[3] Ignatova A M, Artemov A O 2012 Analytical review of modern and promising materials and designs of armor obstacles and defenses against destruction (*Fundamental research* vol 6–1) pp 101–105

[4] Shustov Y S, Kurdenkova A V 2016 Investigation of the physicomechanical properties of fabrics made from Rusar filaments (*Scientific-methodical electronic journal "Concept"* vol 3) pp 36–40

IOP Conf. Series: Materials Science and Engineering **511** (2019) 012031 doi:10.1088/1757-899X/511/1/012031

[5] Shuldeshova P M, Zhelezina G F 2014 Aramid layered woven material for protection against ballistic and shock effects (*Transactions of VIAM: electron. scientific tech. journals* vol 9) Art 06
[6] Safonov P E 2013 Development of optimal technological parameters for the manufacture of aramid fabrics for technical purposes *Author. diss. Cand. tech. sciences (Place of protection: Federal State Educational Institution of Higher Education "Russian State University A N Kosygin)*16 p
[7] Slugin A I, Parfenov O V, Nikolaev N A 2011 The study of the ballistic properties of tissue (*Technology of the textile industry* vol 4 (333) pp 74–78
[8] Zlobina I V, Bekrenev N V 2016 The Influence of Microwave Electromagnetic Field jn Mechanical Properties of Composite Materials (*High technology* vol 17 - 2) pp 25–30
[9] Zlobina I V, Bekrenev N V 2017 Effect of the microwave electromagnetic field on the strength and microstructure of ballistic nonmetallic materials (*Questions of electrical technology* vol 1) pp 40–47
[10] Zlobina I V, Bekrenev N V 2016 Improving the performance characteristics of materials for pipeline valves in a microwave electromagnetic field (*Questions of electrical technology* vol 2) pp 31–36
[11] Zlobina I V, Bekrenev N V 2016 Study of Flexural Strength of 3D Printing Objects from Dielectric Materials Modified in a Microwave Electromagnetic Field *Dynamics of systems, mechanisms and machines* (vol. 1-1) pp 308–312

IOP Conf. Series: Materials Science and Engineering **511** (2019) 012032 doi:10.1088/1757-899X/511/1/012032

Study of the formation features of hard metal composites structure obtained from bimodal powder mixtures

Yu I Gordeev, V B Yasinskiy, N E Anistratenko and A S Binchurov

Siberian Federal University, 79, Svobodny pr., 660041 Krasnoyarsk, Russia

E-mail: mexanixs@mail.ru

Abstract
A new method of forming dispersion-hardening oxide inclusions in the process of thermoreactive synthesis from aluminum nanoparticles introduced into WC-Co powder mixture is proposed. Such multiphase and fragmentarily nanostructured composite is characterized by additional heterogeneity, which is determined by the difference in sizes and elastic properties of the phases. In the structure of WC-Al_2O_3-Co composites a "barrier" effect is provided, the length of the carbide grains contacts is reduced. Carbide grains are separated by a thin cobalt layer with increased strength. In accordance with the results of numerical estimates using finite element analysis (FEA) method, the maximum stress intensity in tungsten carbide is 9.1 GPa (occurring at an external tension of 3 GPa), which is 30% less than the maximum stress intensity in the basic material. The strength improvement predicted by the FEA method in fragmented nanostructured hardmetal composite correlates with experimental results of measurements of fracture toughness according to Palmqvist data. In case of the optimal amount of additives, the increase in crack resistance is equal to 50% compared with the basic material.

Introduction
The tasks of quality improvement of hard metals can be effectively solved through the use of nano-sized tungsten carbide powders. However, it is well-known that when using traditional consolidation methods for obtaining a high density of the sintered composite, a high temperature and holding time during sintering are necessary, what leads to the increase of initial dimensions and the growth of carbide grains [1]. A possible option and effective technical solution to prevent recrystallization is the introduction of nanoparticles, i.e. inhibitors. Doping is carried out by various methods, such as chemical, mechanical (high-energy ball milling) and others [2]. Being dissolved in cobalt, alloying elements affect not only the microstructure, but also have a modifying effect on the consolidation processes and change the sintering kinetics, as well as mechanical and operational properties of the alloy [3-5]. The use of composite carbide-binder powders with simultaneous doping of inhibitors with nanoparticles (oxides, carbides, nitrides) also appears to be a highly promising direction [6-8]. Additional modification of the main mixture of WC, Co micron powders with nanoparticles also provides (besides preventing recrystallization) an improvement of the binder strength and the level increase of physical and mechanical properties of the composite as a whole. Such multiphased and fragmentarily nanostructured composite is characterized by the additional heterogeneity, which is determined by the difference in sizes and elastic properties of the phases [9].

Methods and equipment
For obtaining hard metals with adjustable grain size, structure and improved properties, the modifying

additives of nanoparticles are introduced into bimodal powder mixture. The mechanism of slowing down the growth of a carbide grain during sintering process is the inhibitor segregation at the emerging boundaries of WC-WC or WC-Co (in the volume of cobalt binder). In the course of experimental studies, two methods of introducing aluminum oxides modifying additives into the structure of a hardmetal composite were compared: at the stage of mechanical mixing WC-Co powders with nanoparticles of oxides, carbides (described in [8,9]), as well as a new method of forming dispersion-strengthening oxide inclusions in the process of thermoreactive synthesis from aluminum nanoparticles of W-Co powders introduced into the mixture. Bimodal powder mixtures (WC-Co) were obtained by the joint grinding and mixing of micron tungsten carbide powders to sizes (d_{WC} ~ 2 μm), cobalt (d_{Co} ~ 7 μm) and submicron composite powders (WC-Co) in planetary ball mills in a liquid alcohol medium within 2 - 6 hours. Then modifying additives of aluminum nanoparticles (dAl ~ 0.08 μm with specific surface area about 12 m^2/g) were additionally introduced into the mixture in amount from 0,5 to 2% wt as suspension, and the resulting bimodal mixture was additionally stirred. Due to the size difference and specific surface area of the particles of the basic mixture and additional phase, it is possible to estimate the volume content of the nanophase in the composite - from 6 to 10% of vol. Changes in the structural-phase state of compacts and sintered samples obtained from the powder mixtures of tungsten carbide, cobalt and aluminum nanopowders were monitored by thermogravimetric method (TG), differential thermal analysis (DTA) and differential scanning calorimetry (DSC) using Jupiter STA449C derivatograph («Netzsch», Germany) and SDT Q600 V20.5 («Netzsch», Germany) in combination with X-ray phase analysis on D8 ADVANCE diffractometer (Germany) – figure 1. Additionally, shrinkage was monitored during sintering on DIL 402 dilatometer («Netzsch»). The complex use of these methods allows to establish the temperature-time intervals corresponding to significant changes of the phase state in the volume of composites. The microstructure of samples was examined on the polished surfaces by means of Hitachi TM1000 and JEOL JSM-7500FA scanning electron microscopes. The fractures analyses were carried out using JAMP 9500F microscope.

Results and discussions

In the process of heating during sintering in vacuum, the destruction of oxide films on the surface of aluminum nanoparticles occurs at relatively low temperatures - already at T°C of about 350° C (peak 1, figure 1a). Subsequently, exothermic reduction-oxidation reactions take place between the aluminum and the oxide film on cobalt $2Al + 3CoO = Al_2O_3 + 3Co$, or with atmospheric oxygen absorbed on the surface of the mixture powders. As a result of the local release of additional heat in the microvolumes of the raw billet, a liquid-phase sintering process is carried out between carbide particles at low temperatures - even in the interval of 550 - 700 °C (peak 2, figure1) by the mechanism of mechanical repacking of the carbide phase particles. At the same time, segregation of oxide nanoparticles on the surface and between carbide grains occurs at an early stage of sintering, which in the course of subsequent sintering provides an inhibitory effect. The completeness of thermoreactive synthesis reaction of oxides from aluminum nanoparticles is also confirmed by the results of X-ray phase analysis of the samples when heated to a temperature of 1000 °C (figure 1b).

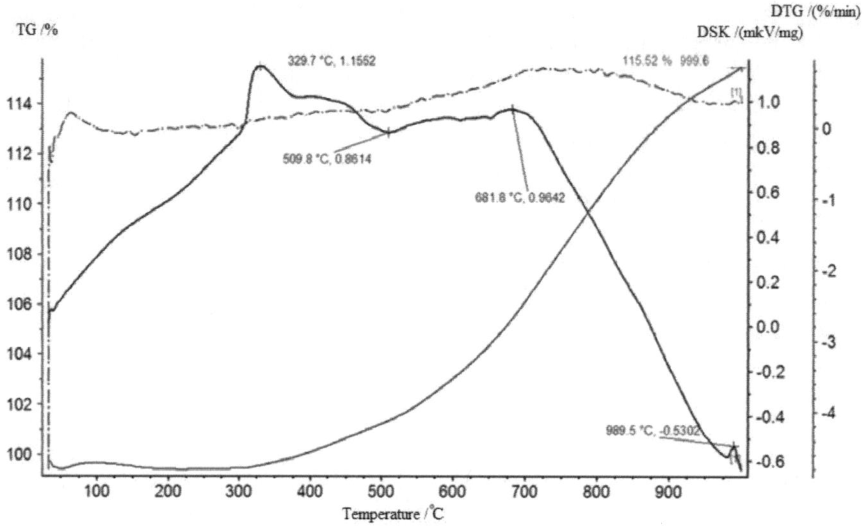

a)

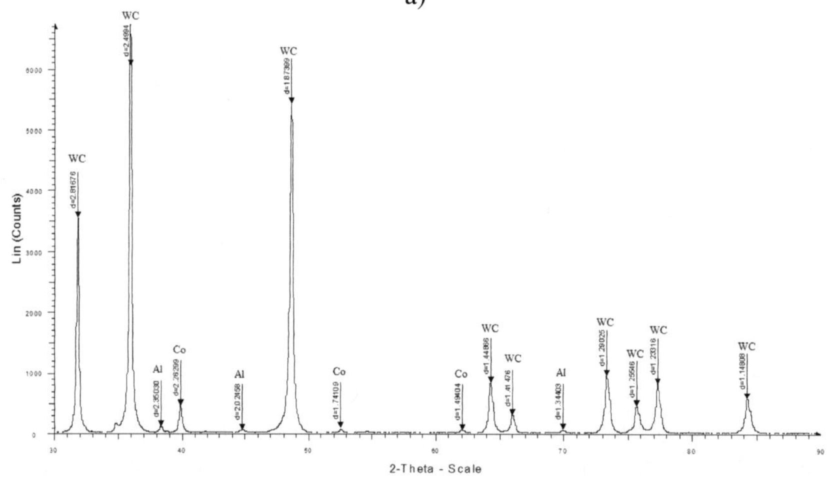

b)

Figure 1. Results of thermal analysis a) and X-ray phase analysis (b) of the hard metal samples from WC-Co-Al bimodal mixtures.

After final sintering at the temperatures about 1370 °C it is possible to form and keep the new type of a fragmentary nanostructured composite WC-Al_2O_3-Co, which is formed by the two interpenetrating scaffolds: carbide and mixed ceramic-metal (metal oxide Co-Al_2O_3). Typical images of the structure of such composites obtained from bimodal mixtures are shown in figure 2 (a, b) for the composition WC-8Co-2Al and in figure 2 (c, d) for the composition WC-8Co-0.5Al.

IOP Conf. Series: Materials Science and Engineering **511** (2019) 012032 doi:10.1088/1757-899X/511/1/012032

Figure 2. Microstructure of composites obtained from bimodal mixtures WC-Co-Al: (a, b) tungsten carbide grains bonded by mixed oxide-cobalt phase - dark background in the form of oxide inclusions; (c, d) the nature of the distribution of nanooxides in the cobalt layer.

The results of the study of the structure parameters basing on typical images (figure 2) using AxioVision 4.6.3, module for the analysis of images Auto Measure on the basis of optical microscope Carl Zeiss Observer.Z1m. (Germany) show that the average size of a carbide grain of hardmetal composites WC-Co, modified by the nanoparticle additives, is equal to 0,8 – 1,2 μm (for WC-8%Co), average thickness λ_{Co} varies from 0,12 to 0,2 μm. (standard hard metals based on WC-Co have the average thickness of λ_{Co} - 0,5-0,6 μm).

Changes in size of the phases are accompanied by a decrease in the contact ability of carbide grains, relative share of the interface borders WC-WC is to 0.3 (in comparison with 0.56 for the base microcrystalline hard metals WC-8%Co). Significant part of carbide grains has submicron size due to the inhibitory effect of oxide inclusions. They are more isolated from each other (compared to base materials). Bimodality by the size of grains of the solid phase (established at the stage of mixture preparation) is relayed in the form of the nanostructural fragmentation of dispersed oxide inclusions in the cobalt interlayer and submicron carbide grains, which is also confirmed by the electron microscopical image of fractures and chemical analysis results shown in figure 3c.

IOP Conf. Series: Materials Science and Engineering **511** (2019) 012032 doi:10.1088/1757-899X/511/1/012032

a)

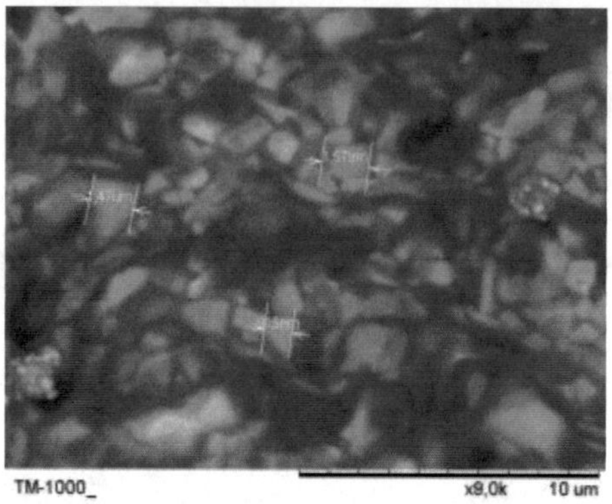

b)

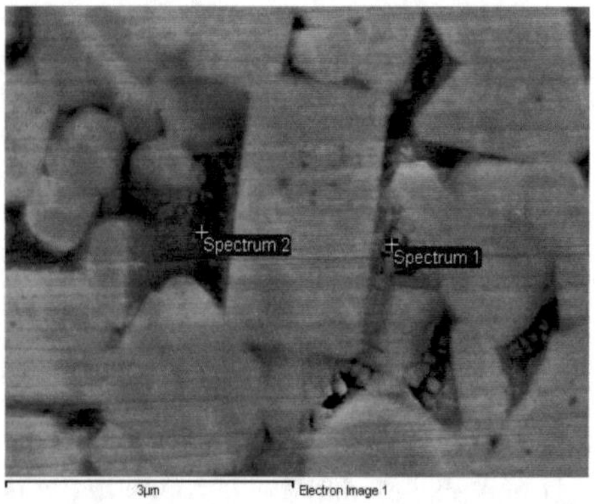

Spectrum	C	Al	Co	W	Total
Spectrum 1	5.53	2.10	32.44	59.93	100.00
Spectrum 2	3.65	2.25	67.10	27.00	100.00

c)

d)

Figure 3. Phase distribution of carbide composites WC-Al_2O_3-Co
a) dispersed inclusions of oxides at the fracture b) the nature of distribution of the cobalt bond (gray background) over the grains of the carbide phase; c) the results of the element-by-element analysis d) the nature of Palmqvist crack opening.

Since the thinner cobalt interlayer is dispersion strengthened, the nature of the stress-strain state in the local volumes (fragments) of Co-Al_2O_3-nano structure also changes. Calculations made using the finite element analysis (FEA) and ANSYS software products allow us to compare the strength of fragmentary nanostructured hardmetal composites with basic (WC-Co). In this case, for the analytical calculations, the microhardness values of the phase components can be used combining with the Palmqvist crack resistance measured data. Since the known microhardness values of tungsten carbide 20,6 GPa and the cobalt phase 7,5 GPa are significantly different, it is necessary to adapt these values to the known expressions for determining the integral parameter, i.e. crack resistance, by one of the known methods [9,10]. Using the finite element analysis (FEA) to estimate the stress state in the microstructure of the basic hard metal WC-8% Co with a carbide “skeleton” showed that the interface boundaries of carbide grains WC are stress concentrators and the weakest structural links in the volume of a hardmetal composite. At a load of 2 GPa, the stress intensity in the composite reaches 10.5 GPa, which leads to a displacement and the formation of microcracks along the boundaries. At the same time, hardmetal composites modified by nanoparticles are characterized by the destruction by another mechanism. The space between carbide grains is filled with cobalt material saturated with nanoparticles, therefore, besides disperse hardening of binder, a new oxide phase blocks the bias of carbide grains, figure 4 a, b, which is also confirmed by the results of numerical modelling of the mutual movement of carbide grains.

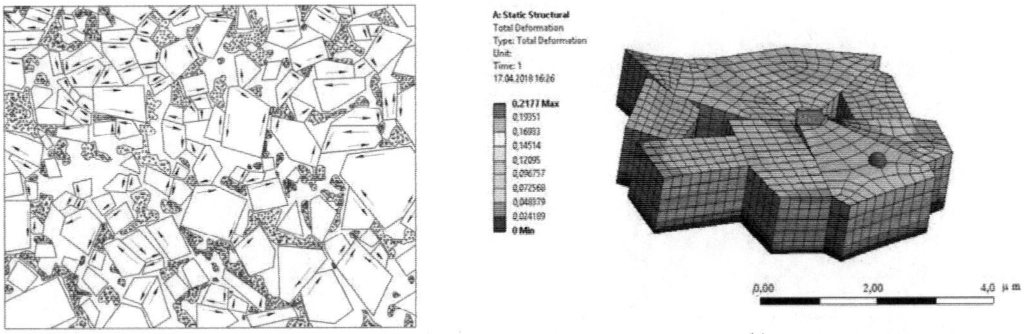

a) b)

Figure 4. Schemes of carbide grains bias as a result of external force impact (a – WC-Co-Al_2O_3-nano; b – bias of carbide grains under the load influence).

It is interesting to compare the experimentally observed effects of transcrystalline destruction of micron carbide grains (polycrystal) with the results of calculations by the FEA method when loading the model of the same structural fragment (figure 5a). Exactly at the maximum stress point the displacement and grain desctruction along crystallographic planes take place (figure 5b).

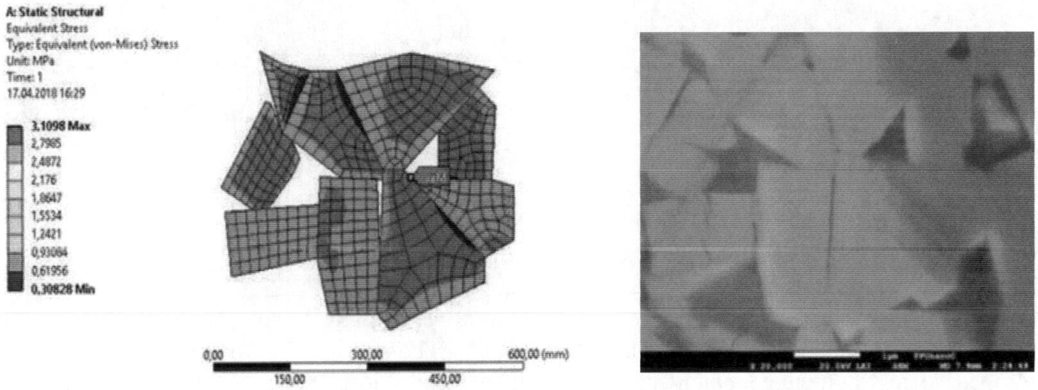

a) b)

Figure 5. Comparison of the results of a numerical experiment (a) and grain destruction in a real fragment of the microstructure (b).

In accordance with the results of numerical estimates, the maximum stress intensity in tungsten carbide is 9.1 GPa (arising under an external tension of 3 GPa) - which is 30% less than the maximum stress intensity in a structure with a continuous carbide “skeleton”. In cobalt, under the same load, the stresses are concentrated in the area with the smallest thickness of the interlayers, their intensity reaches 7,4 GPa, which is very close to the value of their microhardness. It means that the destruction of such microstructure type (figure 6b) should start with plastic deformation of cobalt, and the destruction of tungsten carbide will start with a significantly greater load. Indirect confirmation of this is a morphology change of the indenter imprint and microstructure area adjacent to it. The imprint does not have a strongly pronounced faceting, instead of four Palmqvist cracks in the indentation zone, numerous hidden microcracks are formed. The image of the imprint and its model representation are shown in figure a, b. After the load removal (sometimes with some time delay), the areal structure (figure c) and the traces of

IOP Conf. Series: Materials Science and Engineering **511** (2019) 012032 doi:10.1088/1757-899X/511/1/012032

lateral cracks relieved from residual internal stresses (figure 6d) may be observed on the grinding surface around the imprint along with the "normal" cracks (figure 6a).

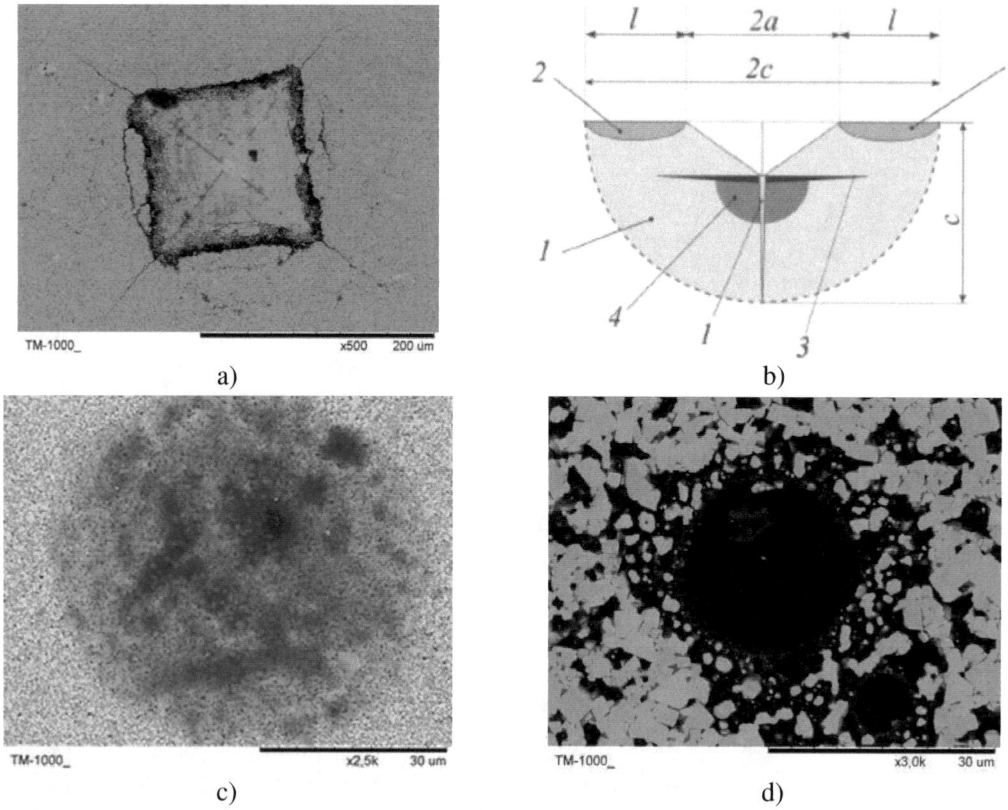

Figure 6. Stress relaxation in the indentation zone on the surface of a fragmentary nanostructured hardmetal composite.

Such effects may be also observed in other composite materials of increased hardness with nanostructural fragmentation of the grain structure [11]. The strength improvement predicted by the FEA method in fragmentary nanostructured hardmetal composite correlates with the experimental results of fracture toughness measurements according to Palmqvist. In case of the optimal amount of additives the fracture toughness growth is 50% in comparison with the basic material. This fact was further confirmed by microhardness numerical values of the individual structural fragments, as well as by the integrated index of Vickers hardness of ~ 1600-1650HV. New type of a hardmetal composite can be previously classified (defined) in the area between tool ceramics and hard metals.

References

[1] C J R González Oliver, E A Álvarez, J L García 2016 Kinetics of densification and grain growth in ultrafine WC-Co composites *Int Journal of Refractory Metals and Hard Materials* vol 59 pp 121–131

[2] Panov, V S, Zaitsev A A 2014 Trends of development the technology ultrafine and nanosized tungsten carbide WC-Co–A review *Proceedings of the universities Powder metallurgy and functional coatings* vol 3 pp 38–48

[3] Rempel' A A , Kurlov A S, and Tsvetkov Yu V 2011 Tungsten carbide nanopowder for WC–Co hard alloys *Proceedings of the Fourth All-Union Conference on Nanomaterials* (Moscow: IMET RAN) p 71

[4] Cholsong P, Zhimeng Ji G 2011 Microstructure and properties of ultrafine WC-10Co composites with chemically doped VC *Rare metals* vol 30 2 pp 183–188
[5] Huang S G 2008 NbC as grain growth inhibitor and carbide in WC–Co hardmetals *International Journal of Refractory Metals & Hard Materials* vol 26 pp 389–395
[6] Sun L , Jia C , Lin Cao R, Lin C 2008 Effects of Cr_3C_2 additions on the densification, grain growth and properties of ultrafine WC-11Co composites by spark plasma sintering *Int J Refract Met Hard Mater* vol 26 pp 357–361
[7] K Essaki, E J Rees, and G T Burstein 2010 Synthesis of Nanoparticulate Tungsten Carbide Under Microwave Irradiation *J Amer Ceram Sos* vol 93 3 pp 692–695
[8] Gordeev Y I, Yasinski V B, Binchurov A S , Anistratenko N E, Nikolaenko I V 2018 Combined Application of Composite Powders WC-Co and Additives of Nanoparticles as an Effective Method of Improving the Properties of Hard Metals *Key Engineering Materials* vol 769 pp 134–140
[9] Gordeev Y I, Abkaryan A K, Binchurov A S, Jasinski V B 2014 Design and Investigation of Hard Metal Composites Modified by Nanoparticles *Advanced Materials Research* 1040 p 13–18
[10] Dvornik M I, Mikhaylenko E A 2009 Study of the strength of VK8 hard metal by the finite element analysis *Chemical physics and mesoscopy* vol 11 4 pp 433–40
[11] Dvilis E S, Khasanov O L, Bikbaeva Z G, Polisadova V V, Struts V K, Omarov G Z, Khasanov A O 2014 Pattern of the B_4C Ceramics Surface Deformation at Local Loading *Advanced Materials Research* vol 872 pp 60–64

IOP Conf. Series: Materials Science and Engineering **511** (2019) 012033 doi:10.1088/1757-899X/511/1/012033

Study of the microstructure of aramid fabrics for structural and protective organoplastics modified in a microwave electromagnetic field

I V Zlobina and N V Bekrenev

Department "Technical mechanics and machine parts", Yuri Gagarin Saratov State Technical University, 77 Politechnicheskaya street, Saratov, 410054, Russian Federation

E-mail: irinka_7_@mail.ru

Abstract. The microstructure of samples of aramid fabric TSVM-DZh, subjected to the influence of the microwave electromagnetic field of an average power level with a frequency of 2450 MHz, was investigated. A significant change in the conditions of contact of individual filaments at the level of the interfibrillary bond of fibers, expressed in the integration of fibrils into agglomerates, filling gaps between the fibers with an increase in the interaction area and in the convergence of individual fibers, which improves the reliability of intermolecular bonds was established. The obtained results can be used as one of the mechanisms for increasing the strength characteristics of structural organoplastics and protective materials from aramid fabrics after microwave exposure.

1. Introduction

1.1. Status of the question

High strength and heat-resistant characteristics of fabrics based on aramid filaments determine their use as not only structural materials in aircraft construction, but also so-called "ballistic" materials used as personal protective equipment of the 1st and 2nd class for equipping EMERCOM employees and personnel working in extreme conditions, when possible emergencies, accompanied by mechanical damage of high-speed objects, exposure to high temperatures or liquids under pressure. One of the first such materials can be attributed to Kevlar, which was also used by the US Army during the Vietnam War. Good protective properties of aramid fabrics in combination with light weight and flexibility led to their widespread use in the army, aviation and navy [1-3]. In Russia, one of the most common ballistic materials based on aramid filaments are TSVM, Rusar and some other materials generally similar in properties [4-5].Compared with metal and ceramic protective agents, aramid fabrics and composites based on them (organoplastics) are significantly lighter and provide flexibility and freedom of movement, can be worn secretly under outerwear, although, of course, they do not provide adequate protection in combat using automatic small arms.

Recently, due to its lightness and durability, organoplastics are finding more and more widespread use in aviation technology as structural materials and means of protection for responsible units (engines, control systems) from damaging factors [6–8].

However, aramid fabrics also have drawbacks associated with weak resistance to low-speed hard drummers [9, 10], weakening of the protective properties when wet, exposure to direct sunlight, significant armor effect, which leads to serious injuries. As a result, the design of elements of protective clothing is complicated by the introduction of water- and light-tight pockets or outer layers of fabric, the use of multilayer (20-30 layers) systems, which makes the protective kit heavier and reduces the mobility of a person.

1.2. Relevance of research

Studies to increase the strength of such fabrics are aimed at improving the technologies for obtaining the initial components and the formation of fabrics by optimizing weaving [11, 12]. This complicates the process, leads to an increase in material, labor and financial costs, and also leads to the need for restructuring or re-equipment of production, which lengthens the cycle of introducing new developments.

The increase in the number of local conflicts, as well as natural and man-made emergencies requires an increase in the protective characteristics of flexible materials and their reliability. Creating a unified methodology for the technological management of the formation of the required characteristics of materials by electromagnetic impact on the finished object without disturbing the existing technological cycle will allow designing optimal serial technological routes with minimum cost and high quality of products, solving the problem of import substitution especially in strategically important industries aimed at ensuring the security of the country and increase the competitiveness of products.

1.3. Statement of the problem

Composites based on aramid fibers are similar to structural glass and carbon plastics in terms of the structuring principle – a polymer matrix and multidirectional reinforcing fibers. Therefore, it is possible to apply the method of microwave exposure to the final product developed by us without modifying the existing technologies [13, 14]. However, this assumption requires experimental verification and theoretical justification. We have conducted experiments to study the effect of microwave electromagnetic field on the breaking strength of aramid filaments [15], during which it was found that microwave processing, with an average specific power, increases the axial breaking load on the TSVM-DJ filament by almost 2 times. It has also been obtained that, after exposure to a single-layer aramid fabric with a microwave electromagnetic field, the resistance of the last piercing by a cone with a corner at apex of 900 increases, both in dry and wet conditions, from 1.5 to 2 times, depending on the applied power density. The best results are achieved when exposed to the microwave electromagnetic field of average power density for 4-8 minutes.

The purpose of the research was to study the specific features of the structure of the TSVM-J fabric and the multilayer composite based on it at different magnifications to reveal the possible mechanism of the effect described above.

2. Research methods and equipment

We have studied the microstructure of the TSBM-DJ monolayer fabric, article 56319A in the initial state (control samples) and after modification in the microwave electromagnetic field. The experiments were carried out on a microwave installation "Zhuk-2-02" (LLC AgroEkoTech, Obninsk, Kaluga region), which allows you to adjust the intensity of the microwave electromagnetic field frequency of 2450 MHz by changing the distance to the object. Modification of the samples was carried out at an average power level with an exposure time of 4 minutes. Previously, in these modes, the best results were obtained for increasing the strength of aramid filaments and fabrics.

The appearance of the surface of the fabric samples was studied using an electron microscope MIRA II LMU (Tescan Ersay Holding, Czech Republic) in the Laboratory of Special-Purpose Materials of the Saratov State University named after N. G. Chernyshevsky.

3. Results and discussion

IOP Conf. Series: Materials Science and Engineering **511** (2019) 012033 doi:10.1088/1757-899X/511/1/012033

Micrographs of samples at different magnifications are shown in figures 1 - 3. It can be seen that with a small magnification (field of view - 2.02 mm) there are no differences in the structure of the woven fabric layer. It may be noted a somewhat dimmer, matt appearance of the treated sample (figure 1). With a larger increase in the field of view of 330.7 microns, separate fibers extending from the filament become noticeable. At the same time, there are significantly more such fibers on the control sample than on the treated sample. This suggests that during the action of the microwave electromagnetic field, a certain consolidation of fibers into bundles in the structure of the filaments occurred. This statement is confirmed by the appearance of agglomerates on the surface of filaments , apparently formed by deformed fibers. With a larger increase in the field of view of 65.9 - 66.13 micrometers (figure 2), the differences in the structure of the filaments and fibers of the fabric become more noticeable. Separate fibers and microfibrill filaments are clearly visible on the surface of the filaments of the control sample, connecting the filaments to each other only at separate points. Also microagglomerates of organic structures are noted, possibly environmental pollution placed on the surface of individual fibers and not simultaneously connected with several fibers. Thus, it is possible to note only the macromechanical connection of the threads among themselves in the process of weaving the fabric, which determines the strength and other operational parameters of the control sample. The surface of the filaments of the treated fabric sample is covered with a sufficiently dense layer of agglomerates, connected simultaneously with several filaments, which are also close to each other, being in close contact. The presence of agglomerates connecting the filaments most likely determines additional points of contact between the filaments and the occurrence of sufficiently strong bonds, which generally contributes to an increase in the interaction regions of the filaments at the intermolecular level. This interaction contributes to the increased strength of the fabric, complementing the purely mechanical contact formed by weaving. The noted features are even more pronounced with a large increase in the surface of the tissue in the field of view of 33.07 μm (figure 4).

On the surface of the filaments of the control sample of tissue there is a large number of randomly located separated misoriented fibers and microfibrils. It is not possible to identify any reliable areas of contact of the fibers of one filament with a neighboring thread. Between the filaments there are significant gaps with sizes greater than 10 microns.

On the surface of the filaments of the treated fabric sample, there are agglomerates of fibers and individual fibers tightly adjacent to it. In this case, the filaments are very close together (5 μm or less), which promotes the contact of the agglomerates with several adjacent filaments. Virtually the entire filament has a coating formed by fibers complexed into agglomerate.

The obtained micrographs can be taken as the rationale for the next mechanism to increase the strength of aramid filaments and fabrics after exposure to a microwave electromagnetic field. Studies [4, 5] found that the structural unit of aramid fibers are rigid macromolecules or their aggregates - fibrils. The high strength and rigidity of such fibers under tension are due to the high degree of orientation of the macromolecules along the fiber axis and the high dissociation energy of chemical bonds.

The difference in the stress state of neighboring structural elements causes shear stresses in the boundary region between the fibers forming the filament. As a result, uniaxial stretching of the fiber is accompanied by the growth of interfibrillary cracks along the direction of the tensile force.

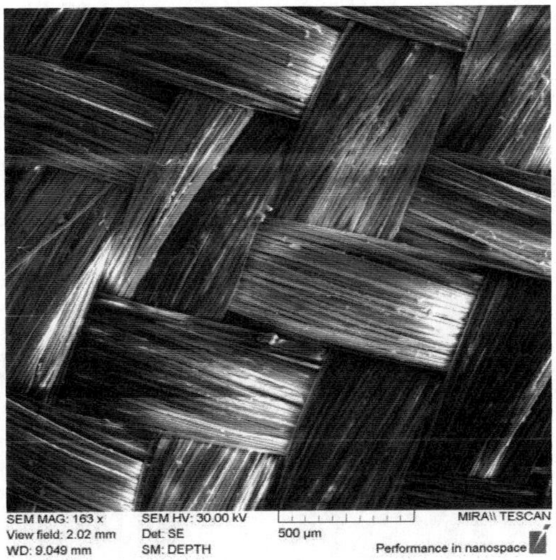

a

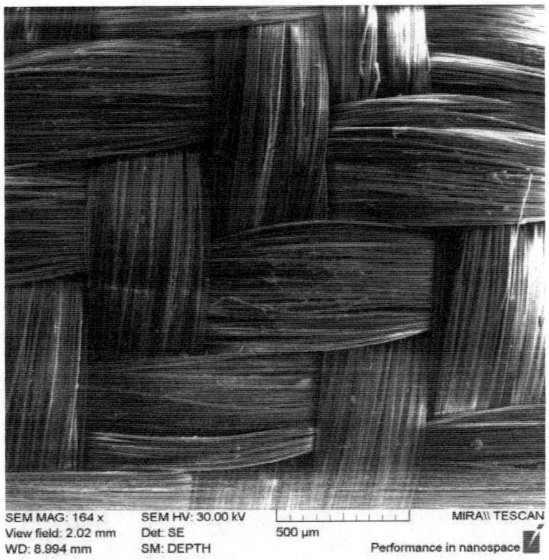

b

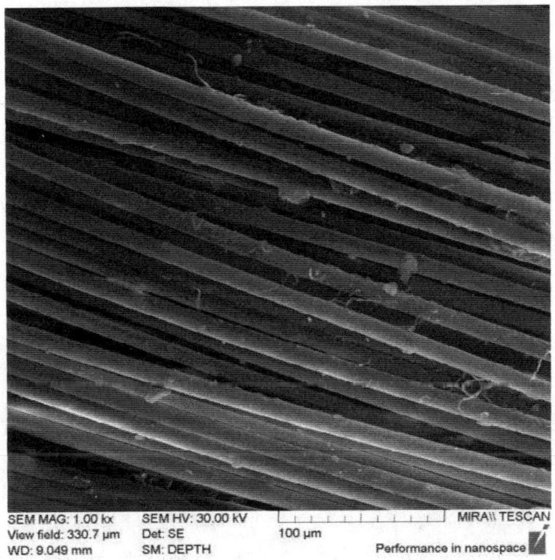

c

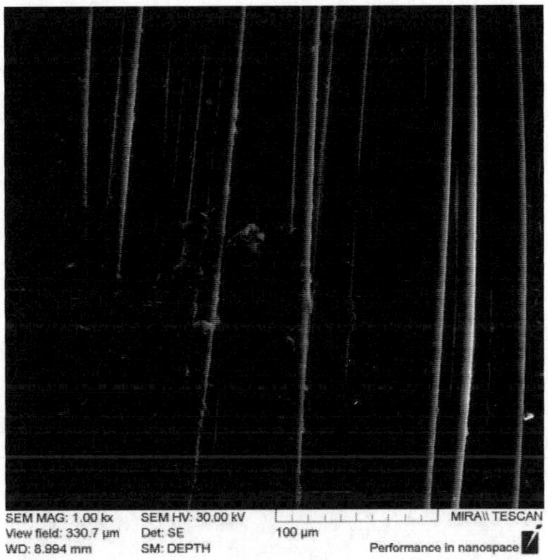

d

Figure 1. The surface of the fabric TSVM-DJ at low magnification. Field of view 2.07 mm: control sample (a), treated sample (b). Field of view 330.7 microns: control sample (c), treated sample (d)

The splitting of the filaments and the separation of the fibers apparently occurs already at the stage of weaving the fabric and is accompanied by the simultaneous breaking of some of the most intense fiber fragments, their exit to the surface of the filament and twisting into spiral-shaped formations under the action of stress relaxation. As a result, the fabric perceives loads with a smaller number of solid filaments and a smaller number of fibers in each filament. Thus it can be concluded that each filament takes an external load individually. As a result, after breaking the first group of fibers during loading, the load is perceived by the next bundle of fibrils, etc., until the remaining fibers are unable to resist the tensile load, and the filament breaks. There is a catastrophic total destruction, accompanied by intensive splitting and fluffing of the filaments and damage to the tissue. The effect of the microwave electromagnetic field most likely contributes to an increase in the number of interfibrillar bonds at the level of macromolecules. As a result, as the load grows, the conditions for joint rather than separate

IOP Conf. Series: Materials Science and Engineering **511** (2019) 012033 doi:10.1088/1757-899X/511/1/012033

operation of the fibers forming the filament are made. The weakening of the structure of the above-mentioned cracks is leveled in this case by the described action of interfiber structures formed in the microwave electromagnetic field and additional fibril bonds, as well as the formation of agglomerates of separated fibers due to their destruction in the electromagnetic field, filling the space between the filaments and entering into surface contact on macromechanical and physical level of intermolecular interaction. The ongoing residual dehydration of the fibers and, as a result, of the filaments leads to their partial shortening and convergence, which facilitates the interaction with the agglomerates of the fibers and contributes to the formation of stronger bonds in the structure of the fabric.

a

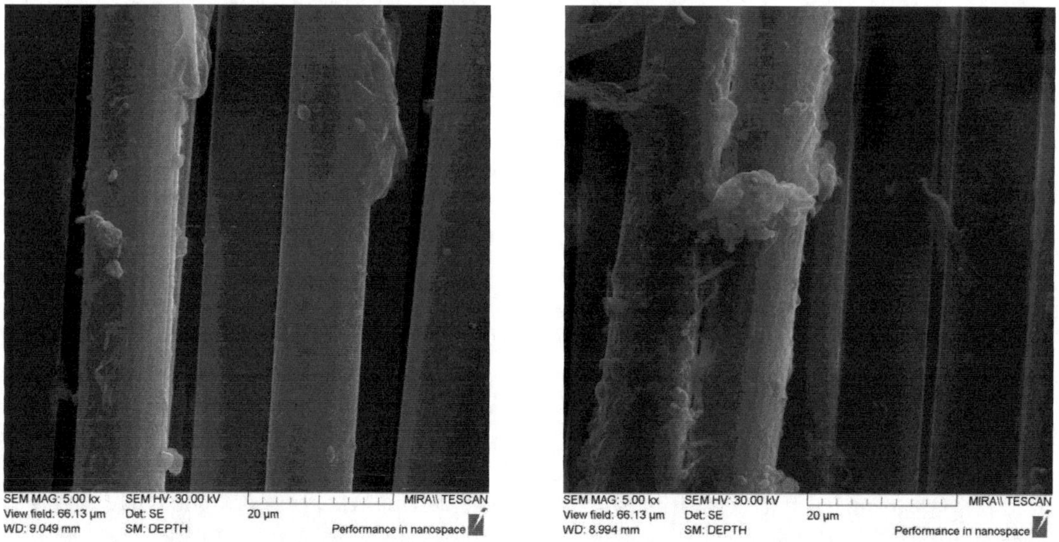

b

Figure 2. The surface of aramid fibers in the field of view is 65.9 - 66.13 microns.Control sample (a), treated sample (b)

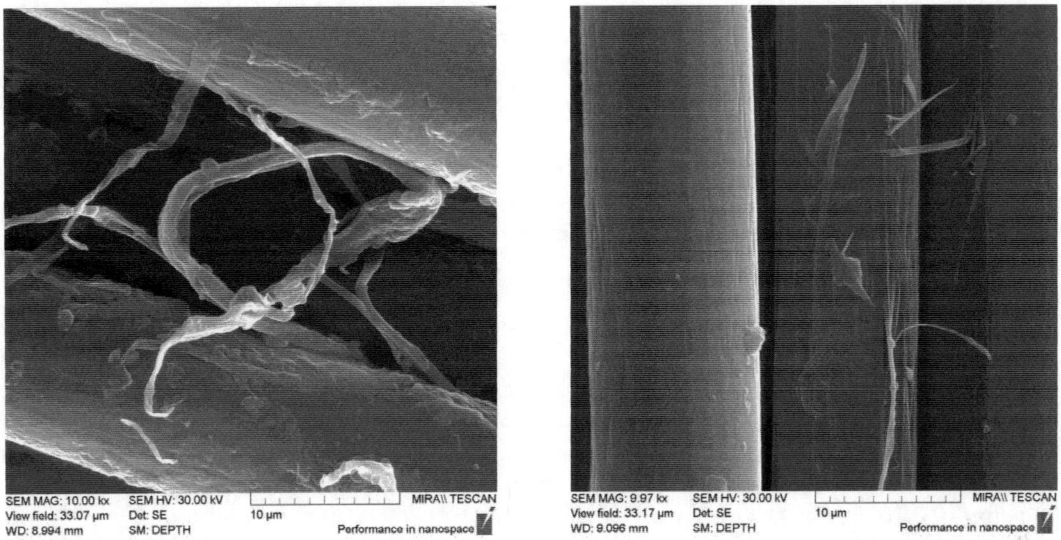

a

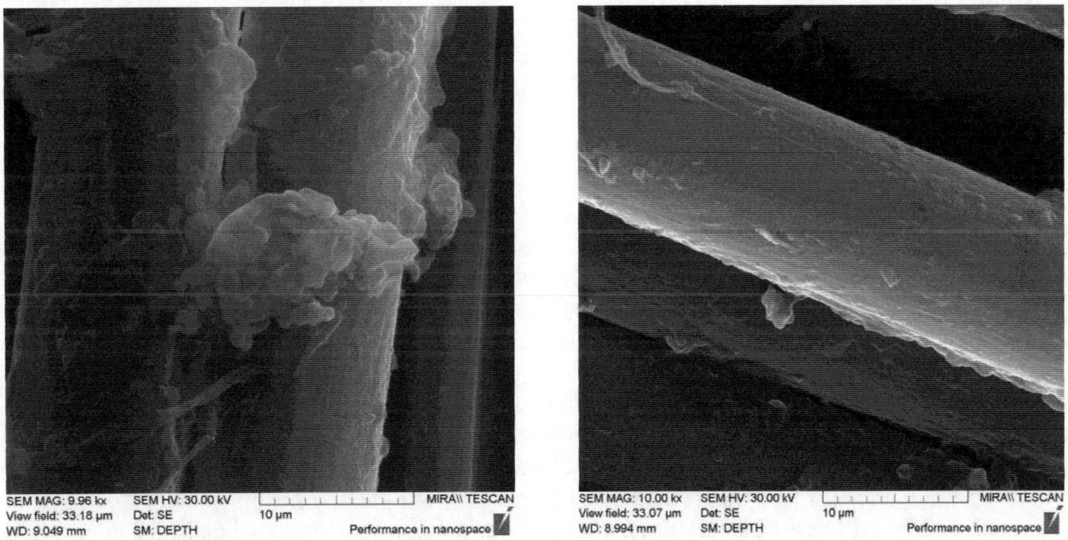

b

Figure 3. The surface of aramid filaments in the field of view is 33.17-33.07 microns. Control sample (a), treated sample (b)

4. Conclusion

Thus, processing in the microwave electromagnetic field causes a certain restructuring of the composite structure based on aramid fabric, which can be called a kind of "quasi-structuring." This change in the structure allows creating conditions for increasing the resilience of external influences, in particular, by increasing the tensile strength of individual filaments and the puncture resistance of the woven structure. Processing of aramid fabrics in the microwave electromagnetic field, therefore, allows to improve the operational, in particular, protective characteristics of the elements of light armor, without making changes of existing technologies for the synthesis of fibers, weaving filaments and the formation of fabric.

Acknowledgments
The studies were carried out with the support of the RFBR grant No. 17-03-00720 "Methodology of optimization micro-design of composite materials for objects of complex shape with increased dynamic strength, which are layer-by-layer formed by electrotechnological methods".

References
[1] Bhatnagar A 2006 Lightweight ballistic composites *Woodhead Publishing Limited (Cambridge England)* 416 p
[2] Chistyakov E 2013 Russian body armor and new armor materials (*Technique and weapons* vol 8) pp 2–6
[3] Ignatova A M, Artemov A O 2012 Analytical review of modern and promising materials and designs of armor obstacles and defenses against destruction (*Fundamental research* vol 6–1) pp 101–105
[4] Shustov Y S, Kurdenkova A V 2016 Investigation of the physicomechanical properties of fabrics made from Rusar filaments (*Scientific-methodical electronic journal "Concept"* vol 3) pp 36–40
[5] Musina T K, Volokhina A V and Schetinin A M 2010 Polyimide and aramid fibers and threads with special properties and products based on them (*In the world of equipment* vol 2) pp 4–8
[6] Shustov Y S, Kurdenkova A V 2016 Investigation of the physicomechanical properties of fabrics made from Rusar filaments (*Scientific-methodical electronic journal "Concept"* vol 3) pp 36–40
[7] Natrusov V I, Shatskaya T E and Kudinov V V 2007 Creation of competitive composite materials based on high-strength high-modulus polyethylene fibers International *Conference "Fundamental and Applied Materials Science" Russia, Barnaul* pp 189–191
[8] Zhelezina G F, Zhelezin I V and Lukin N F 2007 and others Light structural organoplastics resistant to shock and ballistic effects (*All materials. Encyclopedic reference* vol 8) pp 25–37
[9] Murakhovsky V 2015 The winged tissues of Russia *Arsenal of the Fatherland* vol 4
[10] Dolganina N Y 2010 Estimation of the ballistic limit and deflection of multilayer fabric plates upon impact by an indenter *Vestnik Yuzh.-Ural. state un-that* vol 10 pp 2–3
[11] Shuldeshova P M, Zhelezina G F 2014 Aramid layered woven material for protection against ballistic and shock effects (Transactions of VIAM: electron. scientific tech. journals vol 9) Art 06
[12] Safonov P E 2013 Development of optimal technological parameters for the manufacture of aramid fabrics for technical purposes Author. diss. Cand. tech. sciences (Place of protection: Federal State Educational Institution of Higher Education "Russian State University A N Kosygin)16 p
[13] Slugin A I, Parfenov O V, Nikolaev N A 2011 The study of the ballistic properties of tissue (Technology of the textile industry vol 4 (333) pp 74–78
[14] Zlobina I V, Bekrenev N V 2016 The Influence of Electromagnatic Field Microwave on Physical and Mechanical Characteristics of CFRP (Carbon Fiber Reinforced Polymer) Structural *Solid State Phenomena, Materials Science Forum* vol 870 pp 101–106
[15] Zlobina I V, Bekrenev N V 2016 The Influence of Microwave Electromagnetic Field jn Mechanical Properties of Composite Materials (*High technology*) vol 17 - 2 pp 25–30
[16] Zlobina I V, Bekrenev N V 2017 Effect of the microwave electromagnetic field on the strength and microstructure of ballistic nonmetallic materials (*Questions of electrical technology* vol 1) pp 40–47

Study of Welding Properties of Fused Weld Flux Produced by Electric Arc Granulation

M F Kartashev[1,a], S V Naumov[1,b], A N Urchenko[1,c] and M A Sheksheev[2,d]

[1]Perm National Research Polytechnic University
29, Komsomolskyprospekt, 614990, Perm, Russian Federation
[2]Nosov Magnitogorsk State Technical University
38, Lenin Avenue, 455000, Magnitogorsk, Russian Federation

[a]goncharsk@mail.ru, [b]NaumovStanislav@yandex.ru, [c]aleksmto@gmail.com, [d]shecsheev@yandex.ru

Abstract. The article outlines comparative analysis of surfacing under granulated flux cover produced using the developed technology and under AN-348 flux cover. Visual inspection, hardness tests, photoelectric spectral analysis, and studying of geometric characteristics of weld beads are carried out. The results of the comparative analysis and studying of welding properties of fused fluxes are comparable or in some cases identical, and in terms of geometric characteristics of weld beads, outperform the well-known analog of surfacing under flux cover.

1. Introduction

Modern market for arc welding materials offers a large range and variety for welding and surfacing of various metals and alloys. The main directions for increasing competitiveness and quality of welding materials are the development of new flux compositions and improvement of production technology [1-6].

At present, the authors of the article have developed a fundamentally new technology for production of welded fused fluxes using electric arc granulation technology[7], which lies in passing a fine-dispersed charge through a coal arc to fuse and then to solidify it, in a state of weightlessness during falling, into granules of a fused welding flux. However, the welding properties of flux produced using developed technology have not been studied yet and that does not allow fully positioning it as a welding material.

In this connection, the aim of the work is to study the welding properties of a fused welding flux produced using electric arc granulation, namely, mechanical properties, visual inspection results, chemical composition and obtaining the required geometric characteristics of weld beads produced using granulated flux.

2. Experiments

To achieve this goal, a comparative surfacing (figure 1) was carried out under granulated flux cover produced using the developed technology and under AN-348 flux cover (a common Ukrainian welding fused flux widely used by Russian machine-building enterprises and defense industry) in two modes: welding current I = 550 A (mode 1) and 400 A (mode 2), welding voltage U = 34 V, welding speed V = 55 cm/min, diameter of the wire Sv-08G2S 4 mm, material - Steel grade 20 with plate thickness of 12 mm [8]. The only difference in the comparative analysis apart from production method is the different type of flux, AN-348 has manganese-silicate base MS ($CaO + MgO$%, 19, $Al_2O_3 + MnO$%, 47.5, $SiO_2 + MnO$%, 76), while the granulated flux has silicate base ($CaO + MgO$%, 19, $Al_2O_3 + SiO_2$%, 57.5). Granulated flux is produced from hornblendite of Pervouralsk deposit in Ural region as its slag base has shown good efficiency in submerged arc welding [9, 10].

Weld beads obtained from the control surfacing were prepared as samples for studying its geometric characteristics, chemical analysis and hardness tests. The numbering of the prepared samples is shown in table 1.

Table 1. Studied sample lot.

Sample №	Flux	Surfacing mode			Study type
		I (A)	U (V)	V (cm/min)	
111	Granulated flux	400	34	55	geometric characteristics of weld beads
211	AN 348	400	34	55	geometric characteristics of weld beads
121	Granulated flux	550	34	55	geometric characteristics of weld beads
221	AN 348	550	34	55	geometric characteristics of weld beads
112	Granulated flux	400	34	55	chemical analysis and hardness
212	AN 348	400	34	55	chemical analysis and hardness
122	Granulated flux	550	34	55	chemical analysis and hardness
222	AN 348	550	34	55	chemical analysis and hardness

Note: Flux: 1. Granulated flux; 2. AN 348;

Modes: 1. 400 A, 34 V, 55 cm/min; 2. 550 A, 34 V, 55 cm/min;

Study type: 1.geometric characteristics of weld beads (end face); 2. chemical analysis and hardness (weld bead is taken off evenly to base metal).

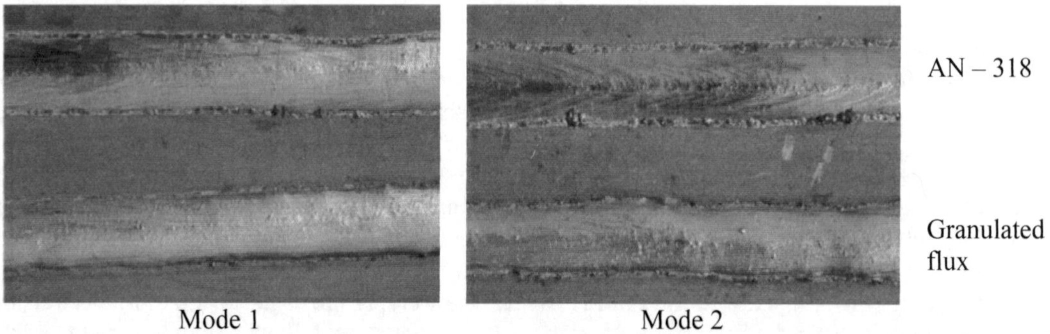

Figure 1. Comparative surfacing with visual inspection, carried out under granulated and AN-348 flux cover.

During visual inspection of weld beads it was established that AN-348 flux and granulated flux produced using the developed technology have equally satisfactory properties of beads formation, separation of slag crust after surfacing, absence of external pores and cracks; at the same time weld beads deposited under granulated flux cover have smaller ripple (table2).

Comparing AN-348 and granulated flux produced using the developed technology slag crusts, it is seen (figure 2) that they have similar density, but granulated flux slag crust has greater porosity, which may indicate a better passing of gases from the weld metal to the slag and promote better slag crust reparability [11].

Table 2. The main parameters of welding properties of granulated flux in accordance with the guidance documents 03-613-03 [12].

Test parameter	Score (1-5)	Parameter brief description
Arc starting	5	Easy. Starting right after touching of electrode and work piece
Arc stability	4	Good. Evenly burning arc with a slight vibration and crunchy noise (sputtering)
Weld formation quality	4	Good. The bead is finely scaly with a few small irregularities in height and small exceedings along weld edges.
Slag crust separability	5	High. Separates after welding without additional mechanical action

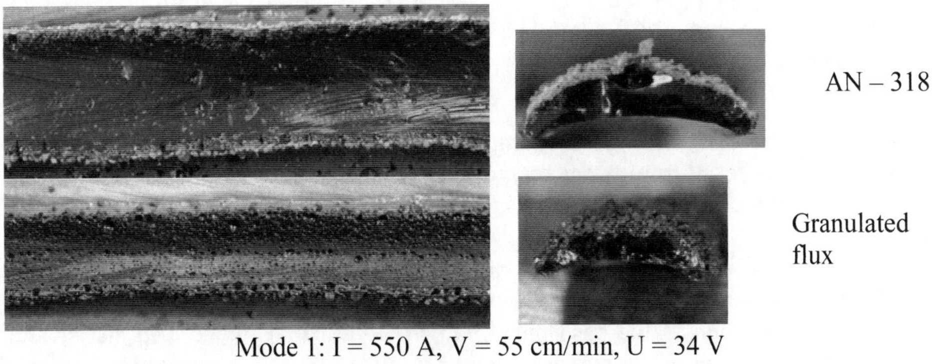

Figure 2. Inner surface of granulated flux and AN-348 flux slag crusts.

2.1. A photoelectric spectral analysis of deposited metal.

Photoelectric spectral method was used to determine the percentage mass fraction of elements in weld beads metal. Photoelectric spectral analysis was performed using Oxford Instruments PMI-Master Pro spectrometer. In this case, bulging of weld beads was milled flush with base metal. Three measurements for each sample were made on the flat area formed after milling and then the average value was found.

Data on chemical composition of the metal was received by means of photoelectric spectral analysis of beads deposited under a granulated flux, as well as under AN-348 flux, that is shown in table 3.

The metal deposited under the AN-348 has a decrease in C and an increase in Mn, whereas a metal deposited under granulated flux has an increase in Si and an insignificant increase in Cr and Cu. No other changes in other elements are observed.

Table 3. Results of photoelectric spectral analysis of deposited metal.

Studied metal	Chemical composition [mass%]									
	Fe	C	Si	Mn	P	S	Cr	Mo	Ni	Cu
Base metal	98.3	0.159	0.196	0.500	<0.005	<0.003	0.084	0.040	0.310	0.244
Bead metal (Sample112)	98.2	0.151	0.471	0.307	<0.005	<0.003	0.133	0.029	0.241	0.355
Bead metal (Sample 212)	98.0	0.119	0.282	0.940	<0.005	<0.003	0.0621	0.027	0.232	0.228
Bead metal (Sample 122)	98.1	0.152	0.503	0.350	<0.005	<0.003	0.125	0.032	0.260	0.346
Bead metal (Sample 222)	98.0	0.116	0.278	0.928	<0.005	<0.003	0.063	0.030	0.242	0.228

The increase in Mn or Si is determined by the type of the flux. It is evident that due to the increased content of manganese oxide in AN-348 flux composition (type MS), its content is also increased in deposited metal. Similarly, with silicate-based granulated flux, the deposited metal produced under this flux has increased Si content. It is found that the chemical analysis of the metal deposited under the

IOP Conf. Series: Materials Science and Engineering **511** (2019) 012034 doi:10.1088/1757-899X/511/1/012034

granulated flux in terms of harmful impurities content presents low S (<0.003 mass%) and P (<0.005% by weight) content. This corresponds to the results obtained during welding under the widespread analog fused welding flux AN-348, which indicates the purity of weld metal.

2.2. The hardness of weld beads.

Measuring the hardness of beads fused under a granulated flux, as well as under an AN-348 flux was carried out using Brinell method, received data on weld metal is shown in table 4.

Table 4. The main parameters of welding properties of granulated flux in accordance with the guidance documents 03-613-03 [12].

Studied metal	HB1	HB2	HB3	$HB_{average}$
Base metal	143	156	143	147.3
Bead metal(Sample 112)	187	196	196	193
Bead metal(Sample 212)	163	170	163	165.3
Bead metal(Sample 122)	187	179	170	178.6
Bead metal(Sample 222)	143	156	146	148.3

The beads produced during surfacing under granulated flux have an increased hardness compared to the base metal, which may result from: an increased Si or Cr content; C content close to the base metal; change in the structure of weld metal; presence of nonmetallic inclusions in weld metal; increased stress in the weld bead.

Bead metal produced from surfacing under AN-348 flux has hardness close in value to hardness of the base metal. The reason for this can be a reduced carbon content in the weld metal of weld beads compared to content of carbon in the base metal, which significantly reduces its hardness, despite welding stress, leading to a significant increase in hardness of weld beads compared to hardness of the base metal.

2.3. Geometric characteristics of weld beads.

Studying geometric characteristics of weld beads was carried out using Olympus SZX16 microscope and the Olympus Stream Motion1.8 program that allows calculating of sectional area of the bead when tracing its contour. Before studying geometric characteristics, micro polished sections of the samples were etched by rubbing with a 4% alcohol solution of nitric acid.

An example of images of micro polished sections with dimensions obtained in Olympus Stream Motion1.8 program is shown in figure 3. The results of geometric characteristics measurements of the samples are given in table 5.

Table 5. The main parameters of welding properties of granulated flux in accordance with the guidance documents 03-613-03 [12]..

Measurement place	Sample 111	Sample 211	Sample 121	Sample 221
Area (mm^2)	57.96	59.81	103.62	98.15
Fusion depth (mm)	4.24	4.41	7.20	5.95
Bead width (mm)	13.33	13.58	16.44	17.19
HAZ (mm)	1.60	1.24	3.12	2.46

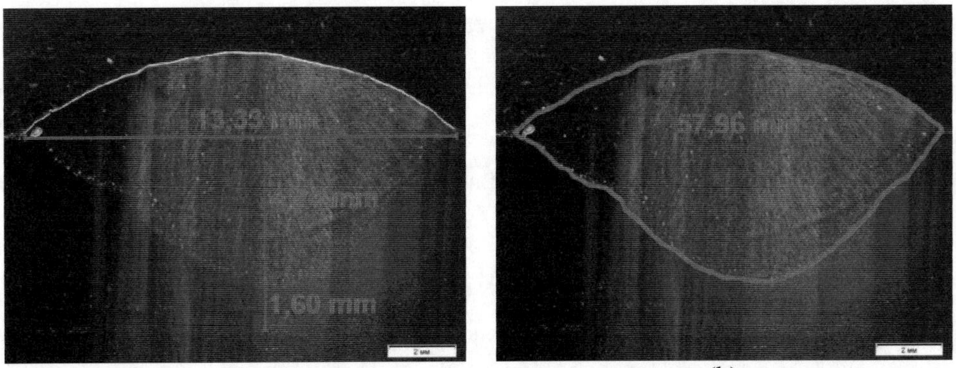

(a) (b)

Figure 3. Geometric parameters of sample 111.(a)bead width, fusion depth, HAZ.(b)area of weld bead.

Analysis of geometric characteristics of the samples makes it possible to make the following observations:

- During surfacing under current of value I = 400A, geometric characteristics of the beads deposited under granulated flux and under AN-348 flux practically do not differ. Current of value I = 400 A is relatively small for selected welding speeds for chosen wire diameter and welding method;
- During surfacing under current of value I = 550 A, beads welded under the granulated flux have an 22% greater fusion depth and 5% smaller width than beads welded under AN-348 flux. It enables to draw a conclusion about better technological properties of granulated flux in comparison with AN-348, since it is possible to obtain a penetration in lower modes and, therefore, lower thermal effect on the base metal;
- bead width and HAZ welded under granulated flux cover is 19.52 mm, which is practically equal to the width value of beads and HAZ welded under the AN-348, that amounts 19.65 mm. Consequently, the width of the zone with a modified structure during surfacing under either flux is practically the same.

3. Conclusion

According to welding and geometric characteristics, chemical composition and hardness of weld beads, deposited under granulated flux comply with quality to those deposited under widespread fused welding flux widely used by Russian machine-building enterprises and defense industry. However, there is a significant decrease in welding properties of granulated flux when using it at relatively low current. Therefore, there is a need to research for optimal welding modes under granulated flux that allows producing high-quality welding and welded joints.

References

[1] Kozyrev N A, Kryukov R E 2017 Welding Fluxes and Additives Based on Metallurgical Wastes *Innovations in Fuel and Energy Complex and Mechanical Engineering* pp 128–133

[2] Golovko V V, Potapov N N 2011 Special features of agglomerated (ceramic) fluxes in welding *Welding International* vol 25 (11) pp 889–893

[3] Kryukov R E et al. 2017 Manufacturing of New Welding Fluxes Using Silicomanganese Slag *IOP Conference Series: Materials Science and Engineering* vol 253

[4] Kartsev D S, Zernin E A 2016 Use of Refractory Nanoparticles as a Component of Welding Materials in Welding and Surfacing with Coated Electrodes and Flux Cored Wires *IOP Conference Series: Materials Science and Engineering* vol 142

[5] Gupta P et al., Effect of B_2O_3 Containing Fluxes on the Microstructure and Mechanical Properties in Submerged Arc Welded Mild Steel Plates 2016 *IOP Conference Series: Materials Science and Engineering* vol 114

[6] Kumar A et al. 2015 Fuzzy Logic Optimization of Weld Properties for SAW Using Silica Based Agglomerated Flux *Procedia Computer Science* vol 57 pp 1140–1148
[7] Naumov S V, Ignatov M N and Sheksheev M A 2017 Technology of mineral raw materials granulation by electric arc for manufacturing of welding fused flux *Solid State Phenomena* vol 265 pp 290–295
[8] Naumov S V 2014 Development of Slag Basis foe Welding Materials from Mineral Raw Material of the Urals (PhD dissertation UrFU, Yekaterinburg)
[9] Ignatova A M, Naumov S V, Ignatov, Pushkin S A and Suslov S B 2010 Assessment of the Suitability and Availability Basaltic M N and Gabbroic Complexes West of the Urals (Perm region) for Welding Materials *Vestnik PNIPU* vol 4 pp 104–116
[10] Naumov S V, Ignatova A M, Ignatov M N Submerged Arc Welding Using Slag Base of West Urals Mineral Raw Resources with Low Detrimental Impurities Content 2017 *Procedia Engineering* vol 206 pp 1355–1359
[11] Kartashev M F, Naumov S V 2017 Detecting and examination of solution mechanisms and oxidation states of sulfur volatiles in silicate melts of fused welding fluxes, which are made from petrurgical raw materials of Ural region *Welding. Renovation.Tribotechnics: abstracts of the 8th Ural Scientific and Practical Conference* (Nizhny Tagil, STI UFU) pp 49–54
[12] Guidance document 03-613-03, Welding Application Procedure for Manufacture, Installation, Repair, and Reconstruction of Technical Devices for Dangerous Production Objects 2003

IOP Conf. Series: Materials Science and Engineering **511** (2019) 012035 doi:10.1088/1757-899X/511/1/012035

Synthesis of non-metallic inorganic radiation-absorbing coatings on the surface of VT1-0 titanium alloy

A K Chubenko, T A Baranova, Yu N Dolgova, V A Mamaeva and A I Mamaev

National Research Tomsk State University, 36 Lenina Ave., Tomsk, 634050, Russia

E-mail: ch@qwertynet.ru

Abstract. The paper is devoted to the microplasma synthesis of radiation-absorbing coatings on VT1-0 titanium alloy surface. This method allows the formation of thin nanostructured non-metallic inorganic films. The process is carried out in an electrolyte solution. Iron and cobalt compounds in forms of potassium hexacyanoferrate and cobalt (II) acetate are added to the solution in the present work. It is shown that the formed coatings have high coefficients of electromagnetic radiation absorption in the terahertz range and in the near-infrared spectrum.

1. Introduction

Radiation-absorbing (RA) materials have a wide range of applications, and the production of materials for RA coatings is one of the most relevant areas in modern material science. RA coatings are used nowadays to solve the problems of electromagnetic compatibility of radioelectronic devices, protect biological objects from electromagnetic radiation and are an instrument of reducing the radar visibility of military equipment [1-2].

It was shown recently that iron-containing oxide coatings with RA properties could be obtained by the method of valve metals surface microplasma oxidation in solutions. Even so, microplasma synthesis is rarely used for RA coatings formation nowadays [3-5].

The core of the method is electrochemical oxidation of metal surfaces under the influence of spark and arc electric discharges developing in the near-surface area. At the same time, various components can be introduced from the solution into the coating formed by this method. As a result, the coating may contain dispersed particles such as metal iron, nickel, cobalt, their oxides and other compounds with ferromagnetic properties. These particles are synthesized during the coating formation process in specific quantities in order to achieve the target magnetic properties of the coating [6-7].

The present work is devoted to the microplasma synthesis of nonmetallic inorganic RA coatings on the surface of VT1-0 titanium alloy in aqueous solutions containing iron and cobalt.

Coating composition and uniformity are important properties at the end of RA coatings synthesis. The microplasma coating formation is carried out in an electrolyte solution, and this feature allows to control and change the coating composition. Combinations of electrolytes define the element composition of the synthesized coating. It should be noted that the coatings formed in the microplasma mode have a number of valuable functional properties: wear resistance, corrosion resistance, thermal stability, etc. The gradient structure, specific porosity and thickness are also the distinctive features of the obtained coatings [8].

High-frequency radiation absorption by the coating occurs due to the superposition of alternating magnetic and electric fields. This fact causes molecular vibrations and microcurrents in the coating that in turn leads to the energy dissipation by transferring it into the heat. Thus, the external energy is absorbed by the coating and dissipated due to either dielectric or magnetic losses. This mechanism is defined as non-resonant absorption. Non-resonant magnetic RA materials as a rule contain magnetic particles (for example, ferrite particles) as well as particles with high electrical conductivity (graphite particles, fullerenes) which are distributed in the polymeric material layer.

Analysis of the literature data shows that the currently used RA materials are formed by the distribution of a magnetically active phase in an organic polymer or an inorganic material. Therefore, the purposeful forming of this phase in the form of magnetically active particles (mixed iron oxides) in the coating layer is the primary task.

The paper [9] shows the positive effect of iron and molybdenum compounds in electrolyte solutions for RA coating microplasma synthesis. Iron and titanium mixed oxides as well as iron and cobalt mixed oxides are proposed in the present work to use as a magnetically active phase. It is known that cobalt-containing spinels have sufficiently high magnetic activity values and their formation in the coating composition can result in an increase of the coating absorption ability.

2. Materials and methods

Widely used in engineering, VT1-0 titanium alloy was chosen in the present work as the base material for the coating formation.

Alkaline solutions containing compounds of iron and cobalt (solution 1 and solution 2 respectively) were developed for the coating formation. **Solution 1** included H_3BO_3, Na_3PO_4, NaOH, $Co(CH_3COO)_2$, Trilon B, $K_3[Fe(CN)_6]$ (5 g/l for each component). The pH value of the solution was stabilized at 8,69 by using a buffer mixture. **Solution 2** included $K_4P_2O_7$ (10 g/l), $(NH_4)_2MoO_4{\cdot}4H_2O$ (1 g/l), $Na_2SiO_3{\cdot}9H_2O$ (30 g/l), $K_3[Fe(CN)_6]$ (15 g/l), $Co(CH_3COO)_2$ (1 g/l) and KOH (1 g/l).

Compositions of these solutions were developed due to the fact that phosphorus ions supported the barrier film formation on the metal surface and also increased the electrochemical polarization and porosity of the formed film. In addition, phosphorus ions increased the hardness of the coating while being injected into it in the form of oxygen-containing ions [10].

Boric acid was introduced into the electrolyte composition in order to control the rate of the coating formation and maintain the solution pH value. High voltage and high temperatures developing in the zone of microplasma discharges led to the formation of solid and heat-resistant boron-containing oxides.

$$4H_3BO_3 \rightarrow B_4O_7^{2-} + 5H_2O + 2H^+,$$

$$Na_2B_4O_7 \xrightarrow{t} 2B_2O_3 + Na_2O,$$

$$Na_2O + H_2O \rightarrow 2NaOH.$$

The presence of silicate ions in the solution composition contributed to the rapid formation of the barrier film and the emergence of silicon oxide in the coating:

$$2SiO_3^{2-} + 4H^+ \rightarrow H_2SiO_3,$$

$$H_2SiO_3 \rightarrow SiO_2 + H_2O.$$

It should be noted that silicate-containing compounds are widely used in the microplasma oxidation process.

Potassium hexacyanoferrate and cobalt (II) acetate were used as additives responsible for the magnetically active phase formation. The advantage of using potassium hexacyanoferrate is its high water solubility as well as the presence of iron in the form of a complex anion that simplified the injection of iron into the coating composition:

$$[Fe\ (CN)_6]^{3-} + 3H_2O \rightarrow 6CN^- + Fe(OH)_3 + 3H^+,$$

$$Fe(OH)_3 \rightarrow 4\ Fe_2O_3 + 3\ H_2O,$$

$$Fe_2O_3 + FeO \rightarrow Fe_3O_4.$$

IOP Conf. Series: Materials Science and Engineering **511** (2019) 012035 doi:10.1088/1757-899X/511/1/012035

Mixed iron and titanium oxides were formed from simple iron and titanium oxides as a result of melting under the high temperature effect.

2.1. Preparation of samples
VT1-0 titanium alloy samples were cleaned from organic impurities by washing the samples in 0,5% NaOH solution at a temperature about 20-25°C.

2.2. Formation of the coatings
The processed samples served as anodes and a water-cooling stainless steel bathtub served as a cathode. The temperature in the bathtub did not exceed 40°C. The electrical impact was applied by voltage pulses with 310 V amplitude and 500 μs duration using the Corund Power Source developed by Sibspark LLC. The coating formation process lasted for 30 minutes. The formed coatings were uniformly dark gray, almost black, rather dense and had sufficient adhesion to the substrates. Figure 1 shows the appearance of the samples with the obtained coatings.

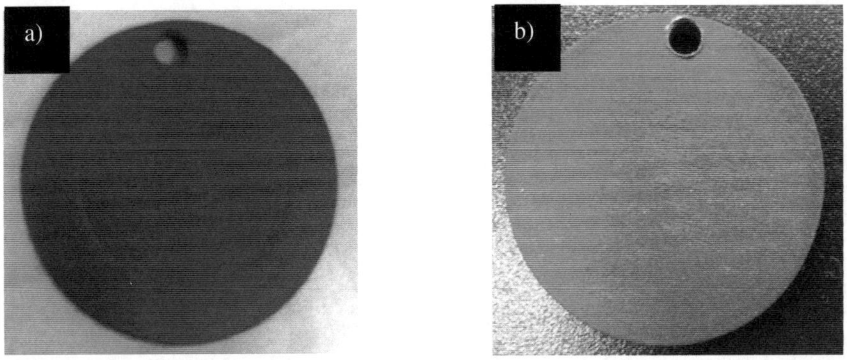

Figure 1. Appearance of the coatings formed on the surface of VT1-0 titanium alloy (a – electrolyte 1, b – electrolyte 2).

3. Results and discussion

The eddy current thickness gauge QuaNix 1500 was used to control the coating thickness. The thickness of the formed coatings was 19 μm for electrolyte 1 and 25 μm for electrolyte 2. The coating morphology analysis by the method of scanning electron microscopy showed the uniformity and high porosity of the obtained coatings at the microscopic level (figure 2). High porosity of the coatings contributed to the increase of RA characteristics [11].

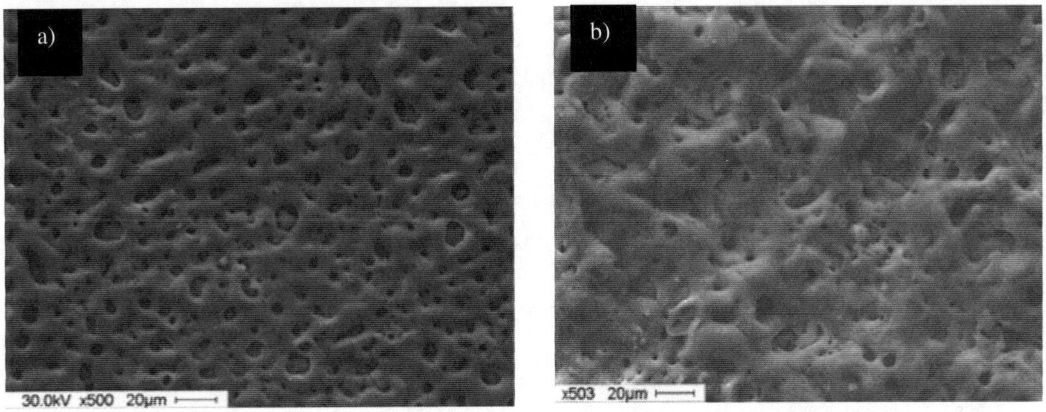

Figure 2. Micrographs of the coatings formed on the surface of VT1-0 titanium alloy (a – electrolyte 1, b – electrolyte 2).

The element composition of the obtained coatings was determined by the X-ray fluorescence method (table 1).

Table 1. Element composition of the coatings formed on the surface of VT1-0 titanium alloy.

Electrolyte 1		Electrolyte 2	
Element	Mass fraction (%)	Element	Mass fraction (%)
O	41.71	O	41.78
Ti	38.02	Al	25.13
P	11.03	Si	17.89
Fe	**6.03**	P	4.28
Co	**2.13**	**Fe**	**5.89**
other	1.08	**Co**	**1.35**
		Mo	0.03
		other	3.65

It should be noted due to the obtained data that the microplasma treatment in the proposed solutions resulted in magnetically active phase formation. The coatings included iron and cobalt which were capable of forming the magnetically active compounds. Iron and cobalt oxide mixtures melted thereby forming crystalline ferrite spinels as a result of exposure to the microplasma discharges under the high temperature effect:

$$Fe_2O_3 + CoO \rightarrow CoFe_2O_4,$$

$$Fe_2O_3 + FeO \rightarrow Fe_3O_4.$$

Some fraction of these oxides also deposited on the metal surface due to the oxide film perforation. The base metal was able to reduce iron and cobalt from their oxides during the process under the high temperature effect thus forming the metallic phase of iron and cobalt:

$$Me_mO_n + mTi \rightarrow TiO_2 + mMe^{n+}.$$

Absorption coefficients for all samples were measured on an STD-21 terahertz spectrometer developed by KDP OJSC in the frequency range from 428 to 973 GHz. The electromagnetic radiation absorption coefficient oscillated from 0.7 to 0.8 within the specified frequency range that was extremely high for thin films with a thickness of less than 30 microns.

Reflection spectra for the near-infrared area were also obtained by means of the Nicolet infrared spectrometer with normal incidence of electromagnetic radiation (figure 3).

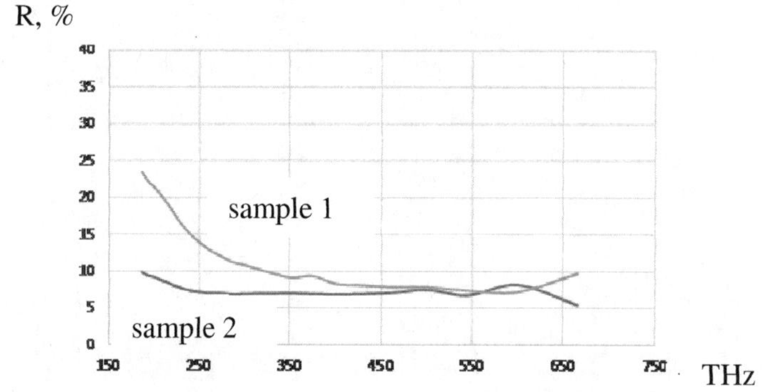

Figure 3. Reflection coefficient (R) dependence on the frequency (VT1-0 titanium alloy).

The obtained dependence demonstrated high absorption ability of the synthesized coatings in the near- infrared spectrum. Differences in the absorption ability of the coatings could be noted only in the frequency range of 250 THz.

Conclusion

Thus, it is shown in the present paper that radiation-absorbing coatings can be formed by the microarc oxidation method on the surface of VT1-0 titanium alloy. Potassium hexacyanoferrate and cobalt (II) acetate are added to the electrolyte composition and form magnetically active compounds. Electromagnetic radiation absorption coefficients of the obtained coatings oscillate from 0,7 to 0.8 within the frequency range from 428 to 973 GHz that is extremely high for thin films. At the same time, such coatings are also effective electromagnetic radiation absorbers in the near-infrared spectrum with absorption coefficients 0.78-0.95.

Acknowledgement

The results were obtained in the framework of the state assignment by the Ministry of Education and Science of Russian Federation, project № 16.9621.2017/8.9

References

[1] Latypova A F and Kalinin J E 2012 Analysis of promising radiation-absorbing materials *Bulletin of the Voronezh State Technical University* **8** No. 6 p 70

[2] Rudnev V S, Adigamova M V, Tkachenko I A, *et al* 2017 Influence of electrolyte components on magnetic and magnetic-resistant characteristics of plasma-electrolytic Fe-containing oxide coatings on titanium *J. Phys. Chem.* **91** No. 3 p 562

[3] Mamaev A I, Dorofeeva T I, Emelyanova E Y, *et al* 2012 Investigation of the properties of layered nanostructured non-metallic inorganic coatings formed by high-energy flows localization at the interface *Izv. Vyssh. Uchebn. Zaved.* **55** No. 7/2 p 107

[4] Emelyanova E Y, Dorofeeva T I, Mamaev A I, *et al* 2011 The formation of layered nanostructured non-metallic inorganic coatings *Izv. Vyssh. Uchebn. Zaved.* **54** No. 10/2 p 131

[5] Baranova T A, Chubenko A K, Ryabikov A E, *et al* 2017 Microarc synthesis of nanostructured radiation-absorbing coatings on aluminum and titanium surfaces *IOP Conference series: materials science and engineering* **286** p 12

[6] Rudnev V S, Sergienko E S, Haritonskij P V, *et al* 2018 Distribution of iron and ferromagnetic characteristics of Fe-containing PEO coatings on aluminum *Prot. Met. Phys. Chem. Surfaces* **54** No. 5 p 488

[7] Vladimirov B V, Krit B L, Lyudin V B, *et al* 2014 Microarc oxidation of magnesium alloys: A review *Surf. Eng. and Appl. Electrochem.* **50** No. 3 p 195

[8] Andreev A S and Snezhko A A 2009 Influence of electrolyte composition on the structure and properties of oxide coatings formed on titanium alloys by microarc oxidation *XV Reshetnev Readings: Conference Proceedings* **1** No. 13 p 307

[9] Mamaev A I, Chubenko A K and Suslyaev V I 2017 The method of obtaining radiation-absorbing coating on valve metals and their alloys and the coating obtained by this method Patent RF No. 2637871

[10] Mamaev A I and Mamaeva V A 2005 *High-Current Processes in Electrolytic Solutions* (Novosibirsk: the Publishing House of the Siberian Branch of the Russian Academy of Sciences) 254 p

[11] Kazmina O V, Dushkina M A, Suslyaev V I and Semukhin B S 2014 Porous material for protection from electromagnetic radiation *AIP Conference Proceedings* **1623** p 214

IOP Conf. Series: Materials Science and Engineering **511** (2019) 012036 doi:10.1088/1757-899X/511/1/012036

The ability of usage various RGD configurations for the biodegradable vascular graft modification

L V Antonova[1], E O Krivkina[1], E A Senokosova[1], V N Silnikov[2], E A Velikanova[1] and T V Glushkova[1]

[1]Research Institute for Complex Issues of Cardiovascular Diseases, Kemerovo, Russian Federation
[2]Institute of Chemical Biology and Fundamental Medicine of the Siberian Branch of the Russian Academy of Sciences, Novosibirsk, Russian Federation

E-mail: leonora92@mail.ru

Abstract. The results of modifying non-woven matrices based on polyhydroxybutyrate / valerate and polycaprolactone with various RGD peptides (peptides: RGDK, AhRGD, c[RGDFK]) attached to the surface of matrices using hexamethylenediamine were studied. The structure of the surface of the matrices before and after the modification was not significantly changed. After 7 days of cultivation, the largest number of adherent and viable human umbilical vein endothelial cells of a person were detected on the surface of matrices with AhRGD or c[RGDFK]. However, cell proliferation was practically absent on the matrices with AhRGD. Thus, the modification of the surface of matrices with the AhRGD peptide is most suitable for products requiring rapid endothelization.

1.Introduction

The surgery needs vascular grafts to be used due to high risk and widespreadness of cardiovascular diseases. Autologic blood vessel grafts are the clinical standard of todays for CABG and small blood vessel reconstruction. But they may be not available. Dacron or ePTFE grafts may parttically solve this problem but with diameter less 5 mm they may cause thrombosis or hyperplasia of intima.

A functionally active biodegradable vascular prosthesis of small diameter, which has high permeability, atrombogenicity and biocompatibility can become an alternative to arteries and veins. Such strategies as immobilization of specific proteins and bioactive peptides can make the polymer surfaces highly biocompatible and non thrombogenic. One of the most perspective method is to immobilize RGD peptides (arginine-glycine-aspargine acid) on the surface of polymer vascular grafts as a sites of cellular adhesion which the most proteins of extracellular matrix have. This RGD peptides is one of the the key ligands for integrines – the receptors that are responsible for cellular adhesion, cellular migration, differentiation and existence. The RGD which contains peptides have simple structure and are more chemically stable than proteins.

Today in literature the usage of wide specter of peptides containing RGD parts is described, but there are no supports of any preferences for exact peptides.

2.Research methods and equipment

The polymer matrices had been produced by Nanon-01A using electrospinnig method from PHBV and PCL composition in chloroform. Hexamethylenediamin, glutaric dialdehyde, ninhydrin, ascorbic acid, cyclic peptides c[RGDFK], RGDK and AhRGD had been used for surface modification. The quality of the modification were measured using ninhydrin test and by arginine-contain peptide finding.

The structure of matrices was studied by SEM before and after modification. Cellular adhesion, viability and proliferation of HUVEC which were cultivated for 7 days on matrices surface with and without RGD were studied using confocal laser scanning microscopy and fluorescence microscopy.

The obtained data was processed using the STATISTICA 6.0. The significance of differences was determined using the non-parametric Mann – Whitney test for disconnected pairs. Differences between groups were statistically significant at $p < 0.05$. The Bonferroni amendment on multiple comparisons was also applied. Data are presented as medians (Me) and 25% and 75% percentiles.

3.Results

The surface modification of non-woven PHBV / PCL matrices with peptides containing the RGD fragment was carried out in two stages: at the first stage, partial aminolysis of the ester bonds of the graft material under the action of diamines, at the second stage, the introduction of the RGD peptide by means of cross-linking reagent.

The presence of peptides on the polymer surface was confirmed using the Sakaguchi test for the presence of arginine [8, 9].

Thus, 3 samples of polymer matrices were obtained, with a different structure of the peptide containing the RGD fragment (Table 1).

Table 1. Samples of polymeric grafts obtained during the modification.

	Peptide 1 (RGDK)	**Peptide 2 (AhRGD)**	**Peptide 3 (c[RGDFK])**
Hexamethylenediamine Amine 1	A1P1	A1P2	A1P3

Scanning electron microscopy showed that non-woven PHBV / PCL matrices, both before and after modification with various RGD configurations, had a homogeneous highly porous structure formed by randomly arranged curved filaments (Figure 1).

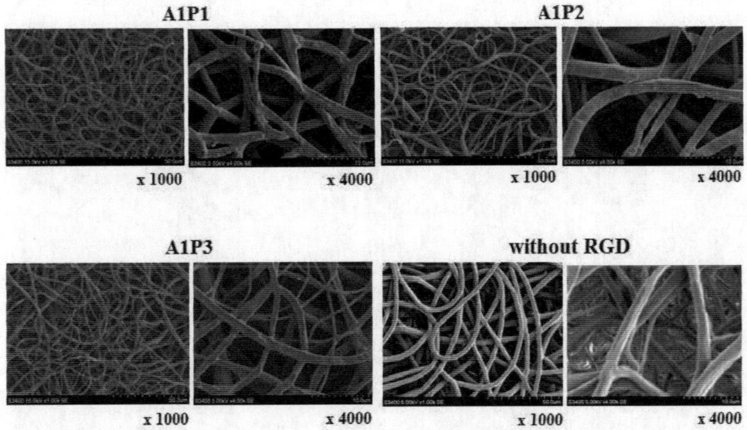

Figure 1. Scanning electron microscopy of the surface of non-woven matrices PHBV / PCL with and without RGD.

Modifying RGD peptides did not lead to structural changes in the surface of the matrices, but unlike unmodified samples, the matrices with RGD had bulges on the polymer filaments that were clearly noticeable at high magnification. These irregularities, presumably, are RGD peptides immobilized on the surface.

There was no adhesion and proliferation of HUVEC on the unmodified PHBV / PCL matrices, due to the lack of cell adhesion sites, which are mandatory for HUVEC adhesion.

The performed matrix modification with RGD peptides ensured the presence of cell adhesion sites on the polymer surface. On all presented types of modified matrices, the adhesion process of HUVEC proceeded with similar dynamics (Figure 2).

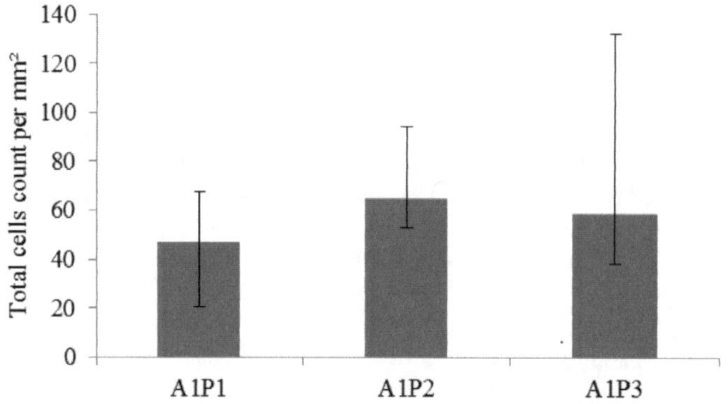

Figure 2. Absolute amount of adhesion of HUVEC on the surface of non-woven matrices PHBV/PCL/RGD.

The following results were obtained by the number of cells adhered on the matrixes: A1P2: 64.7 (52.9; 94.1)> A1P3: 58.8 (38.2; 132.4)> A1P1: 47.1 (20.6; 67.6). There were no statistically significant differences in the total number of cells per 1 mm^2 between the groups.

The assessing the relative number of viable and dead cells on various matrices with RGD showed that A1P2 matrices have more than 50% viable cells (range from 53% to 84%), whereas in A1P1 and A1P3 matrices, on the contrary, more than 50% of dead cells (from 54% to 77%) (Figure 3).

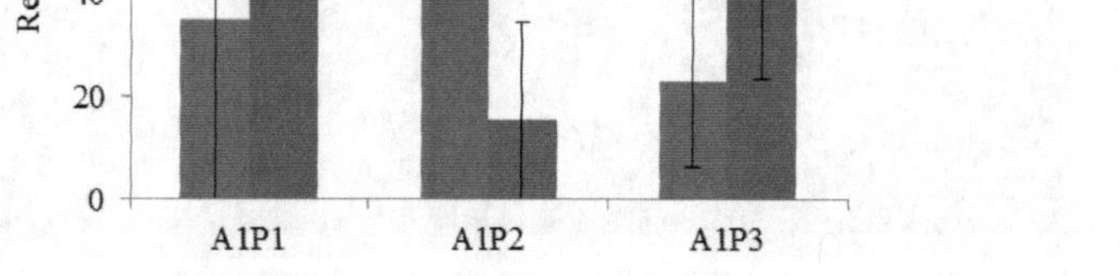

Figure 3. The relative number of viable and dead cells on the surface of non-woven matrices PHBV/PCL/RGD.

The largest relative number of proliferating cells was detected on A1P1 matrices: 11,1 (7,1; 23,1)%. While on the A1P3 matrices this amount was 5.0 (0,0; 22,2)%. The smallest relative number of proliferating cells was found on the A1P2 1.4 (0,6; 11,1)% matrices (Figure 4).

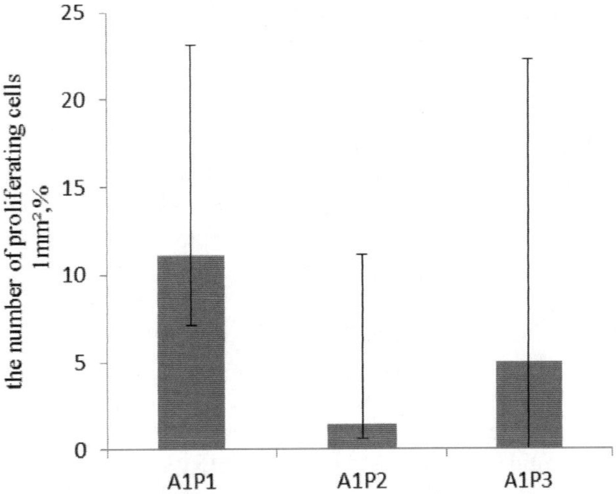

Figure 4. Relative number of proliferating cells on the surface of non-woven matrices PHBV/PCL/RGD.

4.Discussion

The results of the modification of non-woven matrices based on PHBV / PCL with different configurations of RGD peptides were studied: A1P1 - RGDK peptide; A1P2 - AhRGD peptide; A1P3 is a cyclic peptide c [RGDFK] sewn using hexamethylenediamine.

A1P3 matrices showed the lowest results of cell viability, against the background of satisfactory cell adhesion. Dead cells prevailed on the surface of these matrices, and proliferation of endothelial cells was reduced.

Polymer grafts with A1P1 have had high proliferation of endothelial cells on the surface, which may be due to the effective tropism for endothelial cells of a selected configuration of the A1P1 peptide itself.

Matrices with A1P2 showed the highest number of adherent and viable cells in comparison with A1P1 and A1P3. Also, more than 50% of viable cells were found on A1P2 matrices. However, according to the number of proliferating cells, the matrices with this peptide showed the worst result.

5.Conclusion

It can be concluded that the AhRGD peptide is the most favorable for cell adhesion and cell viability: the PHBV / PCL samples modified by it exceeded the matrices with RGDK and c [RGDFK] in the number of adhered and viable cells. The results show that the AhRGD peptide is most suitable for modifying the surface of grafts for their quick endothelialization.

Funding: This study was supported by the Complex Program of Basic Research under the Siberian Branch of the Russian Academy of Sciences within the Basic Research Topic of Research Institute for Complex Issues of Cardiovascular Diseases No. 0546-2015-0011 “Pathogenetic basis for the development of cardiovascular implants from biocompatible materials using patient-oriented approach, mathematical modeling, tissue engineering, and genomic predictors”.

IOP Conf. Series: Materials Science and Engineering **511** (2019) 012036 doi:10.1088/1757-899X/511/1/012036

References

[1] Taggart DP 2013 Current status of arterial grafts for coronary artery bypass grafting *Ann Cardiothorac Surg*. **2(4)** 427
[2] Ren X, Feng Y, Guo J, Wang H, Li Q, Yang J and Li W 2015 Surface modification and endothelialization of biomaterials as potential scaffolds for vascular tissue engineering applications *Chem Soc Rev*. **44(15)** 5680-742
[3] Wang F, Li Y, Shen Y, Wang A, Wang S and Xie T 2013 The functions and applications of RGD in tumor therapy and tissue engineering *Int J Mol Sci*. **14(7)** 13447-62
[4] Harburger DS, Calderwood DA 2009 Integrin signalling at a glance *J Cell Sci*. **122(2)** 159-63
[5] Gabriel M, van Nieuw Amerongen GP, van Hinsbergh VW, van Nieuw Amerongen AV and Zentner A 2006 Direct grafting of RGD-motif-containing peptide on the surface of polycaprolactone films *J Biomater Sci Polym Ed*. **17(5)** 567-77
[6] Gabriel M, Nazmi K, Dahm M, Zentner A, Vahl CF and Strand D 2012 Covalent RGD modification of the inner pore surface of polycaprolactone scaffolds *J Biomater Sci Polym Ed*. **23(7)** 941-53
[7] Hersel U, Dahmen C and Kessler H 2003 RGD modified polymers: biomaterials for stimulated cell adhesion and beyond *Biomaterials*. **24(24)** 4385-415
[8] Lin HB, Sun W, Mosher DF, García-Echeverría C, Schaufelberger K, Lelkes PI and Cooper SL 1994 Synthesis, surface, and cell-adhesion properties of polyurethanes containing covalently grafted RGD-peptides *J Biomed Mater Res*. **28(3)** 329–42
[9] Parniak MA, Lange G and Viswanatha T 1983 Quantitative determination of monosubstituted guanidines: a comparative study of different procedures *J Biochem Biophys Methods*. **7(4)** 267-76.

IOP Conf. Series: Materials Science and Engineering **511** (2019) 012037 doi:10.1088/1757-899X/511/1/012037

The fabrication of patterned metallic master by photolithography and electroplating technique for making PDMS-stamp as a tool for drug delivery system preparation

I A Zykova[1], A A Udalov[1,2], V L Kudryavtseva[1], E V Shesterikov[1,2,3] and S I Tverdokhlebov[1]

[1]National Research Tomsk Polytechnic University, 30, Lenina Avenue, Tomsk, 634050, Russian Federation
[2]V.E. Zuev Institute of Atmospheric Optics, Russian Academy of Science, Siberian Branch, 1, Academician Zuev square, Tomsk, 634055, Russian Federation
[3]Tomsk State University of Control Systems and Radioelectronics, 40, Lenina Avenue, Tomsk, 634050, Russian Federation

E-mail: tverd@tpu.ru

Abstract. In this work, the micropatterned (Ni micropillars) masters were fabricated by combination of photolithography and electroplating technique. The polydimethylsiloxane (PDMS) stamps were successfully made by casting technique on micropatterned masters and used for further preparation of polymeric microchamber arrays film considered to be a drug delivery system. Micropatterned master fabrication is more precise technique, which is necessary for defects elimination.

1. Introduction

Preparation of biodegradable microchamber arrays film which is considered to be a drug delivery system requires using a special polymeric micropatterned stamp as it was demonstrated in previous work [1]. Such stamp is a reverse replica of a master, thus, it is necessary to fabricate patterned master for it. Since the stamp and microchamber patterns directly depending on the master, the technique of master fabrication should be appropriate to provide the desired pattern.

Photolithography is a standard well-spread technique for the pattern fabrication on a metallic substrate by a top-down approach. This technique consists of covering substrate with photoresist; transferring patterns from a photomask to a photoresist by using UV light exposure and its development; etching unwanted parts of the substrate [2]. However, despite the simplicity of standard photolithography technique, it has a limitation in shapes of patterns [3].

Combination of photolithography and electroplating technique named as LIGA (lithography, electroforming and molding) is down to top approach which includes plating of metal instead of etching substrate [4]. Such combination allows getting more precise structures with variety of shapes and dimensions from micro- to nano- sizes. Furthermore, due to low roughness of plated metallic pattern polymeric reverse replica has a high quality [5].

Polydimethylsiloxane (PMDS) is the most common elastic polymer used for stamp making because it ensures a conformal covering over a large area including micro- and nano-nonuniform surfaces, has a quick curing time and has a low cost. Moreover, prepared PDMS-stamp is chemically stable, non-toxic and hydrophobic [6,7]. Traditionally, PDMS-stamp made by the casting of PDMS and curing agent mixture onto a master, wherein it is easy to gain necessary PDMS stamp stiffness by the variation of curing agent concentration.

In this work, the micropatterned master fabrication by combination technique of photolithography and electroplating in order to make micropatterned (MP) PDMS-stamp for the future biodegradable microchamber arrays film preparation was described.

2. Materials and methods

2.1. Materials

Aluminum ceramics plates (Policor, Russia) VK-100 were used as a substrate. Positive photoresist AZ 9260 from the AZ Electronic Materials (Luxembourg) and photomask with pattern diameter of 10 μm was used for MP master fabrication. Chemicals as dimethylformamid (EKOS-1), hydrochloric acid (HCl), sodium hydroxide (NaOH) and nickel electrolyte were used for master fabrication without further purification.

Poly(dimethylsiloxane) (PDMS) (ELASTOSIL RT 601) from Wacker AG, Germany was used for stamp making. Poly(L-lactic) acid (PURASORB PL38) from Purac, Netherlands and chloroform from Fisher Scientific, UK were used for microchamber arrays film preparation.

2.2. Micropatterned master fabrication

The scheme of MP Masters fabricating by photolithography and electroplating of nickel combination technique is shown in figure1.

Ceramic plates were taken as a substrate. A copper (Cu) layer with a thickness of 0.5 μm was deposited on ceramic plates by magnetron sputtering as a conductor for electroplating. For better adhesion between Cu-layer and plates, a tantalum (Ta) layer with thickness of 200 nm was deposited on plates.

For the photolithography process, a positive photoresist AZ 9260 was spin coated on a substrate by a centrifuge at 2000 rpm for 60 s. The photoresist layer having a thickness of 10 μm on the plate was exposed by UV light lamp (380 – 410 nm) through the photomask for 30 seconds. After developing in 1.5% NaOH (m/m) solution for 8 s, the pattern in a photoresist was formed and clean from contaminants in 10% HCl (v/v) solution.

Electroplating was carried out in a bath with Ni electrolyte (table 1) under the current density of $0.1 A\cdot mm^2$, a direct current of 0.5×10^{-1}A. Such direct current and current density were used to avoid detachment of the photoresistive mask. The different electroplating time was chosen: 20 minutes (Master 1) and 10 minutes (Master 2).

At the end of the electroplating, the photoresistive mask was removed in dimethylformamid, and then the masters were washed and dried.

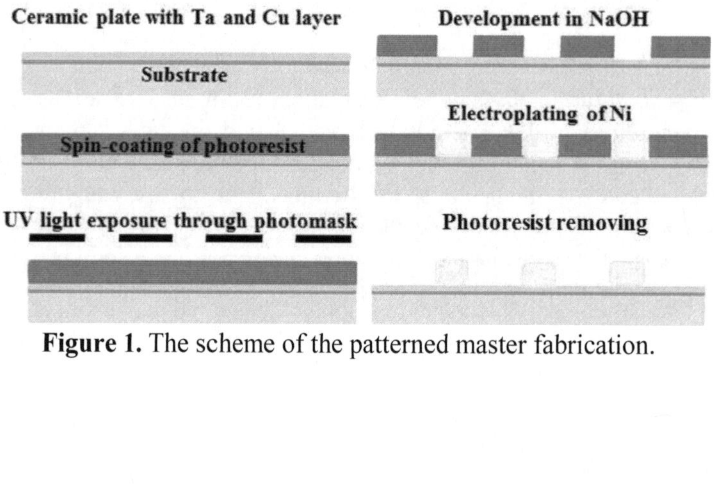

Figure 1. The scheme of the patterned master fabrication.

Table 1. Ni electrolyte composition.

Composition	Quantity
$NiSO_4 \cdot 7H_2O$	150 $g \cdot l^{-1}$
$Na_2SO_4 \cdot 10H_2O$	60 $g \cdot l^{-1}$
HBO_3	24 $g \cdot l^{-1}$
NaCl	6 $g \cdot l^{-1}$

2.3. Micropatterned PDMS-stamp preparation

The MP PDMS-stamp was prepared according to the following method. A liquid PDMS and curing agent were thoroughly mixed in a standard ratio of 10:1 (w/w). The mixture was poured onto a clean master and vacuum degassed for 10 min to remove air bubbles. The PDMS curing was in a heat chamber at 70ºC for 20 min. The cured PDMS was easily peeled off and was subsequently used for polymeric microchamber arrays film preparation.

2.4. Polymer microchamber array film preparation

The MP PDMS-stamp was dip-coated into 1wt.% PLLA solution in chloroform and then it was dried under an ambient condition to full solvent evaporation for receiving PLLA film on it. The MP PDMS-stamp/PLLA was sealed by micro-contact printing on the pre-coated in PLLA solution flat substrate. After that, the MP PDMS-stamp was lifted off and the PLLA film repeating pattern of the PDMS-stamp left on a flat substrate. The model drug was loaded prior MP PDMS-stamp/PLLA sealing.

2.5. Characterisation

Scanning electron microscopy (SEM) on ESEM Quanta 400 FEG (FEI, USA) was used to investigate the morphology and the quality of the masters, the MP PDMS-stamp and polymeric microchamber arrays film. Preliminary the stamp and microchamber arrays film were coated with a conductive thin gold layer *via* magnetron-sputtering system SC7640 (Quorum Technologies Ltd, UK).
The SEM images was analysed with using ImageJ program. For each parameter 100 measurements were carried out.

3. Results and discussion

3.1. Fabricated micro-patterned master

The fabricated MP masters after photolithography development process and Ni electroplating on a substrate are shown in Fig 2. SEM images shows the repeating structures as ordered Ni micropillars with some defects.

The center-to-center distance of micropillars for the both masters is 20 μm. The micropillars height is nonuniform for the both masters. The micropillars height on the Master 1 is in the range of 4.1 to 6.9 μm; on the master 2 is in the range of 1.8 to 3.2 μm. As the electroplating time for the Master 1 was two times longer, its micropillars are two times higher.

On the Master 1, the micropillars have a trapezoid-shaped profile: the diameter of micropillars at the base of the plate is lower than at the top. However, on the Master 2 fabricated under the same condition, but with a less electroplating time, has regular shaped micropillars. A lack of exposure in the lower layers of the photoresist due to a relatively large thickness of the photomask and not complete photoresist development could cause changing in patterns shape. Since residual photoresist remains near the substrate, the contact area for nickel electroplating is less.

As Figure 2 shows some defects on the masters could be observed. The substrates cracks associated with a low-quality of aluminum ceramics plates. The defective pillars on the Master 1 (figure 2a) is a result of over electroplating. Defects can be eliminated by more precise selection of parameters of the photolithography such as time exposure, developer composition and time of development and electrical conditions such as electroplating time and current density which have the influence on a plating speed.

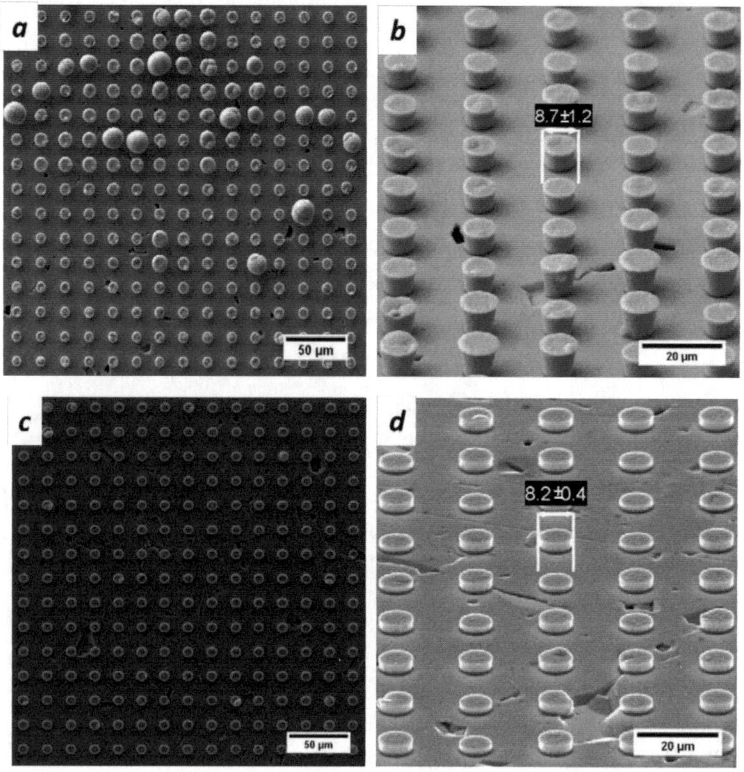

Figure 2. SEM images of the Master 1 (*a, b*) and the Master 2 (*c, d*).

3.2. The PDMS-stamp

The MP PDMS-stamps from the masters are shown in figure 3. The MP stamps from the Master 1 and the Master 2 are named as PDMS-stamp 1 and PDMS-stamp 2, respectively.

The PDMS-stamps have ordered arrays of microwells completely reverse replicate patterns of masters (figure 3 *b*, *d*).

Microwells of PDMS-stamp 1 keeps a trapezoid-shape profile with expansion to the bottom after lifting off due to the elasticity of stamp. The microwells average depth is 6.4±0.3 µm as the height of a metallic pattern. The upper microwells diameters (d_1=6.4±0.3 µm and d_2= 7.4±0.3 µm) are less than the metallic patterns one.

Microwells of PDMS-stamp 2 has a rectangular-shape profile with an average depth of 2.6±0.4 µm. The microwells diameters are the same as for the master 2.

The microwells of PDMS-stamp 2 are the more identical in depth and have a regular shape, so it was decided in further to use just this stamp for microchamber arrays film preparation.

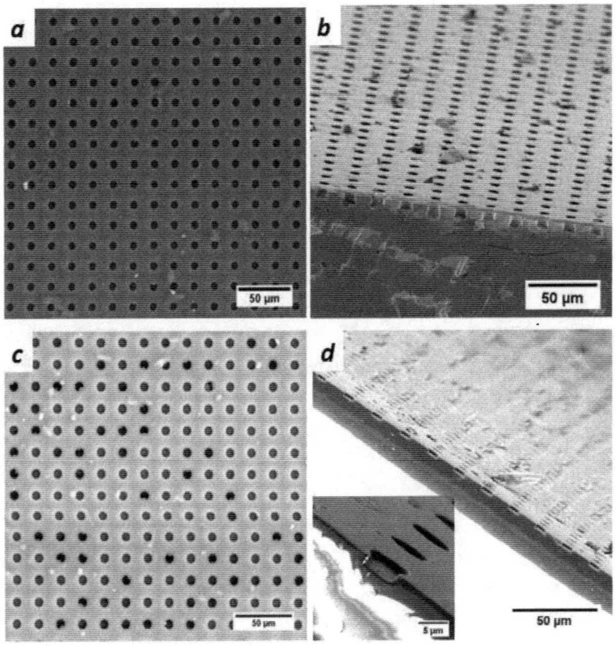

Figure 3. SEM images of the MP PDMS-stamps 1 (*a, b*) and MP PDMS-stamps 2 (*c, d*) on top-view (*a, c*); the tilt-view (*b, d*). The part of image g is magnified in 10 times.

3.3.Polymeric microchamber arrays film

The PDMS-stamp has a low surface energy [6], due to that fact the adhesion of PLLA film to the stamp is a poor that allows easily print it on a flat substrate.

Figure 4 shows the PLLA microchamber arrays film prepared by using the MP PDMS-stamp 2. Microchamber arrays are reverse pattern replica from the PDMS-stamp and a direct replica from the master. The thickness of PLLA film is about 0.5 μm. Microchambers save a free spaces into which drug substance to be loaded. The approximate loading capacity of individual microchamber is 9.12×10^{-8} μl.

It was investigated, that the PDMS-stamp could be swelling in chloroform after being there more than 10 min at 37°C [8]. However, the microchamber arrays film preparation excludes a long contact with the solvent and lasts only 10 s. Thus, the dimensional and shape PDMS-stamp microwells are not changed, so the PDMS-stamp could be used several times.

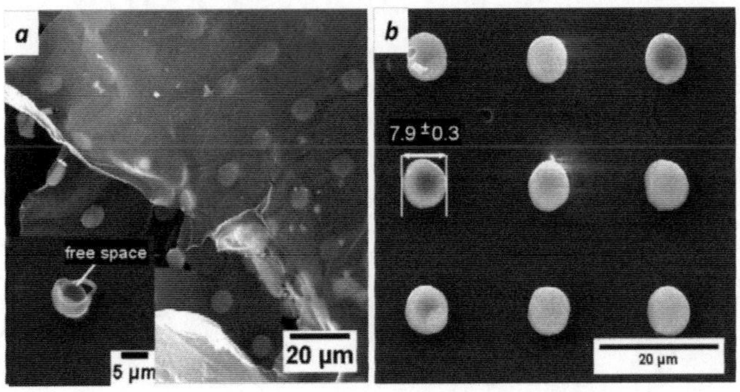

Figure 4. SEM images of PLLA microchamber arrays film prepared with using the PDMS-stamp 2.

4.Conclusion

In this work, the micropatterned metallic masters were fabricated by combination of photolithography and electroplating technique. The micropattern is an ordered Ni micropillars. The increasing electroplating time in two times allows getting masters with the micropillars two times higher.

The PDMS-stamps were successfully made by casting on fabricated Masters and applied for microchamber arrays film. However, it was observed some defects such as cracks, nonuniform height dismersion on both masters and irregular shapes of micropillars which can influence on a quality of PDMS-stamps and microchamber arrays film. It was considered, that for the elimination of defects, more precise choice of fabrication parameters is necessary.

Acknowledgments

This work was financially supported by the Ministry of Education and Science of the Russian Federation, Federal Target Program (agreement #14.575.21.0140, unique identifier RFMEF157517X0140)

References

[1] Gai M, Frueh J, Tao T, Petrov A V., Petrov V V., Shesterikov E V., Tverdokhlebov S I and Sukhorukov G B 2017 Polylactic acid nano- and microchamber arrays for encapsulation of small hydrophilic molecules featuring drug release via high intensity focused ultrasound *Nanoscale* **9** 7063–70

[2] Shelly M, Lee S-I, Suarato G, Meng Y and Pautot S 2017 Photolithography-Based Substrate Microfabrication for Patterning Semaphorin 3A to Study Neuronal Development pp 321–43

[3] Deng T, Wu H, Brittain S T and Whitesides G M 2000 Prototyping of Masks, Masters, and Stamps/Molds for Soft Lithography Using an Office Printer and Photographic Reduction *Analit. Chem.* **72** 3176–80

[4] Lee Y, Ahn S K and Roh Y 2005 Comparison of nanometer-scale gold structures electrodeposited on Au and Pt seed electrode *Surf. Coatings Technol.* **193** 137–41

[5] Hwang S Y, Park H H, Kang S M, Shin H B and Kang H K 2013 Fabrication of nanopatterned metal sheet using photolithography and electroplating *Thin Solid Films* **546** 132–5

[6] Friend J and Yeo L 2010 Fabrication of microfluidic devices using polydimethylsiloxane *Biomicrofluidics* **4** 26502

[7] Mata A, Fleischman A J and Roy S 2005 Characterization of polydimethylsiloxane (PDMS) properties for biomedical micro/nanosystems. *Biomed. Microdevices* **7** 281–93

[8] Lee J N, Park C and Whitesides G M 2003 Solvent Compatibility of Poly(dimethylsiloxane)-Based Microfluidic Devices *Analit. Chem.* **75** 6544–54

IOP Conf. Series: Materials Science and Engineering **511** (2019) 012038 doi:10.1088/1757-899X/511/1/012038

The heat-resistant gas-generating energetic composite materials for needs of the oil industry

L M Ponosova, E F Okhrimenko, A V Kuznetsova and D V Rychkova

JSC "Research Institute of Polymeric Materials", 614101, 16, Chistopolskaya str., Perm, Russia

E-mail: l.m.ponosova@yandex.ru

Abstract. The article shows the development prospects for the producing of the heat-resistant gas-generating energetic composite materials (ECM) recommended as propellant of the downhole devices intended for increase in oil recovery of oil wells. The operational characteristics of series of the versatile heat-resistant propellants for downhole devices of different purpose and the mechanism of action are provided. These versatile propellants are recommended for use in wells with any geothermal conditions and allowing increasing significantly in oil recovery of wells.

1. Introduction

According to officially published data, the average coefficient of oil recovery in the oil industry of Russia is only 0.2-0.4. It should be noted that in USA, for example, the share of secondary and tertiary treatment of oil wells, which increase their efficiency (oil recovery ratio), accounts for more than 80% of the total volume of oil produced [1]. In Russia, unfortunately, despite the efforts of the profile scientific research institutes [2-5], necessary attention was not paid to these methods, and huge number of the so-called "wasted" wells have accumulated, the resources of which are not exhausted. From the positions of the rational use of natural resources and the economic feasibility the depreservation and "revival" of such wells is very perspective direction in the oil industry.

In this regard, the development of new efficient methods for the secondary and tertiary treatment of "wasted" and other low-yield wells is very relevant task, and in recent years, these developments in Russia found "second wind" and, finally, receive due attention of oil producers.

In the oil industry, different methods of increase in efficiency of oil wells are applied [6]:

- treatment with liquids: hydraulic fracturing (HF), hydrochloric acid treatment (HAT), treatment with gas heat-generating solutions (GHGS) and combustible-oxidative mixtures (COM);
- mechanical drainage using cumulative, jet, bullet perforators;
- timely locking of exploration and production wells using sealing devices;
- method of filtration blast waves (MFBW)$
- thermobaro-gas-chemical effect (TBGCE) on the well bottom-hole zone using downhole devices (DD) of the different mechanism of action.

The choice of method is largely defined by the composition and structure of the rock in which the oil layer is located. In some cases, it is advisable to use more simple methods, in others - the combined use of the different methods providing different mechanisms of action.

IOP Conf. Series: Materials Science and Engineering **511** (2019) 012038 doi:10.1088/1757-899X/511/1/012038

The implementation of these methods of influence on the bottom-hole zone is carried out using special downhole devices of the different mechanisms of action:

- devices of thermobaro-gas-chemical effect on the bottom-hole zone, which include the powder pressure generators (PPG), generators with an adjustable pressure impulse (GAPI), downhole pressure accumulators (DPA), etc.;
- devices for mechanical drainage of the well and near-wellbore zone: bullet (BP), cumulative and jet perforators;
- devices of combined (perforation and TGHV) action of type GD, GDK, PG;
- devices of complex action of AKV, MAC;
- well sealing devices (WSD) preventing emergencies and environmental pollution.

It is considered that the methods of thermobaro-gas-chemical effect (TBGCE) are most effective, because they provide a complex effect of several working factors – the baric (pressure), thermal (temperature) and chemical (chemical activity of gaseous combustion products of propellant charge of DD).

2. The baric factor

When burning the powder charge in the range of the productive layer, the pressurized fluid of the formed gases that is there, is displaced into the structure of the layer, expands the natural cracks, pores and channels and creates new cracks of greater length and degree of openness. The increase in the layer structure of similar cracks leads to repeated increase in the conductivity of the layer and, consequently, the productivity of the well.

3. The thermal factor

During the combustion of 1 kg of TN type propellants offered by us for increase the efficiency of the DD, 3344-5434 J of thermal energy is released. The maximum temperature of gases emitted during charge combustion of the downhole device can reach 3500 ° C, but due to the relatively good thermal conductivity of the well column, liquid and rock, the medium temperature at the level of the downhole wall does not exceed 350 C. The heated powder gases, penetrating deep into the layer through the cracks, melt the heavy components of oil (resins, asphaltenes and paraffins) which have dropped out in use. After charge combustion, the pressure in the well decreases, and the powder gases in the layer displace the formation fluids and the melted deposits in the wellbore.

4. The chemical factor

During the propellant combustion in the well occurs chemical effects of aggressive gaseous combustion products on the rock structure and formation fluid. For example, the effect of hydrogen chloride released during the propellant combustion causes the fluidization of hard deposits and fluids due to their partial or complete decomposition. Sometimes, to enhance the effect of chemical factor, for example in carbonate collectors, the powder charge is burned in the medium of hydrochloric acid which is previously pumped to the well that leads to repeated increase in productivity of wells.

The effectiveness of the above factors largely depends on the performance properties of the propellants used in DD as propulsive mass.

The Research Institute of Polymeric Materials (now JSC "NIIPM") is leading enterprise in the Russian Federation in the field of producing of the heat-resistant gas-generating solid propellants. In addition, JSC "NIIPM" is one of the first DD developers for thermobaro-gas-chemical effect (TBGCE) on the bottom-hole zone, which include the powder pressure generators (PPG), generators with an adjustable pressure impulse (GAPI), downhole pressure accumulators (DPA), etc.

The experimental work in this area began at the institute in the early 70s of the twentieth century. The serial production of the powder pressure generators for oil wells was developed by "NIIPM" in 1979. Industrial application under the trade name "DPA - downhole pressure accumulators" found three modifications of the powder generators: DPA-5; DPA-6; DPA-200U.

DPA are intended for processing the bottom-hole zone of oil wells with the dismantled operating equipment. The temperature-time resource of operation of DPA-5, DPA-6 is 3 hours at 100 °C; and DPA-200U is not less than 12 hours at 200 °C and not less than 24 hours at 150 °C. The DPA consists of one igniter, several burning elements (5-7) and the device for assembly, collected in the set on piece of geophysical cable.

The main technical characteristics of serially produced DPA are given in Table 1.

The main difference and advantage of DPA-200U is that the burning elements of DPA-200U are made of high-heat-resistant energy-intensive composite material (propellant).

The heat resistance of combustible composite materials is provided by the use of heat-resistant polymers such as ethylene-propylene, divinyl-styrene copolymers, compounds containing α-oxide cycles in their chain, etc.

The range of the heat-resistant propellants developed in NIIPM is rather wide, which increases possibilities of the choice of propellant in relation to specific DD and conditions of specific wells. Depending on the purpose and application conditions for DD in addition to heat resistance, impose whole range of different requirements (Table 2, 3) to propellants.

For impact on bottom-hole zone of producing wells, it is advisable to use slow-burning propellants in PPG and DPA devices, which provide the long and deepest heating of the zone.

The fast-burning propellants apply in bullet perforators for well drainage, as well as in PPG and DPA for thermobaro-gas-chemical effect on the bottom-hole zone of both the forcing and producing wells. At the same time, the productivity of wells increases, not only due to temperature and gas chemical effects, but also due to the baric influence of shock character (the sharp pressure jump exceeding the ground pressure), creating a system of cracks in zone of the forcing and producing wells and also pressure shock in zone of the forcing wells.

For providing the alternating effect of working pressures from low to high shock pressures, including impulse, in order to build up the thermodynamic system of the oil layer, it is advisable to combine slow and fast-burning elements in one DD.

It is also possible to increase the efficiency of propellants by increasing the role of the chemical influence of the gaseous products of their decomposition, increasing the share of reactive gases and liquids such as chloride and fluoride hydrogen. The use of so-called "acid-forming" propellants is much more effective than the currently used acid treatment of wells, as it provides the complex effect on the bottom-hole layer zone of the different factors (temperature, pressure, gases and acid). For example, hydrogen chloride is the active chemical agent that effectively decomposes different deposits that prevent the outflow of oil from the oil layer. In flooded wells, the hydrogen chloride interacting with the well fluid forms hydrochloric acid. In addition, the thermo-gas-chemical impact apparatus, equipped with propellant, works directly in the productive zone of the oil layer, and the hot hydrochloric acid produced during the combustion process affects the structure of the rock directly in the layer treatment zone, increasing its porosity and permeability. The high temperature, pressure and gases provide deep penetration of the reactive hydrogen chloride into the layer and its effective cleaning.

A large range of the propellants with the wide range of performance characteristics (energy, ballistic, heat resistance, etc) provides the possibility of targeted regulation of the DA operating mode depending on the application conditions. For providing of universal use in the different geothermal conditions, including in deep and ultra-deep wells and in wells with the complicated geothermal conditions, for example, at the Tyumen, Kamchatka, Tajikistan, Uzbekistan, and Ukraine fields, the special requirements are imposed on the thermal stability of propellants (up to 100-300 °C).

The studies carried out in JSC "NIIPM" allowed us to create the series of the heat-resistant, up to 200-300 ° C, gas-generating acid-forming propellants of the TN series for downhole devices with an increased (more than 2 times) content of hydrogen chloride in the combustion products. The acid-forming propellants are more expedient to use in downhole devices for processing of carbonate and terrigenous collectors in wells with the increased water content. In addition, taking into account their high heat resistance, they can be successfully used in wells with the complicated geothermal conditions.

The latest studies of JSC "NIIPM" are directed to development of the new more effective, universal and low-price propellants. As a result, in production are implemented the several new brands of solid

IOP Conf. Series: Materials Science and Engineering **511** (2019) 012038 doi:10.1088/1757-899X/511/1/012038

propellants like TN-15/5E with the improved operational characteristics for production of the wide nomenclature of charges for equipment of the modern downhole devices.

These propellants have the following advantages:

- increased amount of the hydrochloric acid emitted during the charge operation;
- high, up to 150-200 °C heat resistance;
- possibility of long stay of charges in the well without changing the performance characteristics: at 150 °C - not less than 24 hours; at 200 °C - not less than 12 hours;
- large gassing;
- low sensitivity to mechanical influences and electric spark;
- simple and low-price technology that provides the rapid transition of mass production to any standard sizes of charges.

The main operational properties of the new composite propellants are given in Table 4. The developed heat-resistant propellants are universal and can be recommended as propulsive mass of the most different downhole devices (accumulators and powder pressure generators of wells, bullet perforators, soil intake devices and soil samplers such as DPA, PPG etc.) for operation practically in any geothermal conditions.

5. Conclusion.

The actuality and development prospects for the producing of the heat-resistant gas-generating propellants for needs of the oil industry are shown.

The series of compositions of the versatile heat-resistant propellants for downhole devices for various purposes and the mechanism of action that meets all the requirements defined by their operating conditions is developed.

The performance characteristics of the various heat-resistant propellants developed at JSC «NIIPM», which increase the versatility of the use of downhole devices and their efficiency are presented.

Table 1. The main characteristics of DPA produced by JSC «NIIPM».

Designation	Serial			New modifications	
	DPA-5	DPA-6	DPA-200U	DPA-5	DPA-6
Outside diameter, mm	110	110	65-110	70-110	102
Passage diameter, mm	-	22	20	-	24
Length of powder charge, mm	1200	1200	1000	900	1060
Mass of powder charge, kg	16	14	7	5	14
Quantity of charges in one set, pcs.	6	6	6	5	6
Mass of the pressure accumulator, kg	94	84	42	25	84
Ignition method	Electric minimum current of operation 1,5A				
Maximum application temperature, °C	100	100	200	100	100
Permissible time of stay in the well at the maximum temperature, h	3	3	12	3	3
Minimum hydrostatic pressure in the well treatment interval, MPa	3	3	3	3	3

IOP Conf. Series: Materials Science and Engineering **511** (2019) 012038 doi:10.1088/1757-899X/511/1/012038

Table 2. Apparatus and heat-resistant propellants to increase the productivity of oil wells using the TBGCE method.

THERMOBARO-GAS-CHEMICAL EFFECT (TBGCE) ON THE BOTTOM-HOLE ZONE

DOWNHOLE PRESSURE ACCUMULATORS (DPA) AND PRESSURE GENERATORS (PPG)

SLOW-BURNING SOLID PROPELLANT		
For production wells		
Provide the heating and gas-chemical fluidization of high-viscosity oils, deposits and formation fluids in zone of production wells		
Require-ments	It is reached	It is perspective
Heat resis-tance ↑	100…200°C	150…270°C
Burning velocity mm/s ↓	1,5...10	‹1,5

FAST – BURNING SOLID PROPELLANT		
For forcing and production wells		
Provide the physical effect of shock character for creating: - system of cracks in zone of the forcing and producing wells - pressure shock in zone of the forcing wells		
Require ments	It is reached	It is perspective
Heat resis-tance ↑	200...250°C	270°C
Burning velocity mm/s ↓	10...25	›30

PULSING SOLID PROPELLANT		
For forcing wells		
Provide the alternating effect of slow- and fast-burning propellants in order to build up the thermodynamic system of the oil layer around the well and its driving		
Require ments	It is rea-ched	It is pers-pective.
Heat resis-tance ↑	100...200°C	270°C
Burning velocity mm/s ↓	2...25	2...30

ACID-FORMING PROPELLANTS		
Provide the fluidization of high-viscosity oils, decomposition of oil deposits and formation fluids in structure of cracks and macropores of layer		
Require-ments	It is rea-ched	It is perspe-ctive.
Volume of gases, l/kg ↑	900…1000	1100
Amount of 12% acid (HCl and HF), kg/kg	2,0	3,1

Table 3. Apparatus and heat-resistant powders and propellants to increase the productivity of oil wells by methods of mechanical drainage of the bottom zone and timely blocking of wells.

MECHANICAL DRAINAGE OF DOWNHOLE ZONE

CUMULATIVE, JET, BULLET PERFORATORS

HEAT-RESISTANT FINE-GRAINED POWDERS		
Provide the drainage of an upsetting column and creation of network of cracks in structure of layer for increase in its permeability		
Requirements	It is reached	It is perspective
Heat resistance ↑	290°C	300...500°C
Force ↑	80-115	›115

TIMELY LOCKING OF EXPLORATION AND PRODUCTION WELLS

WELL SEALING DEVICES (WSD)

HEAT-RESISTANT LOW-GAS SOLID PROPELLANTS		
Provide the rational use of minerals, prevent emergencies and environmental pollution		
Requirements	It is reached	It is perspective
Heat resistance ↑	200°C	270°C
Burning velocity 5...10 mm/s	5	10…20

IOP Conf. Series: Materials Science and Engineering **511** (2019) 012038 doi:10.1088/1757-899X/511/1/012038

Table 4. The main operational properties of propellants of the TN series.

Composition brand	Characteristics with P = 50 MPa, V = const				Density, g/cm3	Characteristics with P = 50 MPa, V = const	
	Energy-capacity, kgf · m/kg	**Volume of gases, l/kg**	**Temperature of gases, K**	**Amount of 12% acid (HCl), kg/kg**		**Temperature of the beginning of decomposition, Tnir, °C**	**Temperature and time operation resource**
TN-15/5	106400	1029	2790	1.9	1.68	≥240	200° C≥ 12 h
TN-18/5K	80200	735	2860	0.2	1.84	≥300	200° C≥ 24 h 250° C≥ 12 h
TN-15/5E.1	110670	979	2576	1.9	1.74	≥240	150° C ≥ 24 h
TN-15/5E.2	109500	1028	2814	2.0	1.64	≥240	160° C ≥ 24 h
TN-15/5E.2V	93647	872	2845	0.9	1.77	≥300	180° C ≥ 6 h
TN-15/5E.2G	72215	687	2930	0.2	1.92	≥300	180°C ≥ 6 h
TN-15/5E.2D	76100	8326	2176	1.4	1.71	≥300	180° C ≥ 6 h
TN-15/5E.2E	64630	739	2300	0.6	1.83	≥300	180°C ≥ 6 h
TN-15/5E/4	102880	904	3009	3.1	1.87	≥300	150° C ≥ 24 h

Notes:

1. The velocity of the gas input can be adjusted over the wide range by the surface of the charge burning.
2. The experimental testing of the high-heat-resistant propellants is carried out (up to 250-300 °C and above).

References

[1] Kostochko A V, Smola E B, Vahitova T T 1995 *Conversion of organizations and enterprises of special chemistry and special technology* (Kazan: Collection of scientific papers) p.16

[2] Ponosova L M, Sharina T P, Znamenskaya L B 1997 *Heat-resistant compositions for downhole devices* (Perm: Collection of reports of the 2nd Ural Conference *High technology polymers and dual technologies of technical chemistry*) p.160

[3] Alikin V N, Kuzmitsky G E, Zabelin L V, Klyachkin Yu S, Fedchenko N N 1999 *Conversion of special technical chemistry. Powders. Propellants. Charges* (Perm: Perm Scientific Center of URORAN) p 176

[4] Kostochko A V, Kostochko A A, Ibragimov R A, Khramova E V 2012 *The problematic issues of the disposal of powders and some of their applications in the national economy* (Kazan: Bulletin of Kazan Technological University Vol. 15 № 10)

[5] Ponosova L M., Golubev A E 2016 *Heat-resistant polymeric materials for the needs of the oil industry* (Ivanovo: Collection of theses of the VI Russian Scientific Conference with international participation *Physic chemistry of polymer processing processes*) p.109

[6] Ponosova L M., Golubev A E 2017 *Application possibilities of the developments of JSC NIIPM in the apparatus of the oil industry* (№ 7. - ISSN 2077-542) p 52-55

IOP Conf. Series: Materials Science and Engineering **511** (2019) 012039 doi:10.1088/1757-899X/511/1/012039

The obtaining of biologically active substance by mechanical dispersion of selenium particles in water

V N Kolokoltsev, G E Folmanis and M A Fedotov

A A Baykov Institute of Metallurgy and Material Science Russian Academy of Sciences, 49, Leninsky prospect, 119334, Moscow, Russia

E-mail: mfedotov@imet.ac.ru

Abstract. Selenium colloidal solutions with particles ranging from 20 to 100 nm were obtained by mechanical dispersion of selenium pellets in water at frequency of 100 Hz. It is shown that the properties of selenium colloidal solutions prepared by mechanical dispersion depend on the material of the vessel in which they are obtained. The selenium solutions obtained in plastic vessels have acidic properties, while in glass vessels they are alkaline. Upon evaporation of colloidal solutions, highly anisotropic dendritic structures are formed, which may indicate anisotropy of the formed nanoparticles. The aqueous colloidal system of nanoscale selenium obtained by mechanical dispersion in concentrations physiologically significant for plants has a positive effect on the seeds of agricultural crops. Selenium colloidal solutions have greater antioxidant activity compared to sodium selenite with lower toxicity values.

1. Introduction

In the early 60s of the 20th century, the establishment of the antioxidant properties of selenium made it possible to consider it as a trace element for human and animal life. One of the main functions of selenium is to enhance the immune defense of the body [1]. Selenium basically enters the body with plant origin food. The use of inorganic selenium compounds for the preparation of functional foods is undesirable, since its particles are highly toxic [2]. The use of selenium nanoparticles has proven to be a promising process. Unlike ionic forms, selenium nanoparticles are less toxic, are easily absorbed by the body, and have a prolonged effect [3]. Selenium also plays an important role in the regulation of plant life [4].

Medications generally are prepared on the basis of colloidal solutions of biologically active substances [5]. For the preparation of colloidal solutions, various methods of dispersion are used, allowing one way or another to grind up the initial material to a nanoscale state. The laser ablation method turned out to be very effective, making it possible to obtain Se nanoparticles [6, 7]. However, this technique is still expensive, which limits its widespread use. An alternative is the method of mechanical dispersion in distilled water as an environmentally friendly dispersion medium [8]. The preparation of a substance in a nanoscale state in mechanical plants occurs by the shock interaction between macroscopic particles of a substance.

The aim of the work is to obtain an aqueous colloidal solution of biologically active selenium by mechanically dispersion its granules at a sufficiently low frequency (100 Hz).

IOP Conf. Series: Materials Science and Engineering **511** (2019) 012039 doi:10.1088/1757-899X/511/1/012039

2. Materials and Methods

Chemical glass vessels and clear polystyrene vessels were installed on a vibrating polishing machine Metapolan-2 with a vibration frequency of 100 Hz. Bidistilled water with pH value of 6.32 and pellets of high-purity selenium was poured into the vessels. Mechanical dispersion was carried out for 25–30 hours at a temperature of 300 K. After dispersion, the aqueous solution was settled together with the granules until clarification.

The concentration of selenium in the solution was determined on an atomic emission spectrometer with inductively coupled plasma ULTIMA-2 instrument (Horiba Jobin-Yvon, France). The size distribution and ζ-potential of selenium particles was measured on a Zetasizer Nano ZS instrument (Malvern, United Kingdom). It is known that as a result of the thermal motion the particles of the dispersed phase in colloidal systems perform random motions, changing the direction about 10^{14} times per second, therefore several successive measurements were on the instrument. Determination of the solution pH was carried out on a S220 Seven Compact device (Mettler Toledo, Switzerland).

The clarified selenium solution was evaporated at a temperature of 50–70 °C in glass vessels with a volume of 20–30 ml (figure 1). The upper narrow glass plate with size 1.5x15x50 mm prevented the entrainment of selenium particles with water vapor when the vessel was heated. The lower glass plate with a size of 1.5x10x25 mm was used to control the precipitation of selenium particles from a colloidal solution on its front and back sides. The face side was considered the side of the plate facing the solution in the vessel. The solution was evaporated on an electric heater. The surface temperature of the heater was controlled by a copper-constantan thermocouple. At the temperature of 60-70 °C, the process of complete evaporation of the solution in air lasted about 60 hours. After the solution was completely evaporated, the glass plates were taken out for investigation on an optical microscope LENOVO (Germany).

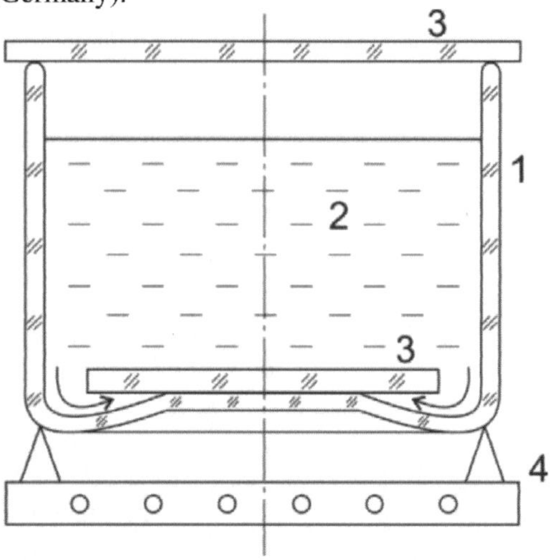

Figure 1. Evaporation process scheme: 1 – glass vessel; 2 – selenium solution; 3 – glass plate; 4 – electric heater.

3. Results

The figure 2 shows the distribution of selenium particles by size and zeta potential in a solution obtained by mechanical dispersion in a plastic vessel for 27 hours. The sample was taken with a syringe from the upper layer of liquid, which was settled in the vessel with selenium pellets for more than five days until complete clarification at 300 K. The selenium concentration was 1.8 mg/l.

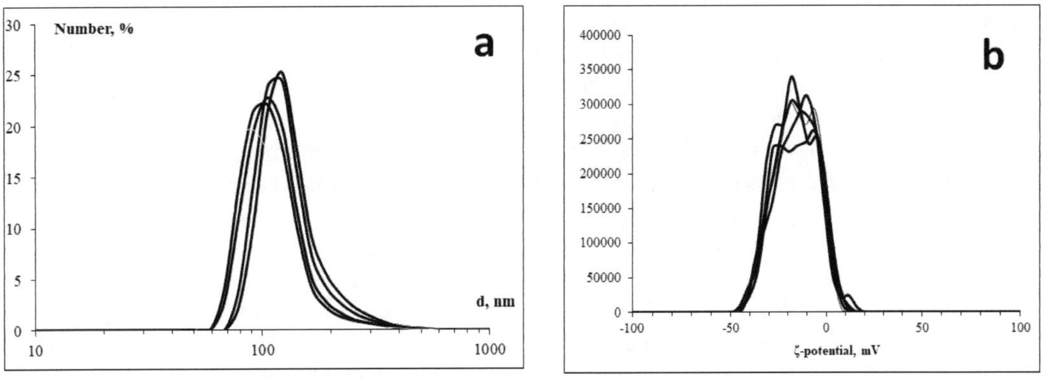

Figure 2. Selenium particle size distribution (a) and particles ζ-potential (b) in an aqueous solution in a plastic vessel after mechanical dispersion of the granules for 27 h. Particle distribution was measured with a time interval of 2 minutes.

The figure 2 shows that the maximum of the particle size distribution is in the region of about 100 nm. The average value of the particles ζ-potential in the solution obtained in the plastic vessel is in the range from -15.4 to -17.0 mV.

The distribution of selenium particles and the ζ-potential of a solution in a glass vessel are shown on figure 3, the dispersion time is ~27 h. Sampling was carried out with a syringe near the surface of the granules. The concentration of selenium in the solution was 10 mg/l. As it is shown on figure 3, particle sizes are grouped around discrete values of ~40, 100, and 500 nm. Thus, a system with a multimodal particle size distribution is formed. Unlike the solution obtained in a plastic vessel, in a glass vessel the particles ζ-potential can be both positive and negative. Its value changes from negative values of about -5 mV to positive +20 mV. In this case, significant changes in time are observed, both in the size distribution of particles and in the values of the ζ-potential.

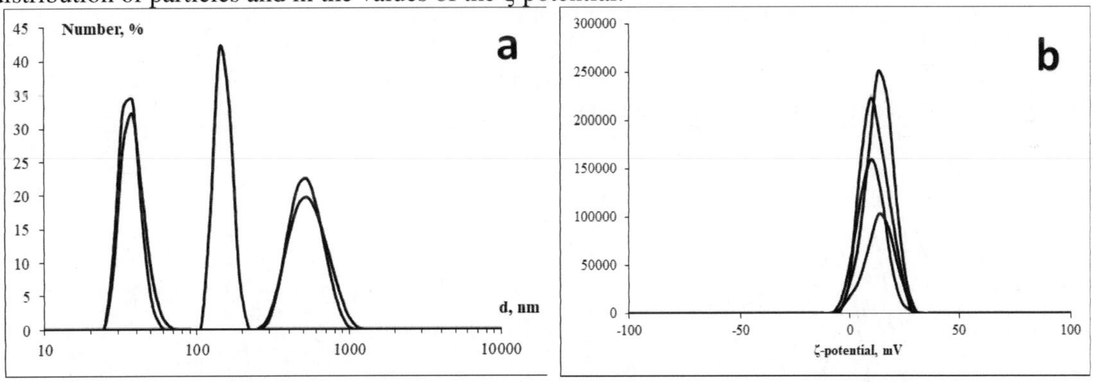

Figure 3. Selenium particle size distribution (a) and particles ζ-potential (b) in an aqueous solution in a glass vessel after mechanical dispersion of the granules for 27 h. Particle distribution was measured with a time interval of 2 minutes.

The results indicate that the process of dispersing the selenium granules proceeds differently in the plastic and glass vessels. This fact is confirmed by the formation of a red strip of amorphous selenium on the wall of the plastic vessel, equal in height to the thickness of the layer of selenium pellets; this strip is not observed on the wall of the glass tube. This can be explained by the difference in the charge of particles formed in the glass and plastic vessels, and the charging of the surface of the plastic wall. When large granules of selenium collide with the wall of a plastic container, a positive charge accumulates on the surface, which binds small particles of amorphous selenium. On the surface of glass wall, the charge does not accumulate, possibly because of its rather high electrical conductivity. In addition, it is noticed that in the plastic vessel the granules of gray selenium become shiny during the dispersion process, and in the glass vessel the surface of the granules remains matt gray. On the surface of the granules, there is always a gray selenium oxide – SeO_2, which is formed in air at the melting point of selenium, when the granules are obtained by the drip method. In water, selenium oxide dissolves to form weak selenous acid H_2SeO_3. It is possible that negative SeO_3^- ions are bound by positive charges on the wall of the plastic container. In this case, SeO_2 is carried away from the surface, and it becomes shiny since pure selenium is practically insoluble in water. It is possible that there are other mechanisms in the plastic vessel that prevent the formation of an oxide film on the surface of the granules. If the solution is removed from the plastic vessel, then after a while the granules will become gray in the air again. In the glass vessel selenium granules always remain gray under the same experimental conditions.

After settling of selenium solution with granules for a long time (several months), the other particle distributions and their ζ-potential values were observed. Thus, the concentration of selenium in the upper layers of the solution decreases, and increases above the surface of the granules. Under the same conditions of preparation process, the selenium concentration above the surface of the granules in colloidal solution in a glass vessel reaches 8–10 mg/l, and in the plastic vessel the concentration of selenium is 2 mg/l.

Depending on the material of the vessel, the colloidal solution may be acidic or alkaline. In glass vessels, the alkaline nature of the solutions increases with time (from 6.94 to 7.23). In plastic vessels, an acidic environment is usually observed, the acidity of which also increases with time (from 6.28 to 5.96). The increase in acidity can be explained by the fact that some of the SeO_2 can enter the colloidal solution from the surface of the plastic wall over time. The other part of SeO_2 probably comes from selenium granules located in the inner layers of the granulate backfill. In our case, the height of the granule layer was 8-10 mm. As for the glass vessel, here it can be assumed that in the presence of weak selenous acid alkali metal atoms contained in glass are transferred into the solution. An excess of alkali metal ions leads to the alkaline nature of the solutions in glass vessels.

Figure 4 shows the distribution of particles in the selenium solution obtained in a plastic vessel with mechanical dispersion for 27 h and clarified during the year without granules. The data on the ζ-potential of the particles are also presented here. It can be seen that the relative proportion of small nanoparticles with a size of 20–40 nm has increased, but the solution is not stable: significant changes in the spectrum are observed during the measurement time. This may be due to a significant decrease in the concentration of particles due to the Ostwald ripening.

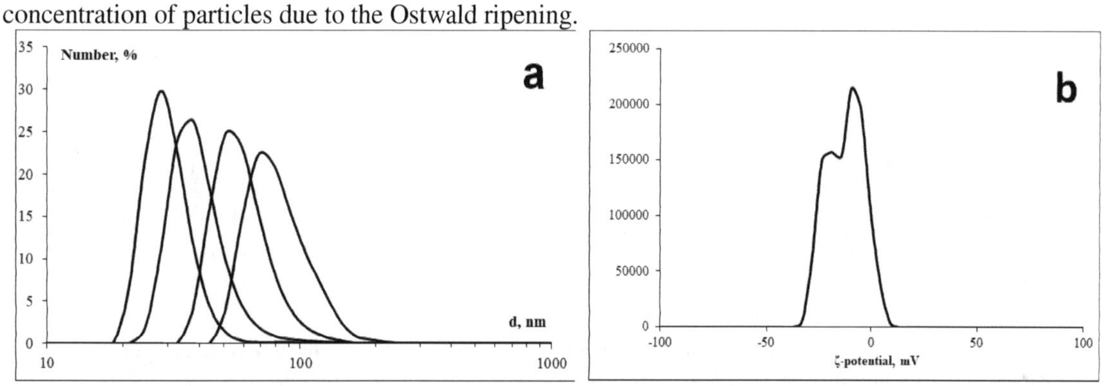

Figure 4. Selenium particle size distribution (a) and particles ζ-potential (b) in an aqueous solution obtained in a plastic vessel by mechanical dispersion for 27 h. Clarification of the solution - 1 year. The concentration of selenium in the solution is 0.17 mg/l. Particle distribution was measured with a time interval of 2 minutes.

When the selenium solutions is evaporate on glass plates, the structures of a dendritic type are also formed, whose dimensions and structure depend on the concentration of selenium in the solution. Figure 5 shows some of the characteristic types of structures observed on the surface of a glass plate turned to the bottom of the vessel. The concentration of selenium in the solution was 0.25 mg/l.

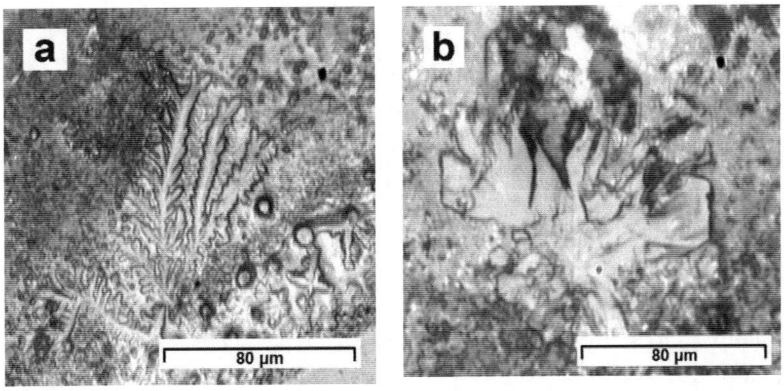

Figure 5. Fractal structures on a glass plate turned with the front surface to the bottom of the vessel (a) and on the front side (b). The concentration of selenium in the solution was 0.25 mg/l. The evaporation temperature of the solution was 60 °C.

It was established that when the solution is evaporated on the upper plate (position 3, figure 1), small accumulations of selenium particles are observed, i.e. there is an entrainment of selenium particles with water vapor.

The obtained colloidal selenium solutions were used to treat the seeds of garden crops. It is established that it has a growth-promoting effect. When comparing the obtained suspension with selenium salts, it was noted that the content of ascorbic acid in radish seedling seedlings is reduced from 35.3 mg/100 g (control sample) to 20.1 mg/100 g in the case of sodium selenite and to 15.3 mg/100 g in the case of colloidal selenium. Selenium colloidal solutions have an antioxidant effect. Antioxidant activity is expressed in mg-equivalents of gallic acid per gram of dry mass (mg-eq. GK/g d.m.) Seeds treated with colloidal selenium have values of 20.1 mg-eq. GK/g d.m., while for sodium selenite this value is 6.9 mg-eq. GK/g d.m. Antioxidant activity of the control sample was 16.3 mg-eq. GK/g d.m.

4. Discussions

By mechanical dispersion of selenium granules in water at a low frequency (100 Hz), colloidal solutions of selenium nanoparticles 20–100 nm in size can be obtained; however, their relative fraction is small compared to larger particles (500 nm or more). With a sufficiently long exposure of colloidal aqueous solutions with selenium granules at 300 K, it is possible to increase the concentration of selenium in the solution near the surface of the granules to 10 mg/l. However, during long-term storage of solutions in the absence of granules, a decrease in selenium concentration to values of 0.1 mg/l or less is observed. At the same time, a precipitate of red amorphous selenium is observed at the bottom of the vessel, which indicates that the nanoparticles interact with each other to form large particles.

When the selenium solutions are evaporated, different fractal structures form on the surface of the glass plates, which indicates the more complex nature of the colloidal solution with selenium nanoparticles than just water and a suspension of particles. It can be assumed that the colloidal solution contains a weak selenous acid, which is formed when the selenium dioxide is dissolved in water. It is known that selenium granules are always covered with a thin layer of SeO_2, which is highly soluble in water. The mechanical dispersion of the granules of the selenium in glass and plastic vessels showed that the material of the vessel significantly affects the chemical composition of the solutions and their pH. This means that upon receipt of selenium solutions, it is necessary to select carefully the vessel material. Preliminary studies have shown that pure quartz is the most suitable material. An interesting feature of the mechanical dispersion of selenium granules is the change in the optical properties of the surface of the granules depending on the material of the vessel. In a plastic vessel, selenium granules in the solution become shiny, while in a glass vessel the surface of the granules remains matt gray. This indicates that we are dealing with a different chemical environment in the studied solutions, the nature of which is currently not clear. It also follows from the presented results that a part of selenium atoms are included in various molecular compounds, therefore measuring the concentration as a whole does not give a true distribution of selenium nanoparticles in solution. For example, in a glass vessel, part of the selenium may be bound in the form of alkali metal selenates. Therefore, the given concentration values cannot be attributed only to nanoparticles. It follows from the above that the known antioxidant and immune effect of colloidal solutions of selenium nanoparticles may also be due to various molecular compounds.

5. Conclusion

By mechanical dispersion of selenium granules at a frequency of 100 Hz, colloidal solutions of selenium nanoparticles ranging in size from 20 to 100 nm were obtained. The concentration of selenium in colloidal solutions depends on the material of the vessel. The selenium solutions obtained in plastic vessels have acidic properties, while in glass vessels they are alkaline. Upon evaporation of colloidal solutions, highly anisotropic dendritic structures are formed, which may indicate anisotropy of the formed nanoparticles. Selenium colloidal solutions have greater antioxidant activity compared to sodium selenite with lower toxicity values.

Acknowledgement

The authors are grateful for the help in the work and discussion of the research results for the staff of IMET RAS: prof. L.V. Kovalenko, V.A. Volchenkova and S.A. Maslyaev. The work was carried out according to the state assignment # 075-00746-19-00. Part of the experiments was carried out using the equipment of the Center for the collective use of physical methods of research of the A.N. Frumkin Institute of Physical Chemistry and Electrochemistry Russian Academy of Sciences.

References

[1] Santi C and Bagnoli L 2017 Celebrating two centuries of research in selenium chemistry: state of the art and new prospective *Molecules* 2017 **22** (12) pp 2124-2128
[2] Dzahmisheva Z A and Dzahmisheva I S 2014 Functional foods for human purposes *Fundamental research* **9** pp 2048-2051 (in Russian)
[3] Zhang J S, Gao X Y, Zhang L D and Bao Y P 2001 Biological effects of nano red elemental selenium *BioFactors* **15** pp 27-38
[4] Nikonov I N, Folmanis J G, Kovalenko L V, Laptev G Y, Folmanis G E, Egorov I A, Fisinin V I and Tananaev I G 2012 Biological activity of nanoscale colloidal selenium *Doklady Biochemistry and Biophysics* **447** (1) pp 297-299
[5] Hint I A 1980 About the fourth technology component *Scientific and information collection SKTB "Disintegrator"* (Tallin, Valgus) (in Russian)
[6] Li L, Hong M, Schmidt M, Zhong M, Malshe A, Huis in'tVeld B and Kovalenko V 2011 Laser nano-manufacturing—state of the art and challenges *CIRP Annals – Manufacturing Technology* **60** pp 735-755
[7] Fedotov M A, Folmanis G E, Kovalenko L V, Nasakina E O and Senchikhin I N 2018 Colloidal selenium solutions prepared by laser ablation and ultrasonic dispersion *Journal of Physics: Conference Series* **1134** 012015
[8] Folmanis G E, Kolokoltsev V N, Kovalenko L V and Fedotov M A 2017 Destruction of selenium granules in the aquatic environment under the action of mechanical vibration *Conf. Proc. "Deformation and destruction of materials and nanomaterials"* (Moscow, IMET RAS) (in Russian)

The relation between the inductive capacity of an oxide surface structure of hard alloys and their wear resistance

V P Nesterenko, A A Kondratyuk, K P Arefyev, P A Chazov, A A Kononova and E Ya Sokolova

National Research Tomsk Polytechnic University, 30, Lenin Ave., Tomsk 634050, Russia

E-mail: nest2004@bk.ru

Abstract. The purpose of this research is to determine the value of specific inductive capacity of oxide structures formed on the contact surfaces of carbide cutting tools. This paper also deals with the study of the effect of specific inductive capacity of polyoxide structures on the wear resistance of cutting tools exposed to intensive adhesive wear. It was established that the decrease of specific inductive capacity of the oxide film mass formed on the contact surfaces of a wedge leads to the increase of the wear resistance of cutting tools.

1. Introduction

The wear resistance magnitude of cutting tools and the variation of its parameters depend considerably on the properties of the oxide film structures formed on contact surfaces. Thin oxide films, formed on the contact surfaces of a wedge, affect such operating characteristics of cutting tools as their wear resistance, durability and reliability. They reduce the intensity of adhesive processes in contact areas by shielding of intermolecular interaction with a work material. It was found that the intensity of the intermolecular interaction between tooling and work materials depends mainly on the specific inductive capacity of oxide structures formed in contact areas. The wear resistance of cutting tools used in processing of materials which cause intensive adhesive wear tends to increase under the conditions when the specific inductive capacity magnitude of oxide structures decreases. Steel which belongs to austenitic, austenitic- martensite, austenitic- ferritic classes and chrome-nickel alloys are work materials causing accelerated adhesive wear of a wedge. These materials are processed with carbide cutting tools which usage class is K [1]. Simultaneously with the wear resistance increase and the decrease of specific inductive capacity magnitude of the oxide surface structure formed on contact surfaces of cutting tools there is a reduction of the variation coefficient for a particular performance characteristic.

2. Experimental technique

The study of wear resistance of carbide cutter plates belonging to the class K was performed on a screw-cutting. 12X18H10T steel which causes intensive adhesion wear was used as a work material. Carbide plates of the engineering grade BK8 belonging to the above mentioned class were used as a cutting tool. When testing it was necessary to fix the cutting speed at 70 m/min, the cutting depth to 1,5 mm and the cutting rate at 0,21 mm/r. The wear resistance of a cutting tool was measured in minutes till the specified criterion at which the cutting tool lost its cutting properties. The back surface wear of a wedge equaling to 0,6 mm was assumed as the criterion of cutting properties loss.

After testing the carbide cutters were firstly cleaned in a special solution, then dried and finally oxidized in an electric furnace with atmospheric air free access to the heating area. The oxidation temperature of carbide cutters in a furnace was approximately equal to the temperature in contact areas of tooling and work materials, and was equal to 850°C. After that the cutting plates were taken from the furnace, the polyoxide structure formed in the process of oxidation was removed, crushed into powder and compounds with the same weight of 10-20 gr were prepared from this powder. These compounds were placed in a cell equipped with the electrode made of copper. The cell, serving as a condenser with a polyoxide mass, was placed in a special compressor between insulating plates. A dynamometer was used to record the pressure acting on the poly-oxide mass. This pressure was equal to $0{,}02 \pm 0{,}005\ kg/mm^2$. The cell was connected to a measuring a.c. circuit to record the capacitance of a polyoxide mass. A special device E7-8 designed to meter the capacity, inductance and ohmic resistance at a.c. frequency of 1000 Hz was used to measure this capacitance.

The magnitude of specific inductive capacity of polyoxide samples was determined by the known equation [2] using the magnitude of capacitance obtained by its measuring:

$$\varepsilon = \frac{C \cdot H}{\varepsilon_0 \cdot S}, \qquad (1)$$

where C is the capacitance of a polyoxide structure of investigated samples. The capacitance was measured in picofarads using the E7-8 device.

H is the layer thinness in millimeters of a polyoxide structure between copper electrodes.

ε_0 is the electric constant.

S is the electrodes area, mm^2.

The thickness of a compressed mass in a cell was determined after metering its electric capacitance when removed from the device. A tool length indicator was used to measure the thickness of the compressed polyoxide mass. The accuracy of this measurement was 0,005 mm.

We constructed the dependence diagram of the wear resistance magnitude expressed in nonfailure operating time of a cutting tool till the specified dulling criterion on the magnitude of specific inductive capacity of a polyoxide with the "wear resistance-specific inductive capacity" structure. This dependence diagram was constructed based on the data obtained from the wear resistance determination of cutting tools belonging to K class by processing chromium-nickel steel of austenitic category causing intensive adhesive wear and on the data from electric capacitance obtained by measuring with subsequent calculation of the specific inductive capacity magnitude.

3. Results and discussion

Upon the research results it was found that there is a close correlation between the wear resistance of carbide cutting tools belonging to the class K and the magnitude of specific inductive capacity of oxide surface structures formed when heated. Figure 1 represents the dependence of wear resistance variation of cutting plates made from hard alloy BK8 belonging to hard alloys of the class K on the magnitude of specific inductive capacity of oxide mass formed on the surface when heated at high temperature in an electric furnace. Chromium-nickel austenitic steel of the class 12X18H10T is used as a work material. As it can be seen from the diagram with the decease of specific conductivity magnitude of the oxide surface mass there is a stable tendency to the increase of cutting tools wear resistance.

Adhesive wear is the major destruction of carbide cutting tools belonging to class K occurring when they are used to process austenitic chromium-nickel austenitic steel [2]. The decrease of the specific inductive capacity magnitude of a structure located in an intercontact space leads to more reliable shielding of intermolecular interaction between tooling and work materials. The durability of micro dot connections in the contact zone does not reach the cutoff value. The contact areas do not develop into the seat of destruction by the disconnection of a cutting wedge and work material.

Close relation between the wear resistance of hard-alloy cutting instruments and the specific inductive capacity of polyoxide structures obtained by oxidation of cutting tools in a muffle electric furnace points out the fact of formation and operation of ultra-thin polyoxide structures on the contact surfaces of a cutting wedge and confirms its great influence on all characteristics of the cutting operation. From the received data it is obvious that polyoxide structures formed in the contact areas are identical to physical and chemical properties of structures formed by oxidizing heating of hard-alloy samples in an electric muffle furnace.

It can be stated that if the conditions of thin oxide film formation in the contact areas approximate the conditions of polyoxide compound formation by heating hard-alloy plates in an electric furnace, the closeness of correlations between the wear resistance and specific inductive capacity will also increase. To create favorable conditions for polyoxide structure formation in an electric furnace identical to those arising by cutting, it is necessary to have profound knowledge of all the processes taking place in interfacial spaces.

Hydrogen, oxygen and other controlled and uncontrolled impurities being a part of the formed polyoxide structure affect considerably the magnitude of specific inductive capacity. The properties of polyoxide structures in the contact surface areas of cutting and work materials are formed both due to the oxidation and simultaneous reduction of formed oxides.

The oxidation process of contact surfaces of a cutting tool in the areas of their interaction with the work material takes place due to the oxygen entering the intercontact space from the surrounding gaseous atmosphere and the oxygen coming from the hard-alloy structure.

Both the oxides reduction process and the formation of lower oxides and other oxide structures formed in the oxide mass are carried out by hydrogen accumulated by a hard alloy with its further entering the intercontact spaces from composite volumetric structure caused by its diffusion. Oxygen penetrates into the inner structure of the hard alloy components at their preparation stage. These stages include tungsten, carbide and cobalt powder preparation from deoxidization (oxide reduction) and powder grinding. While the oxide restorative process is proceeding there is some residual oxygen concentration in the optimal form, the formed tungsten carbide as well as in the metallic cobalt. The content of oxygen in components of a future hard alloy structure can significantly increase as the result of further powder activation when grinding. The saturation of the component structure by atmospheric oxygen in all cases is caused by adsorptive processes [3].

The initial hydrogen accumulation in a hard alloy structure takes place at the preparation stage of its components when sintered in a hydrogenous medium. The introduction of hydrogen in the forming thin-film oxide structure of interfacial space causes significant changes of its properties including chemical, physical and mechanical.

Hydrogen affects both the formation of the surface and near-surface polyoxide structure of a hard alloy and processes developing in the interaction zones between the cutting medium and work material. Hydrogen absorbed by hard alloys components due to its high penetration power is also accumulated on the surface of samples and in their volume. The hydrogen molecules are firstly absorbed by more chemically active parts of the surface and near-surface areas where hard alloy is molded. Surface relief areas belong to carbide granules, cobalt layers and interphase boundaries. Diffusion makes hydrogen penetrate into the near-surface layers of a sintering compound. The area of interacting and closed pores and the surfaces at which boundaries a sharp change of structure properties can be observed (including density, porosity, acting inner stresses) belong to the components of surface and near-face structures of sintering molding where a considerable hydrogen molecules adsorption takes place. The highest concentration of pores and cracks is in the structure of cobalt granules and the interphase space. The cobalt layer has mainly high concentration of short branched cracks. The porous system in a carbide structure is formed at high-temperature processing of tungsten powder while its carbonaceous component is being saturated. The crack system of carbide and cobalt particles is formed at the stage of their grinding in a ball mill. The process of hydrogen saturation when sintering is accompanied by the formation of new systems of pores and cracks and their annihilation [4].

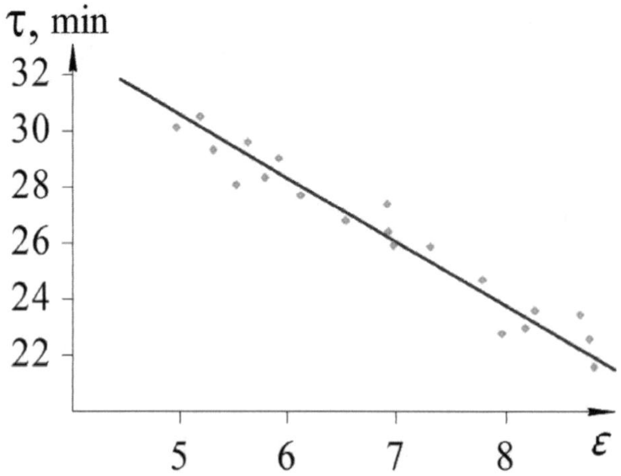

Figure 1. Wear resistance change of cutting plates produced from the hard alloy BK8 by cutting the steel 12X18H10T depending on the specific inductive capacity magnitude of a polyoxide structure formed on the contact surfaces of a cutting tool.

The real impact of the intercontact poly-oxide structure properties on the cutting operation, and hence, actual wear resistance parameters of cutting tools are determined by the hydrogen concentration penetrating into the polyoxide mass and the level of reaction activity between oxides and hydrogen which may result in the formation of dissipative compounds with different frictional characteristics. The impurities included in the hard-alloy structures which are exposed to oxidation also affect considerably the formation of properties in the intercontact polyoxide structures.

The shielding property of a thin-film mass against the intermolecular interaction of contact surfaces of tooling and work materials is caused by improving the dielectric characteristics of intercontact poly-oxide structures associated with the decrease of their specific inductive capacity. The wear resistance of hard-alloy cutting instruments of the class K tends to increase when used for material cutting causing intensive adhesive wear.

The lowest specific inductive capacities have polyoxide structures in the contact areas, if they consist of oxides with higher oxidation level of tungsten and cobalt. The compounds of the contact areas with a stoichiometric composition are formed under the most favorable and optimal conditions [5].

The probability of forming structures in intercontact spaces with the above specified properties will increase if the content of hydrogen and hydrogenous compounds in a polyoxide mass decreases.

The maximum wear resistance of hard-alloy cutting tools of class K used to process materials causing intensive adhesive wear will be achieve if the content of a polyoxide mass concentrated in an intercontact space is as specified above. This can be explained by high dielectric characteristics of a polyoxide structure which shields the intermolecular interaction of contact surfaces and functions as a solid lubricant. In this case dielectric losses and the magnitude of polyoxide specific inductive capacity will reach the lowest magnitude.

Specific inductive capacity of a polyoxide film formed in the intercontact areas is determined by its degree of polarization. The decrease of polyoxide mass polarization rate leads to the decrease of its inductive capacity and the improvement of its dielectric characteristics. On the contrary, the increase of the polarization rate of a polyoxide mass leads to the increase of its inductive capacity and deterioration of its dielectric characteristics. The latter results in the degradation of a polyoxide structure which enhances the chances of intermolecular interaction of cutting and working surfaces. The polarization increase can lead to the increase of dielectric losses, electric conductance and the decrease of thermal capacity which in turn results in the reduction of wear resistance of carbide cutting tools used for processing materials causing intensive adhesive wear [6].

The hydrogen leaving hard-alloy structures penetrates into the contact areas and forms solid solutions and different complex compounds with oxides of film surface structures. Hence, the polarization increases and the degradation of dielectric characteristics of the intercontact dissipative poly-oxide mass is observed. This causes the degradation of shielding properties of a polyoxide mass against the intermolecular interaction of contact surfaces. There is also degradation of polyoxide structure properties as a solid lubricant caused by the reduction and redistribution of atomic bond strength between planes in a crystallographic oxide structure. The polarization process induced by electric and mechanical fields is an example of structural instability formation near the polyoxide composition. Instability takes place if the post oxidation and the renewal processes might develop similarly. The development of the post oxidation processes results in increase of the oxide stoichiometry, while the renewal processes lead to the decrease of oxidation level of a polyoxide structure. In this case we can observe the origination of some areas on the surface or in the whole polyoxide volume with a very high probability for intermolecular interaction of tooling and working materials.

Temperature, contact stress, the availability of reducing and oxidizing medium affect the polarization processes of a polyoxide mass in the contact areas [7]. The electron shells of polyoxide compositions of a polyoxide mass are mixed in the external force field. With the increase of displacement there is also the increase of polarization level. This kind of polarization can be classified as an electron polarization. The higher the polarization, the lower the resistance of the shielding property of a film polyoxide mass to the intermolecular interaction of contact surfaces is. There is a decrease of electron polarization level with the temperature rise till the temperature reaches its optimum or is higher than the cutting temperature. The volume increase of a polyoxide mass and the reduction of particle numbers per unit volume is caused by the density reduction of a substance.

If the poly-oxide mass exposed to cutting operation experiences spontaneous polarization caused by several external actions like elastic or plastic deformation, friction and sudden temperature rise, there is an instant reduction of shielding properties of dielectric medium and increase of intermolecular interaction force in certain local points of contact surfaces. Hence, there is intensification of a micro-welding process between contact surfaces with the further destructive disconnection of welding joints which results in gradual micro- or macro - surface destruction of a cutting tool. Gradual and continuous accumulation of defects and reduction of hard alloy service life are caused by cyclic increase of intermolecular interaction between contact surfaces.

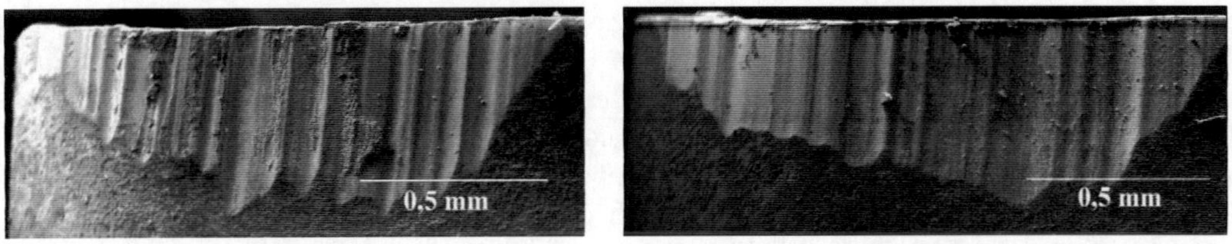

a) b)

Figure 2. Worn out surfaces of cutting blade (cutter plate) made of BK8 a) with a specific inductive capacity of a polyoxide structure $\varepsilon=10$, b) with a specific inductive capacity of a polyoxide structure $\varepsilon=6$.

Hydrogen, which is a constituent part of polyoxide structure, creates some concentration of various electron defects in an oxide lattice at high temperatures developing by the use of cutting tools. Therefore, the electron shells are additionally exposed to the displacement under the action of external force fields. This kind of polarization can be described as electron - relaxational. This polarization is likely to develop in polyoxides with serious structural defects caused by the This type of polarization developed in a polyoxide mass indicates the additional increase of its dielectric properties and eliminates the intermolecular interaction in the contact areas.

Besides the electron nature of the chemical bond between metallic elements and oxygen, the polyoxide mass of the intercontact space has ionic bonds. Hydrogen penetrating into the tungsten and cobalt structure facilitates the formation of an ionic bond. Therefore, we can observe another type of electric instability of an oxide mass which can be defined as the polarization of ions. It develops by a relative displacement of ions in a lattice. The displacement of electron shells of these compounds will differ due to the unequal composition of the oxide mass and different magnitude of the forbidden zone. As a result, the polyoxide film structure has a heterogeneous electric structure that distinguishes by the quantity of reduced and enhanced charges in a volume. This fact contributes to the increased probability of intermolecular interaction between contact surfaces.

The total hydrogen concentration in an intercontact oxide mass will decrease if high temperature is established in contact areas of tooling and working materials. It is possible since the concentration of dessorbing hydrogenous compounds leaving contact zones exceeds the hydrogen concentration which enters the intercontact spaces of the bulk hard alloy structure. This fact leads to the decrease of compound concentration with an ionic bond of a polyoxide mass and hence, its polarization reduction. This can result in improving the dielectric properties of a polyoxide mass and reducing the specific inductive capacity. In this case the screening capacity of the intercontact structure to shield the intermolecular interaction between tooling and working materials will increase.

Polyoxide molecules form large complex compounds with polar radicals by their interaction with hydrogen. Ordering process of polar radicals in a force field represents a dipolar-relaxation polarization. Whereby there is a certain alignment of polar macro-molecules. The probability of molecular alignment is simplified with the temperature rise which leads to the increased intensification of the given dipolar-relaxation polarization and sharp rise in specific inductive capacity. However, alongside with the cutting temperature rise there is a destruction of complex oxide-hydrate compounds in an intercontact space. The latter causes the reduction of the growing polarization and hence, the decrease of specific inductive capacity of a polyoxide film structure.

The polyoxide mass, formed on the contact surfaces of a cutting tool, has a layered structure consisting of tungsten and cobalt oxides and can be exposed to the interlayer polarization. However, this type of possible polarization is not associated with the degradation of dielectric behavior of a polyoxide. This is due to the postoxidation and the volume growth of a polyoxide mass in contact areas caused by the temperature rise by heating and cutting. In this case, there is the stabilization of specific inductive capacity and even its decrease.

The polarization processes of a polyoxide mass in the contact areas can proceed due to its elastic and plastic deformation as well as due to the friction. The highest magnitude of polarization, so called electric structural instability of a polyoxide, can be achieved at one certain temperature or various temperatures.

Under the impact of elastic and plastic deformation, friction and high temperatures the formed and destructive polyoxide film structure can change its composition and aggregate state, and consequently it can change the level of intermolecular interaction between contact surfaces and work materials.

Volumetric layers of a hard alloy contain high concentration of free and dissolved oxygen [8]. Oxygen diffusion from inner layers of a volumetric carbide structure into the contact areas is able to stabilize the oxidation process of the interfacial polyoxide mass and improve its dielectric properties. Specific inductive capacity will decrease. Thus, shielding properties of the intecontact polyoxide structure from intermolecular interaction between contact areas will improve.

Other controlled and uncontrolled impurities found in hard alloys and exposed to diffusion at high temperatures during the use of cutting tools can increase or decrease the level of specific inductive capacity of an intercontact polyoxide mass.

IOP Conf. Series: Materials Science and Engineering **511** (2019) 012040 doi:10.1088/1757-899X/511/1/012040

As it has been established that oxides of phosphorous, sulphur, germanium, niobium, calcium, nickel, iron and other chemical elements as impurities enter into the composition of surface polyoxides. These impurities can considerably affect the magnitude of specific inductive capacity.

Figure 2 presents the images of cutting tools wear. From this figure it's obvious that the hard-alloyed cutting plates, whose poly-oxide structures have low specific conductivity and are less exposed to adhesive effect from the material being processed, are more resistive to wear.

At sufficiently high temperatures at which the cutting tools operate, the part of the poly-oxide mass located in interfacial surfaces is exposed to sublimation and converts from a solid to gaseous state. The gaseous state has lower dielectric characteristics compared to those of a solid poly-oxide mass. Gaseous poly-oxide distinguishes by high dielectric losses and high magnitude of inductive capacity. The gaseous oxide substance formed in the intercontact space leads to the increase of intermolecular interaction between the cutting tool and work material and hence, to the increase of adhesive wear and the reduction of wear resistance.

The transition of the polyoxide of an intercontact space to gaseous state when it is used for steel and alloy cutting is possible at high temperatures. The process is periodically repeated after poly-oxide mass formation and its sublimation.

The gaseous poly-oxide mass can take part in a convection heat exchange. Hydrogen contributes to the development of this process. The complex hydride compounds and hydroxyle groups will also participate in the solid lubrication and contact areas cooling. It should be noted that the effect of these processes on the reduction of wear rate at high temperatures is negligibly small.

The efficiency of sublimated poly-oxide gaseous environment is not high enough when it performs as a shield against intermolecular interaction.

The whole process of the poly-oxide mass formation, its sublimation, partial condensation and the execution of its lubricating and shielding functions against intermolecular interaction of contact surfaces are continuous. However, each next oxidation of carbide contact surfaces depending on the carbide structure state and external factors can be restored at some certain temperature range. With the decrease of this range the cutting stability of hard-alloys will increase.

Thermal pretreatment and some types of radiation exposure of hard alloys affect differently the oxidation processes and inductive capacity of the poly-oxide mass formed when exposed to heating. There is a wide range of magnitude variations and its values do not always correspond to the changes of wear resistance magnitude of carbide cutting tools. These pretreatment types affect not only the nature of the oxidation processes but also other composition properties. Simultaneously with the change of the oxidation processes behavior, the bonds on the interphase boundary of carbide granules and the cobalt layer as well as the chemical bond in carbide granules are exposed to destruction.

Thus, it can be concluded, that the decrease of inductive capacity of poly-oxide structures leads to the increase of wear resistance of cutting tools exposed to intensive adhesive wear.

References

[1] State Standard 3882 – 74 (DIN ISO 513-2014)

[2] Trent E M 1980 *Metalcutting* (Moscow: Mechanical Engineering) p 283

[3] Adamson A 1979 *Physical chemistry of surfaces* (Moscow: Mir) p 568

[4] Vert Ch 1981 Hydrogen Capture in Metals *Hydrogen in Metals* vol 2 (Edited by G Alefeld and J Volkl Moscow: Mir) pp 362–407

[5] Nesterenko V P, Arefyev K P, Tyurin Yu I and Efremenkov A B 2014 Oxygen Effect on Wear Resistance of Hard Alloys from Tungsten-Cobalt Group *Physics. Bulletin of Universities* vol 57 11/2 pp.170–175.

[6] Nesterenko V P, Arefyev K P, Tyurin Yu I and Lychagin D V 2013 The relation between the wear resistance of hard alloys and their saturation with oxygen when sintering *Physics. Bulletin of Universities* vol 56 4/2 pp 221–224.

[7] Gubkin A N 1971 Physics of dielectrics (Moscow: High School) vol 1 p 270

[8] Tretyakov V I 1962 Cerametallic hard alloys (Moscow: Metallurgy) p 592

IOP Conf. Series: Materials Science and Engineering **511** (2019) 012041 doi:10.1088/1757-899X/511/1/012041

The study of the ontogenesis of crystals of the Ural type diamonds by the method of IR-spectrometry

Y Nefedov[1], O Kochneva[1] and G Heide[2]

[1]Oil and Gas Geology Department, St. Petersburg Mining University
[2]Institute of Mineralogy, Freiberg

E-mail: yurijnefedov@yandex.ru, kochnevaoe@mail.ru, gerhard.heide@tu-freiberg.de

Abstract. The collection of the Ural, Anabar and Brazilian diamonds conducted research by infrared spectrometry method. It was found that a diamond with a low degree of aggregation of nitrogen defects composing the crystals of the later generation from mantle eclogite xenoliths and outer zones of the greater part of the crystals from deposits of kimberlite-lamproite type is missing or is poorly developed in the crystals of the Ural type. Conclusions about the thermal conditions of the Ural diamonds formation and their possible affinity to the specification and type of original sources.

1. Introduction

The Ural type diamonds, combining rounded crystals of this mineral, for which flat-cut forms of the most diamonds from the deposits of kimberlite-lamproite type are not characteristic, attracted attention by specific morphological features and high yield of jewelry differences.

The most important characteristic of the Ural type diamond deposits is the lack of a reliable set of the original sources of placers, despite the fact that many of them are being developed more than 100 years. A range of assumptions about the possible types of original sources of these placers is rather great: from ancient completely eroded kimberlite bodies to the modern diamondiferous tuffisites. In the absence of reliable information of original sources and the rarity of finds of minerals-satellites in these placers, the diamonds can be the most important sources of information about the genetic characteristics of potential original sources. Therefore, studies based on the analysis of the crystal-chemical characteristics of this mineral can contribute to the understanding of the formation processes and question decision of possible indigenous sources of the Ural type diamonds.

One of the most important crystal-chemical diamond characteristics is the presence of structural impurities of nitrogen. To identify the concentration and location of nitrogen in diamond crystals the method of IR-spectroscopy has been successfully used. Scientists have made a great contribution to the development of spectroscopic methods of the study of diamond crystals: Kaminskii F V, Sobolev E V, Klyuev Y A, Boyd S R, Taylor W R, Woods G S. A significant amount of research of diamond crystals using the methods of IR-spectroscopy was made by Bogush I N, Vasiliev E A, Kopchikov M B, Palazhchenko O V, Petrovsky V A, Khachatryan G K, Laiginhas F and others, but the Ural type diamonds were insufficiently studied by this method. Therefore, the method of IR-spectroscopy was chosen as the primary method in this study because it allows to identify patterns of occurrence of nitrogen defects in diamond crystals without destroying mineral individuals quantitatively and with a high enough productivity to determine the concentration of nitrogen in the diamond, its shapes in the

IOP Conf. Series: Materials Science and Engineering **511** (2019) 012041 doi:10.1088/1757-899X/511/1/012041

form of A and B1 defects. In the present work under the patterns of the nitrogen defects in diamond crystals refers to the quantitative determination of the nitrogen concentration in diamond, in various forms of its presence, zone identification of their placement within the crystals, the setting of statistical regularities of allocation of diamond crystals in their content of various nitrogen defects.

Conducted research of extensive and representative collection of the Ural type diamonds from the different regions have allowed to reveal regularities of occurrence of the nitrogen defects in the studied crystals and to give them the correct genetic interpretation. These results are the basis for a more balanced approach to the analysis of the possible genetic types of indigenous sources of placers of the Ural type diamonds, which will contribute the choice of optimal search methods or make a reasonable decision about inexpediency of identifying the indigenous sources as potential industrial fields.

The work purpose is to identify regularities of occurrence of nitrogen defects in the crystals of the Ural type diamonds from the various regions.

2. Research techniques

The materials were obtained in the course of studying of the diamond collections of Mining Museum, Department of Mineralogy of St. Petersburg University, Komi Research center of the Ural branch of RAS, the research materials of the Ural diamonds collection of All-Russian scientific research geological Institute and information from published sources. On the basis of the spectroscopy laboratory of St. Petersburg Mining University has been studied the collections of the Ural diamond crystals - 173 pcs., Anabaro-Olenek interfluves - 164 pcs., Brazil - 179 pcs.

The determination of the type and content of nitrogen defects in diamond crystals was carried out by infrared spectroscopy using a FTIR Vertex 70 with IR-microscope Hyperion 1000. The photo luminescence spectra were recorded on a modular Fluorolog 3 spectrofluorimeter by Horiba – Jobin Ivon, equipped with Olympus microscope. Study with the method of optical spectroscopy was carried out on two-beam spectrophotometer UV-2550PC Shimadzu. Processing the results of the analyses the statistical methods have been used (Microsoft Excel, Statistica 7.0). Processing and interpretation of the spectra was carried out with "Spector Examination".

In diamond, the absorption in the range of 1500-3500 cm^{-1} can be used as an internal standard in the study of crystals of small sizes or irregular shapes for the purpose of calculating their thickness and considering the reflection. Spectra of optical density have been normalized at its own two-phonon absorption. The absorption coefficients were taken as internal standard at frequencies of 1973 cm^{-1} and 2500 cm^{-1}, accordingly k_{1973}= 12.5 cm^{-1}, k_{2500} = 4.9 cm^{-1}[2]. Please note that the method of spectra normalization at the optical density in two-phonon absorption now is common and does not require preliminary thickness measurement of the studied sample, which significantly reduces and simplifies the measurement procedure.

The concentration of nitrogen was calculated with the program of visual selection on the standard spectra absorption A, B1 and C, according to the certain coefficients of proportionality [3], [4]. The relative errors of determination of A and B1 defects concentration do not exceed 10% [5], depending on the shape, thickness, presence of inclusions, surface quality of the sample.

3. Research techniques

Ontogenic studies of Beskrovanov V V, who for the first time systematically studied their internal structure on a large and diverse sample of naturally oriented polished plates cut from diamond crystals, allowed us to consider the history of diamond formation within the ontogenic cycle – a set of mineralogical events, including the origin, growth and subsequent changes in crystals [6].

In the diamond crystals, he identified three main ontogenetic areas: central, intermediate and peripheral, which had specific constitutional and anatomical features. Beskrovanov found that for diamonds formed at the intermediate stage, a characteristic feature of anatomy was the appearance of zoning often with sequence of dissolution and regeneration zones. As the most important features of the diamonds of the Ural placers, he considered the curved forms of crystals and the intensive B2 system in IR absorption and assumed that the internal morphology (anatomy) of the Ural diamonds probably does not contain a peripheral region.

The analysis of the literature data on the peculiarities of the appearance of nitrogen defects in diamond, composing the peripheral ontogenic region, indicates the predominance of A-defects in it with a decrease of the nitrogen content in the form of B1 defects to 10-20%. The late generation crystals from eclogitic mantle xenoliths, which are also characterized by a low degree of aggregation of nitrogen defects [9][8], can be considered as a standard of diamond that is typical for this region [8]. IR spectroscopic studies of crystals belonging to different generations from eclogitic xenoliths of the Udachnaya pipe revealed their significant differences in the peculiarities of nitrogen defects (figure 1).

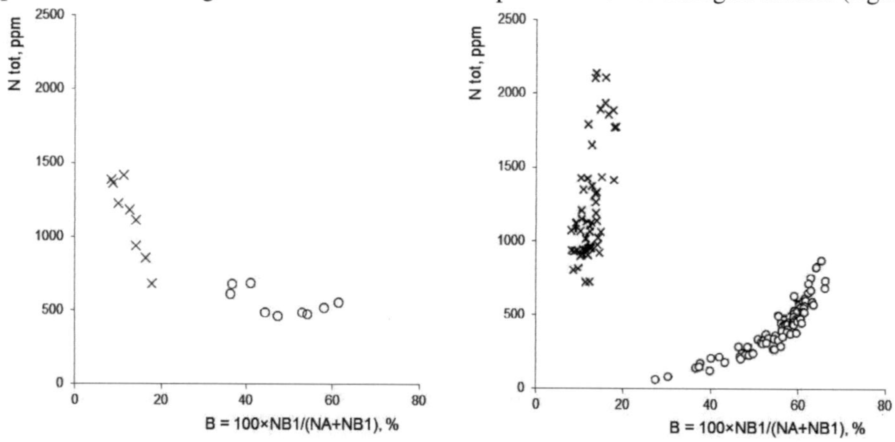

a b

Figure 1. The ratio of nitrogen content and the degree of aggregation of nitrogen: ○– for the first, ×– for the second diamond generation according to table 1 from the work of Spetsius, et al. 2012 (a); from eclogitic xenolith UD 111/02 of the Udachnaya pipe according to Stepanov et al. 2007 (b).

The results obtained by us on the peculiarities of the appearance of nitrogen defects in diamond crystals of the Ural type revealed for internal (figure 2, a) and external (figure 2, b) crystal zones from the deposits of the Urals, Brazil and the Anabar-Olenek interfluve [1] were compared with the results of the determination of these parameters in crystals of two generations from eclogitic xenoliths of the Udachnaya pipe, the late generation of which is considered as a standard for diamond composing the peripheral ontogenic region (figure 2). From the analysis of these diagrams it becomes obvious that the fundamental difference of the whole set of crystals of the Ural type is the absence of crystals or separate zones in it composed by a diamond corresponding to the parameters of nitrogen defects of the peripheral ontogenic region. Consequently, this area was either not presented at all, or it was substantially dissolved at the final stage of their ontogenesis.

The parameters of nitrogen defects in the crystals of the first generation of eclogites fall into the region typical for diamonds of the Ural type (figure 2), which confirms the assumption made by Beskrovanov about the predominance of the intermediate ontogenic region in the internal structure of the Ural diamonds.

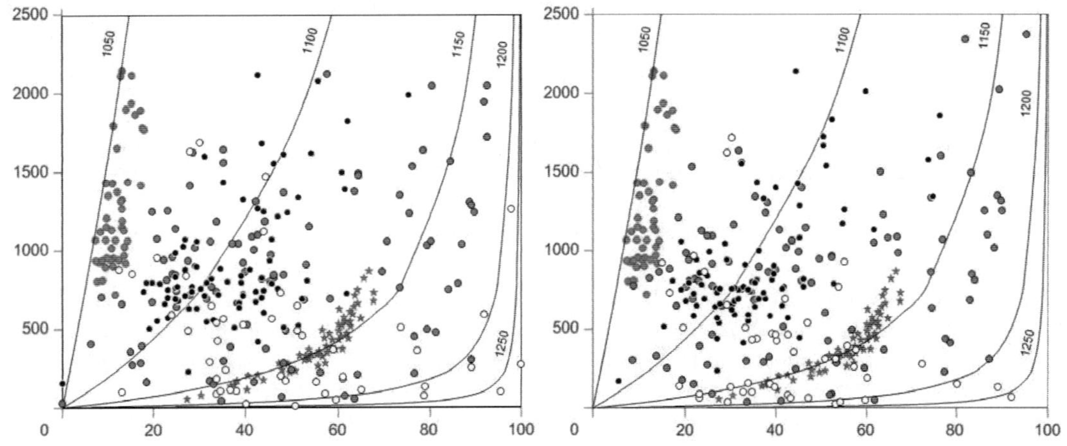

a b

Figure 2. Comparison of parameters of nitrogen defects in diamonds of the 1st (⋆) and 2nd(●) generations from mantle eclogite xenolith (according to Stepanov et al. 2007) and diamonds of the studied placers: • - the Urals; ○ – Brazil ● - Anabar-Olenek interfluve.

Comparison of diamonds of the Ural type with crystals from kimberlite bodies for which the peripheral zone is typical (figure 3), allows assuming that this zone nevertheless was shown in diamonds of the Ural type, but at the final stage of their formation was dissolved. The repeated appearance of dissolution and regeneration process of these diamonds is clearly reflected in their anatomy [6] and morphology [7].

The conducted researches of various material of the Ural type diamonds covering placer deposits of the Urals, Brazil and Anabar-Olenek interfluve confirmed the assumption made earlier about crystal prevalence among them in which the peripheral octahedral zone is absent or poorly shown, but which is the main on a number of fields in kimberlites where composes the most high-quality part of crystals of diamond.

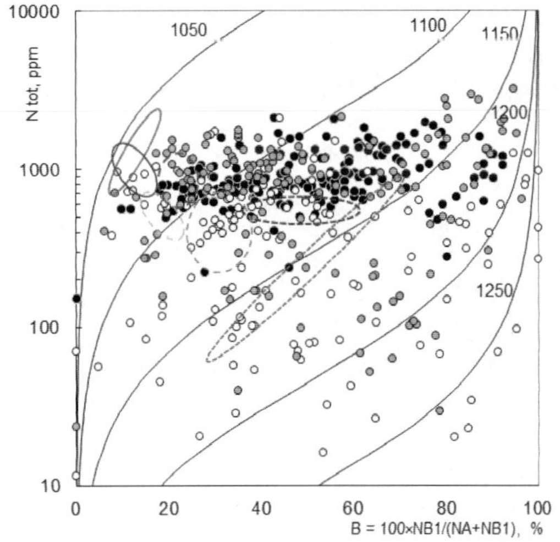

Figure 3. Comparison of the appearance parameters of nitrogen defects of diamonds from different regions: • – the Urals; ○ – Brazil; ● – Anabaro-Olenek interfluve and indigenous sources: brown dotted line – the Arkhangelsk province; yellow dotted line – Malo-Botuobinsky district; green dotted line – Daldyn-Alakit district; red dotted line – diamonds of the first generation; red line – diamonds of the second generation of the eclogite xenolith [9]; blue dotted line – diamonds of the first generation; blue line - diamonds of the second generation of the eclogite xenolith [8].

IOP Conf. Series: Materials Science and Engineering **511** (2019) 012041 doi:10.1088/1757-899X/511/1/012041

4. Conclusions

The comparative analysis of morphological features of diamonds of the Urals, Anabar-Olenek interfluve and Brazil allows to carry them to the same Ural type. They are also united by a wide range of the degree of aggregation of nitrogen defects with a predominance of diamonds with a close ratio of A and B1 defects. The Ural diamonds have a certain specificity, in which there are practically no crystals with a low (less than 500 ppm) nitrogen content; Brazilian diamonds, including crystals containing no A-defects and diamonds of Anabar-Olenek interfluve, some of them have abnormally high (2600 ppm) nitrogen content. Among the Brazilian and Anabar diamonds there are crystals that do not contain B2-defects, which is associated with their destruction at the highest temperature annealing, which is not observed for the Ural diamonds.

A comparison of various indigenous diamond deposits indicates a broad general range as their nitrogen content and degree of aggregation of nitrogen defects, but for the specific indigenous diamondiferous objects the values of these parameters vary over a much smaller range. Therefore, a very wide range of the degree of aggregation of nitrogen defects (0-100%) for crystals of the Ural type does not allow to tie placers containing them with individual indigenous sources of kimberlite-lamproite type. It is necessary to consider the possibility of feeding these placers from many sources, which can be both intermediate reservoirs and various indigenous objects of kimberlite-lamproite type. However, it is difficult to assume the presence in a particular specific and limited area of the alluvial region of the entire spectrum of kimberlite-lamproite bodies, which can provide a wide range of content and degree of aggregation of nitrogen defects typical for diamonds of the considered deposits, and, therefore, it is impossible to exclude the presence of a specific and yet undetected type of indigenous sources, diamonds of which have a very wide range of content and degree of aggregation of nitrogen defects. It is revealed that the principal difference of the whole set of crystals of the Ural type is the absence of crystals or separate zones in them, composed by diamond, corresponding to the parameters of nitrogen defects of the peripheral ontogenic region allocated by Beskrovanov. Consequently, this area was either not presented at all, or it was substantially dissolved at the final stage of ontogenesis. This feature should be taken into account in the development of geological and genetic models of the formation of the Ural type diamonds.

References

[1] Vasiliev E A, Kozlov A V, Nefedov Y V 2013 *A comparative analysis of Anabar, Brazil and the Urals diamonds by method of infrared spectroscopy* (The notes of the Mining Institute) T 200 pp 167-171

[2] Klyuev Y A 1971 *The Intensity of stripes in the IR spectrum of absorption of natural diamonds* (Diamonds) vol 6 pp 9-12

[3] Boyd S R, Kiflawi I, Woods G S 1995 *Infrared absorption by the B nitrogen aggregate in diamond* (Phil Mag B) vol 72 pp 351-361

[4] Boyd S R, Kiflawi I, Woods G S 1994 *The relationship between infrared absorption and the A defect concentration in diamond* (Phil Mag B) vol 69 pp 1149-1153

[5] Khachatryan G K 2003 *An improved method of estimating concentrations of nitrogen in diamond and its application* (Geological aspects of the mineral resource base of "ALROSA". Modern state and prospects of solution) pp 319-322

[6] Lavrenko S A, Korolev I A 2018 *Analysis of Cambrian clay cutting during Saint-Petersburg subway construction* (Gornyi Zhurnal) vol 2 pp 53 – 58

[7] Rakin V I 2013 *Morphology of the Ural type diamonds* (Ekaterinburg) p 396

[8] Spetsius Z V, Kovalchuck O E, Bogush I N 2012 *Properties of diamonds in xenoliths from kimberlites of Yakutia: implication to their origin and exploration* (Materials of 10th International Kimberlite Conference Bangalore) p 5

[9] Stepanov A S, Shatsky V S, Zedgenizov D A, Sobolev N V 2007 *The causes of diversity of morphology and impurities of diamonds from eclogites of the Udachnaya pipe* (Geology and Geophysics) vol 48 № 9 pp 974-988

The dynamic identification of the technical condition of pipelines on the basis of the analysis of the temporal characteristics of electromagnetic-acoustic signal

M G Bashirov, E M Bashirova and I G Khusnutdinova

Ufa State Petroleum Technological University, Branch of the University in the City of Salavat, Russian Federation

E-mail: ilvina011@mail.ru

Abstract. To assess the resource of safe operation of process of technological pipelines, it is proposed to control the technical condition of the pipeline metal using the method of dynamic identification. The method is based on the analysis of the coordinates of the roots of the denominator of the transfer function on the complex plane. The transfer function of the system "object of control-electromagnetic-acoustic transducer" is an integral quantitative parameter characterizing the state of the metal.

Introduction

Ensuring the safety of technological processes of the oil and gas industry is determined by the reduction of failures of technological pipelines, which make up a third of all equipment failures at the enterprises of the industry. During the entire period of operation, the process of technological pipelines are under the influence of aggressive conditions and mechanical stresses, which leads to a change in their technical condition [1], [2].

At the moment, the assessment of the technical condition and service life of technological pipelines is made on the basis of the results of mechanical tests of metal samples cut from the existing equipment, or using physical methods of non-destructive testing. The actual problem of evaluation is the ability to identify defects at an early stage of their development. [3-9]. Existing methods for assessing the stress-strain state have their advantages and disadvantages. A promising direction in the assessment of stress-strain state of equipment and structures is the use of electromagnetic-acoustic effect. Currently, the EMA effect is used for thickness measurement and developed detection of defects such as discontinuities. Separate works are devoted to the detection of degradation of metal properties at an early stage and tensometry using the EMA effect [10], [11]. The questions of assessing the resource of safe operation of technological pipelines using the EMA effect are poorly understood and studied, which necessitates further research.

Materials and methods

Experimental studies were conducted in the laboratory. Experimental studies have allowed to study the degradation processes, the impact of operational loading on the change of electrophysical properties of metal elements of technological pipelines. Since the elements of the process pipeline operate under different loading schemes, the experimental studies were made in two directions: under static loading

IOP Conf. Series: Materials Science and Engineering **511** (2019) 012042 doi:10.1088/1757-899X/511/1/012042

(tension) and under cyclic loading. As the objects of research we used samples of steel grades 10, St3 and 09G2S, the most widely used in the manufacture of process of technological pipelines in the oil and gas industry. To test the samples for low-cycle fatigue at pure bending (figure 2, c), the installation developed by the Professor of the Department "Equipment of petrochemical and oil refining enterprises" of the Ufa State Oil Technical University Gaziev R. R. (patent No. 2262682 of the Russian Federation) was used [12]. The basic parameters of the inductor and the design of the electromagnetic acoustic transducer (EMAP) (figure 1), a General view of the experimental facilities (figure 2). The signals received at the output of the system "control sample-electromagnetic-acoustic transducer" are shown in figure 3. In figure 3a shows the signal at the output of the EMAT, in figure 3, b EMAT signal passed through the filter.

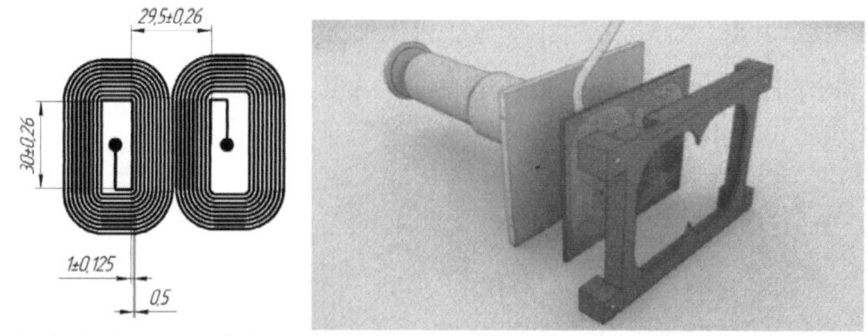

Figure 1. EMA Inductor of the transducer and electromagnetic acoustic transducer (EMAP).

a) b) c)

(a) the static tensile tests using a test machine Walter + Bai; (b) the installation for the study of complex stress-strain state of the pipeline model in the form of a thin-walled cylindrical vessel; (c) the external view of the installation for test of samples for cyclic loads

Figure 2. General view of experimental facilities.

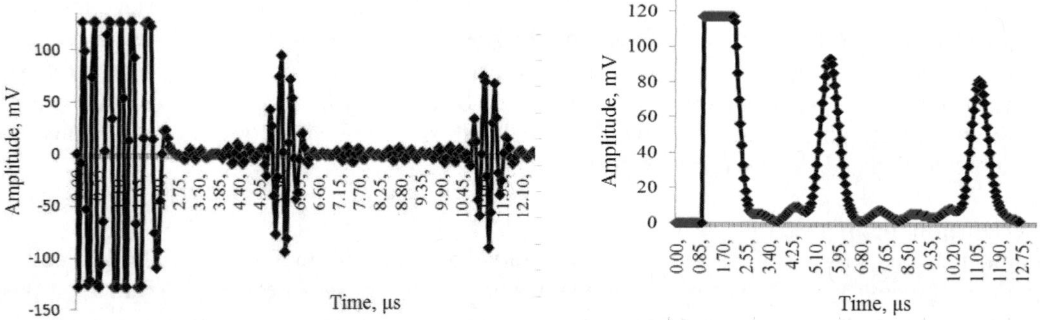

a) b)

(a) sounding and bottom signals at the output of the EMAT; (b) bottom probe and EMAT signals after passing through a coherent filter

Figure 3. Time sweep of EMA signals.

IOP Conf. Series: Materials Science and Engineering **511** (2019) 012042 doi:10.1088/1757-899X/511/1/012042

Experimental research

The developed software processes the signal of the electromagnetic-acoustic transducer, using the first reflected (bottom) signal, which is the most informative and easy to detect. The temporary diagram of the electromagnetic acoustic signal of the transducer is shown in figure 4. In figure 4, the first reflected signal is located in the range from 5.55 to 7.25 μs.

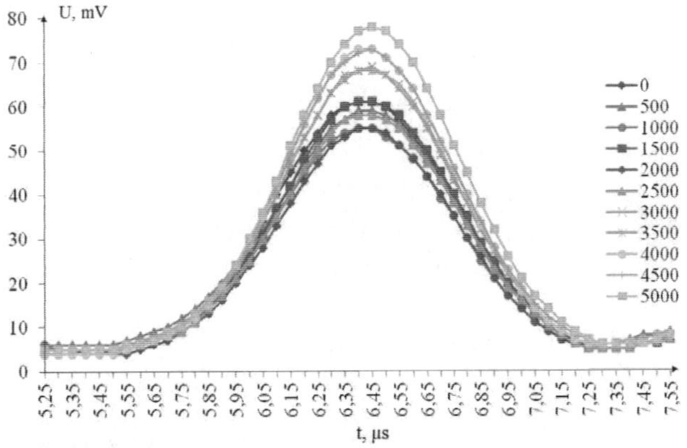

Figure 4. Changing the bottom signal with increasing load on the sample from 0 to 5000 kilogram-force 09G2S grade steel.

The developed program by the method of areas identifies the transfer function of the system "object of control-electromagnetic-acoustic transducer" calculates the roots of the characteristic equation of the system-the polynomial of the denominator of the transfer function [13], [14]. In General, the transfer function W(p) is defined as the ratio of the image of the output signal y(p) to the image of the input signal x (p). For the system" EMAP – object of control " the signal of external action x(P) is a joint action on the object of control of the alternating magnetic field of the current in the inductor and the magnetic field of the permanent magnet, the response signal of the object Y (P) is the EMF induced in the EMAP inductor [14]. To determine the transfer function of the system and search for the roots of the characteristic polynomial, the TAU 2.1 program developed at the Department of "Automation of technological processes and production" of the Ufa State Oil Technical University was previously used. For a more precise definition of the transfer function the calculation was made by the method of moments in the program developed by Professor G. K. Ayasan.

Results

After a series of tests by loading steel samples, the dependence between the applied force and the values of the roots of the characteristic equation was revealed. The obtained dependence is shown in figure 5. The real parts of the roots of the characteristic equation are deposited along the abscissa axis, the imaginary parts of the roots of the characteristic equation are deposited along the ordinate axis. It is possible to estimate the current technical condition and resource of safe operation of the process of technological pipeline of a hazardous production facility by the location of the roots of the characteristic polynomial (denominator) of the transfer function on the complex plane (figure 6). When the state of the metal changes, the coordinates of the roots of the characteristic equation of the transfer function also change.

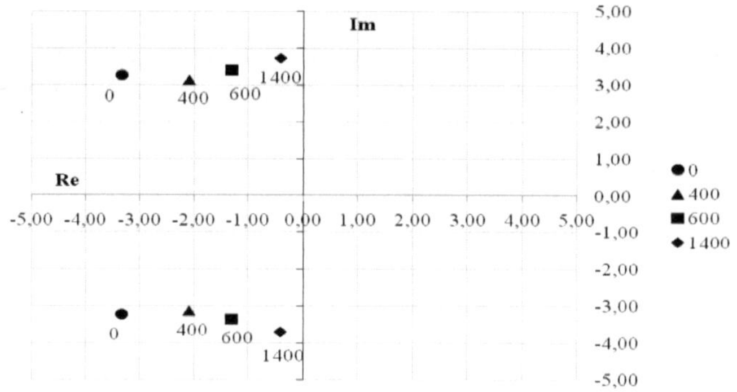

Figure 5. The location of the roots of the characteristic equation under the action of cyclic loading (09G2S, 0-1400 cycles).

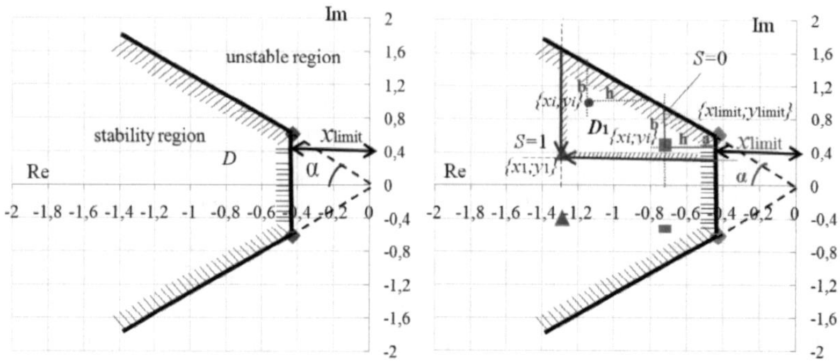

a) b)

a-the area of stability of the pipeline D; b-the initial area of the operational state of the pipeline D1

Figure 6. The assessment of the current technical condition and resource of the pipeline by the coordinates of the roots of the characteristic polynomial.

Conclusion

The studies established the relationship between the quantitative values of the roots of the polynomial transfer function of the system "object of control - electromagnetic acoustic transducer" and between the level of the stress-strain state of metal [14]. The results of the research can be used to develop a methodology for assessing the service life of the technological pipelines safe operation to ensure industrial safety.

References

[1] Sun B, Wang J, Li Z-L, Qin Y, Liu S-Z and Tomker D 2017 Modified modal strain energy method for analyzing the dynamic damping behavior of constrained viscoelastic structures *Journal of Engineering Science and Technology Review.* vol **10 5** (Kavala, Greece: Thrace Institute of Technology) pp 174–180

[2] Bashirov M G and Mironova I S 2015 Integral criteria development for technical state and lifespan assessment of machine assembly in oil and gas industry *SOCAR Proceedings.* No **1** (Baku, Azerbaijan: OilGas Scientific Research Project) pp 46–55

[3] Bashirov M G, Kuzeev M I, Kuzeev I R and Yunkin A I 2004 Electromagnetic method of diagnosis in problems of operation safety assurance and evaluating the service life of equipment at petrochemical and petroleum-processing plants *Bezopastnost' Truda v Promyshlennosti.* (Moscow: Industrial safety) pp 49–54

[4] Ibragimov I G and Vildanov R G 2007 Measuring stresses in welded joints by the remagnetisation loss method *Welding International* vol **21** **2** (United Kingdom: Woodhead Publishing Limited) pp 139–141
[5] Ibragimov I G and Vildanov R G 2005 Measurement of stresses in metallic structures by method of reversal magnetization losses *Zavodskaya Laboratoriya. Diagnostika Materialov* vol **71** **2** (Moscow: Test-zl) pp 29–31
[6] Vil'danov R G 2004 A study of a remagnetization loss sensor *Measurement Techniques* vol **47** **3** (Moscow: Springer New York Consultants Bureau) pp 259–263
[7] Poyarkova E V, Kuzeyev I R 2017 Multifractal parametrization of welded joints structure as an instrument of diagnostics and detection of defects *Solid State Phenomena* vol **265** SSP (Switzerland: Trans Tech Publications Ltd) pp 60–64
[8] Kovshova Y S, Kuzeyev I R, Naumkin E A, Makhutov N A, Gadenin M M 2015 The influence of quasi-static loading regimes on the strength of vessels operating under pressure *Inorganic Materials* vol **51** **15**. (Moscow: Pleiades Publishing, Ltd.) pp 1502–07
[9] Kuzeyev I R, Naumkin E A, Tepsayev A N, Samigullin A V, Gafarova V A 2015 Generation of the response electromagnetic signal in the process of destruction of shell structures *SOCAR Proceedings* (Baku, Azerbaijan: OilGas Scientific Research Project) pp 75–80
[10] Strizhak V A, Pryakhin A V, Balabanov E N, Obukhov S A and Efremov A B 2012 Features of the construction of the sensor for the emission and reception of acoustic pulses of electromagnetic-acoustic method for controlling stress-strain state of the wheel rim *Magnetic phenomena* (Izhevsk: IzhSTU) pp 164–177
[11] Muravyev V V and Balabanov E N 2012 Features of the acoustic strain measurement method in metal products *Proceedings of the XIII All-Russian school-seminar on the physics of condensed matter* (Ekaterinburg: IMP UroRAN) p 190
[12] Gaziyev R R, Zakharov N M, Kuzeyev I R and Nasibullin R I Patent No. 2262682 of the Russian Federation. Installation for testing samples for low-cycle fatigue at pure bending. Declared. 03.02.2004; publ. 20.10.2005
[13] Bashirova E M 2005 Evaluation of the limiting state of the metal equipment for processing of hydrocarbonic raw materials with application of an electromagnetic method of control *Dissertation ... candidate of technical Sciences*: 05.26.03, 05.02.01. (Ufa) p 140
[14] Khusnutdinova I G and Bashirov M G 2017 The use of electromagnetic-acoustic method for estimating the stress-strain state of the metallic elements of power equipment *Key Engineering Materials* vol **743** (Switzerland: Trans Tech Publications Ltd.) pp 463–467

Ultrasonic testing method for quality control of mold castings

I O Bolotina[1,a], D A Sednev[1,b] and V A Portenko[2,c]

[1]634050, Russia, Tomsk, Lenin Ave., 30, Tomsk Polytechnic University
[2]634050, Russia, Tomsk, Lenin Ave., 36, Tomsk State University

[a]bolotina@tpu.ru, [b]sednev@tpu.ru, [c]viktoria98@sibmail.com

Abstract. The paper presents the review of one of the promising testing methods for quality control of mold castings that is ultrasonic inspection. Pulse-echo method, through transmission method and phased array technique are considered. The advantages and limitations for the analyzed methods are discussed. The produced by foreign and national companies aperture is reviewed.

1. Introduction

The share of acoustic methods in the volume of nondestructive testing comprises more than 60% in relation to other testing techniques [1]. This is conditioned by a number of advantages pertaining to these methods such as: high speed, accuracy of measurement, high sensitivity, inspection with one side access to the object, safety for the operator`s health, option to use portable ultrasonic testing equipment, in-service testing during technological processes, capacity to transmit ultrasonic waves into moving objects avoiding their direct contact with transducer.

The analysis of research published [2-4] demonstrates that the systems based on application of ultrasonic testing method are more promising among the nondestructive testing systems for mold castings as these systems possess the highest sensitivity and are highly self-descriptive when it comes to characteristic technological and operational defects in metal. The obtained using this testing method data about the defect`s size, square and spacing enables quality visualization for subsequent assessment of capability and service life of a metal part. The obvious advantage of ultrasonic testing application is the potential to build automated systems applying quite acknowledged in practice solutions [5-7]. In this respect it is possible to state that acoustic methods and systems are the most promising for nondestructive inspection of metal parts quality.

However, in spite of sufficiently wide application of acoustic methods for nondestructive testing, they possess limitations such as low detectability of single volume defects, low efficiency of testing materials with great oscillation damping.

From a wide range of acoustic methods available, pulse-echo method, through transmission method and phased array technique are chosen for testing metal parts.

2. Pulse-echo method

The principle of the pulse-echo method is that during electric pulse generation the ultrasonic transducer generates oscillations transmitted to the object under control. The same transducer receives echo-signals reflected from defects [8]. This method is widely used in industry due to its simplicity as only one transducer is needed to perform testing of an object. In this case there is no necessity to use specific

IOP Conf. Series: Materials Science and Engineering **511** (2019) 012043 doi:10.1088/1757-899X/511/1/012043

constructions for location and alignment of acoustic axes which is required if two transducers are to be used [9]. The ultrasonic nondestructive method also allows quite accurate detection of location, depth of occurrence and spacing of the defect in the object under control in relation to the transducer. The method can be used with one side access to the object under control.

To the main characteristics of pulse-echo method pertain the following: the maximum penetration depth, the minimal dead zone, high sensitivity, resolution, location accuracy and productivity.

In addition to the advantage of one side access, the pulse-echo method possesses the highest sensitivity to detect internal defects as well as reliable accuracy for defects` spacing detection. Judging by the amplitude value of the signal reflected it is possible to detect the defect size, while judging by the spectral composition of the impulse reflected it is possible to detect the defect type and shape.

The diagram of ultrasonic wave front passing along the depth of the object under control is given in Figure.1. Figure. 1, a demonstrates the zone without the defect, while Figure 1,b demonstrates the zone with the defect which re-echoes the signal.

The ultrasonic impulse path occurs from the source to the reflector and backwards and constitutes:

$$2D = C^*t_з, \quad (1)$$

where $t_з$ – the delay time of the echo pulse arrival in relation to the initial one; D – the distance from the defect to the source; C – the ultrasonic wave speed in the object under control.

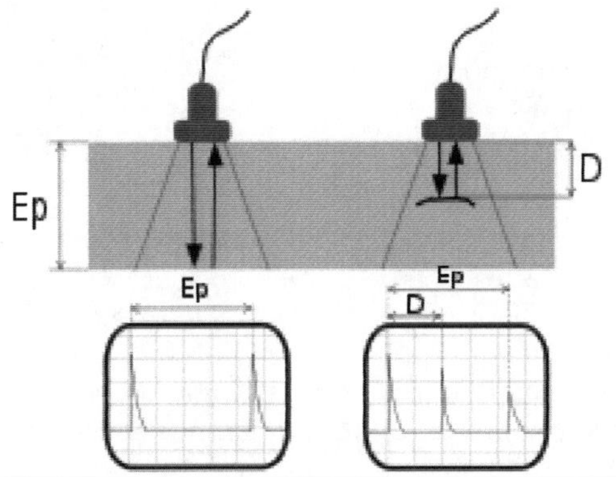

Figure 1. The diagram of the ultrasonic wave passing through the object under control, pulse-echo method used: a) without the defect; b) with the defect.

Pulse-echo method equipment enables to determine the defect behavior, classify defects according to their size, shape and spacing.

The leading world manufacturers of ultrasonic equipment produce instruments for metal parts testing within their product range. *Sonotron NDT* produces ultrasonic defectoscope *ISonic AUT 16* which allows for metal parts testing (Figure 2) [10].

The flaw detector is the portable multichannel instrument intended for high speed automated ultrasonic testing machines and monitoring systems. The instrument contains 16 independent channels, each channel comprising its own pulser-receiver, and 100 MHz sampling rate signal digitizer. Each channel may drive either single element probe or probe pair of phased array.

The scanning zone visualization can be formed in A-, B-,C- and D-Scan, 3D-View modes, etc. Significant increase of scanning speed may be achieved due to various specific probe pairs of phased arrays; each of these is toggled via its separate gate to one control module (Figure 3).

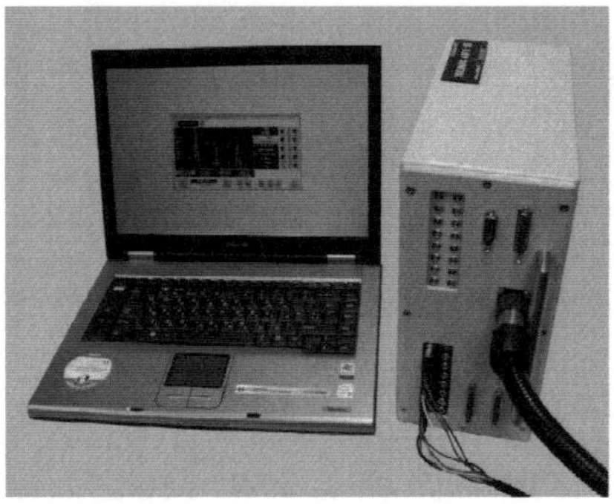

Figure 2. Ultrasonic defectoscope *Sonotron ISonic AUT 16.*

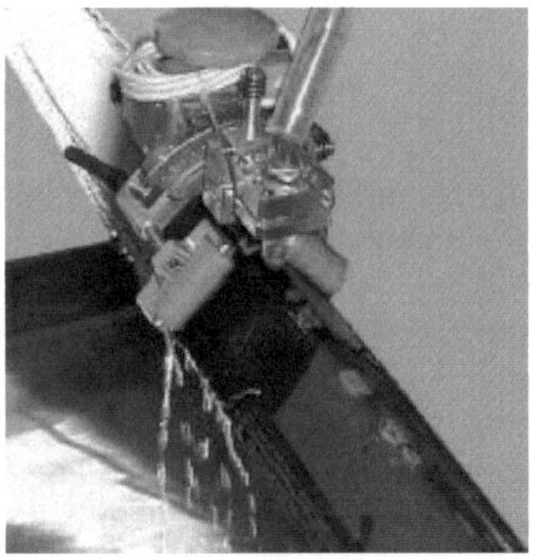

Figure 3. Ultrasonic defectoscope *Sonotron ISonic AUT 16* in service.

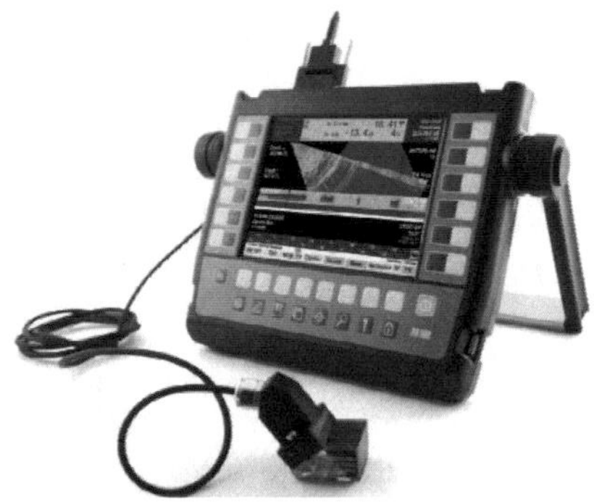

Figure 4. Ultrasonic defectoscope *DIO 1000 PA.*

STARMANS electronics, s.r.o. produces ultrasonic defectoscope *DIO 1000 PA* (Figure 4). Image visualization is formed in various scan modes (A, B, C, S and W - scans) in this defectoscope. Scanning speed increases significantly when using phased array. Phased arrays may be used in manual as well as automated mode (data array storage output in the editing C-scan mode) using the scanners. The application for this instrument is testing products of metals, composites and plastic.

3. Through transmission method

Through transmission method of nondestructive testing is the method where two ultrasonic transducers 2 and 3 are used (Figure 5). The transducers are located on both sides of the object under control 1 along the same axis. Transducer 2 excites ultrasonic waves while transducer 3 receives them. If the defect 4 is present, the amplitude of the ultrasonic signal received reduces significantly or the ultrasonic signal is completely lost due to the defect casting acoustic shade.

Through transmission method for testing metal materials enables detecting more fine defects throughout great thickness of parts in comparison to pulse-echo method. It is based on caused by defects attenuation of elastic oscillations of ultrasonic frequency [11]. Through transmission method is applied to determine layering, foreign inclusions, holes and other defects.

For through transmission method to function efficiently there has to be two side access provided to the object under control. The method limitation is its complexity. First of all, it is related to the problem of two transducers orientation in relation to their central beams of the directional pattern, impossibility for accurate assessment of defects location and low sensitivity (up to 20 times) in comparison with echo methods.

To the advantages of the through transmission method pertain the following: minor dependence of signal amplitude on the material discontinuity location, high noise immunity and absence of dead zone.

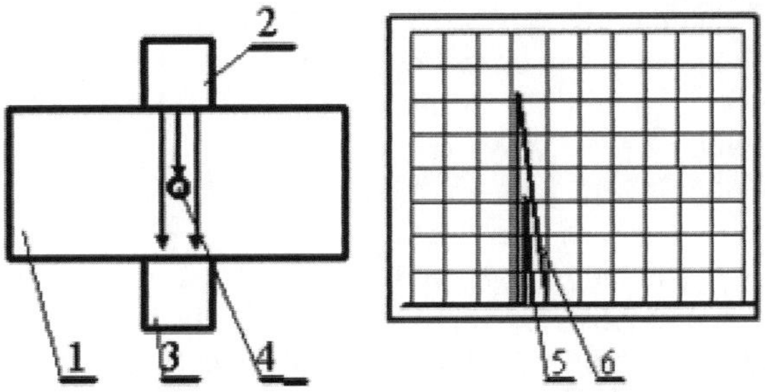

Figure 5. The diagram of ultrasonic wave passing along the thickness of the object under control when using through transmission method:
1– the object under control; 2, 3 – transducers: the source and receiver
4 – the defect; 5, 6 – bed signals with and without the defect.

Figure 6. Ultrasonic defectoscope *DryScan 410* in service, through transmission method is used for carbon products inspection.

Based on the through transmission method a number of low frequency ultrasonic defectoscopes produced in Russia are available: UD4-76 [12], UD-22 UM [13], UD2H-PM [14], as well as defectoscopes produced abroad are available: USN-60 [15], RapidScan2 [16], DryScan 410 (Figure 6) [17], and MasterScan 380M [18]. These defectoscopes use special probes: low frequency direct single probes, dual-mode probes, inclined single, single-dual and immersion probes.

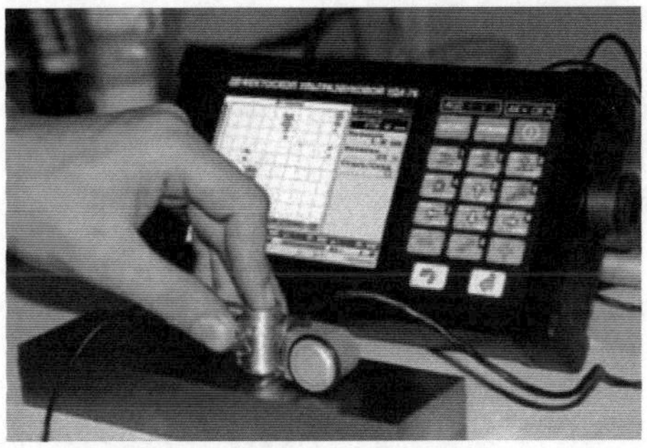

Figure 7. Ultrasonic defectoscope *UD4-76*.

National industry also produces a range of ultrasonic defectoscopes of UD series. For instance, the main purpose of ultrasonic defectoscope UD4-76 (Figure 7) is to detect failures in uniformity and continuity of materials in end products, green parts and welded (soldered) joints.

IOP Conf. Series: Materials Science and Engineering **511** (2019) 012043 doi:10.1088/1757-899X/511/1/012043

UD4-76 enables visualizing and storage of testing results in B-scans mode in relation to the transducer location during scanning.

In spite of through transmission method` theoretical studies and practical applications in ultrasonic testing being known for quite a while [19], its application was limited by only single acoustic path – single source and single receiver.

4. Phased array technique

The main advantage of this technique is high productivity of testing in comparison to single element transducer. The ultrasonic tomography image reconstruction is based on imaging due to reflection of elastic waves from material discontinuity which is equivalent to obtaining C-scan. Application of tomographic reconstruction using equally spaced linear phased arrays allows to determine spacing of several defects located in the inspection zone, with sufficiently high resolution.

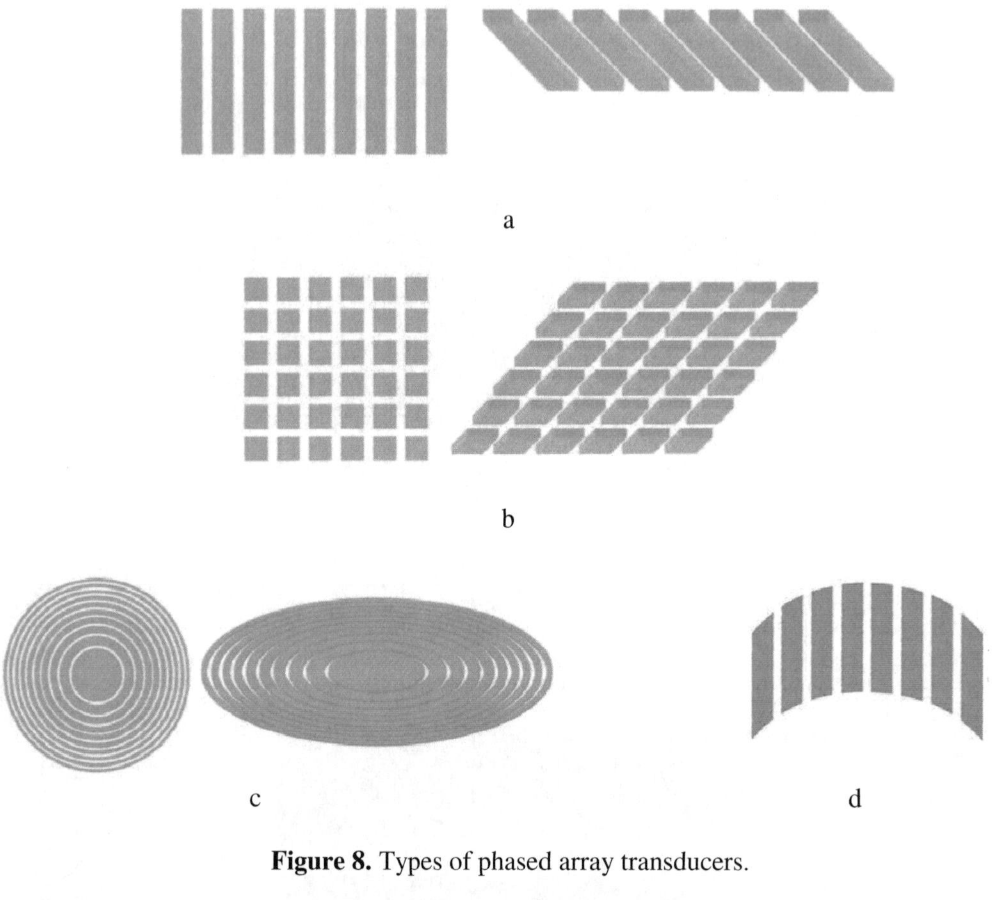

Figure 8. Types of phased array transducers.

Application of phased arrays allows to determine material discontinuity variously oriented to the acoustic axis. Single element transducer is highly likely to miss defects which are located aside of the acoustic beam under great angle to the acoustic axis of the transducer [20].

Depending on the number and location of piezoceramic elements, phased arrays fall into the following types:

1. Linear arrays – a single sequence of elements, usually formed by cutting a large cuboid piezoceramic plate. The beam is steered in single dimension (Figure 8,a).
2. Matrix arrays – the elements are placed as lattice which also can be curved. The beam is steered in three dimensions (Figure 8,b).

3. Circular arrays – the elements represent circles typically formed by cutting from piezoceramic plate (Figure 8,c).
4. Sector (convex, concave) arrays – a single sequence of elements which is curved to form the required beam shape or in compliance with the geometry of the object under control. The beam is steered in single dimension (Figure 8,d).

There are three main methods to steer the phased array beam:

- Electron scanning: single focal law is switching within the group of elements. Scanning is performed under the constant angle and along the phased array larger lobe. This process is equivalent to mechanical movement of traditional single element transducer.
- Dynamic focusing along the depth (along the acoustic axis): scanning is performed by changing the focal length. When applied the same impulse is used for irradiation while the array is refocusing subsequently to various depth sizes in the receiving mode.
- Sector scanning (also called azimuth scanning or angular scanning): irradiation is performed by the same group of elements keeping the same focal length, however subsequently under various angles.

The instruments of Acoustic Control Systems – ACS Group (Moscow) can be named to illustrate this method application, the company produces a range of ultrasonic tomographs for various applications:

✓ *A1040 MIRA* low frequency ultrasonic tomograph [21];
✓ *A1050 PlaneScan* low frequency ultrasonic scanner-topographer [22];
✓ *A1550 IntroVisor* high frequency ultrasonic flaw detector-tomograph [23].

For instance, *A1550 IntroVisor* is ultrasonic flaw detector-tomograph with digital phased array focusing and tomographic reconstruction of the visualized zone. It is designed for defects detection in metals and plastic. The instrument provides real-time section imaging.

To the advantages of ultrasonic tomography pertain the following:

- Real-time flaw detection in material discontinuity in metals, plastic and composite objects with efficient visualization of the received data;
- Velocity and efficiency;
- Real-time imaging in tomograph mode with high information refresh rate;
- Greater phased array aperture enabling weld control without shear scanning;
- High frame rate, at that high scanning speed along welded joint up to 50 mm/sec is enabled.

A1550 IntroVisor ultrasonic flaw detector-tomograph parameters are given in table 1 below.

Table 1. The main *A1550 IntroVisor* ultrasonic flaw detector-tomograph parameters.

Parameter Name	Value
Permissible number of phased array elements	16 ÷ 64
Curvature	transversal, longitudinal
Image reconstruction rate	15 ÷ 35 frames per second
Tomograph size, pixels	256 × 256
Section reconstruction size, mm	±256 × 256 horizontally, in depth
Readings discreteness, mm	0,1 - 1
Operating frequency, MHz	1,0; 1,2; 1,5; 1,8; 2,0; 2,5; 3,0; 4,0; 5,0
Permissible ultrasonic sound velocity, m/s	from 300 to 15000
Display type	TFT SVGA 640x480
Operating temperatures	from –10°C to +45°C
Autonomous work duration	up to 8 hours

All acoustic tomographs of Acoustic Control Systems – ACS Group operate in pulse-echo mode only thus enabling tomograms based on echo signals only.

At present there is a range of produced abroad automated systems to perform ultrasonic testing of metal structures, namely, *Tecnatom* equipment (Spain) which uses echo method with ultrasonic wave input using coupling fluid (Figure 9).

a b

Figure 9. *Tecnatom* equipment [24]: a –through transmission method; b – echo method

In spite of high accuracy and high quality control the equipment possesses some limitations:

- insufficient self-descriptiveness of the testing results to detect potential defect-related risks (during in-service inspection);
- applied, as a rule, with elastic vibrations input using coupling fluid.

5. Conclusion

The existing sufficient range of testing techniques for defects detection in metal parts allows to perform practically complete inspection of products. One of the most promising methods is ultrasonic testing method which possesses a number of advantages in comparison with other methods. However, a significant limitation of the available on the market equipment for ultrasonic testing is great labor costs due to low testing speed. Hence, development of the equipment which would integrate high resolution and high scanning speed to perform inspection is quite relevant nowadays.

It is expedient to use automated systems for acoustic testing method to perform testing of metal mold castings during production as well as at all stages of product life cycle. Application of phased array technique enables to ensure high detectability of defects, high self-descriptiveness of the testing results and identification of the visualized flaws topology.

Acknowledgements

The study is financially supported by the Ministry of Science and Higher Education of the Russian Federation within the Federal Target Program entitled "R&D in Priority Areas of Development of Russia's Scientific and Technological Complex for 2014–2020" (unique project ID: RFMEFI57817X0251).

References

[1] Aleshin N P, Grigoriev M V and Schipakov N A. 2015 Modern Equipment and Technologies of Nondestructive Testing of Composite Polymer Materials Engineers Bulletin vol 1 pp 532-538

[2] Budadin O N, Kuryutin V Yu and Borisenko V V 2007 Automated Ultrasonic Nondestructive Testing of Polymer Products Control Diagnostics Vol 4 pp 19-23

[3] Kablov E, Murashov V and Rumyantsev A 2006 Diagnostics of Polymer Composites by Acoustic Methods *ULTRAGARSAS* vol 2(59) pp 7–10

IOP Conf. Series: Materials Science and Engineering **511** (2019) 012043 doi:10.1088/1757-899X/511/1/012043

[4] Murashov V V, Kosarina E I and Generalov A S 2013 Quality Inspection of Avionics Parts from Polymer Composites and Multi-layer Sandwich Constructions Aircraft Materials and Technologies vol 3pp 65–70
[5] Berke M and Büchler J 2006 Ultrasonic Imaging in Automatic and Manual Testing (Proc. Int. Conf. ECNDT– We.3.1.5)
[6] Bolotina I O, Borikov V N, Ivanova V S, Mertins K V and Uchaykin S V 2018 Application of Phased Antenna Arrays for Pipeline Leak Detection Journal of Petroleum Science and Engineering vol 161 pp 497-505
[7] Bolotina I O, Bulavinov A N, Pinchuk R A and Salchak Y A 2016 Principles of Construction of Ultrasonic Tomographs for Solution of Problems of Nondestructive Testing in Mechanical Engineering (IOP Conf. Series: Materials Science and Engineering) vol 124 012149 pp 1-7
[8] Samokrutov A A and Shevaldyukin V G 2007 Ultrasonic Pulse-echo Tomography of Metal Constructions. State and Tendencies Plant Laboratory. Materials Diagnostics vol 73 1 pp 50-59
[9] Kretov E F 2007 Ultrasonic Defectoscopy in Power Plant Engineering (Saint Petersburg. Sven) 296 p
[10] Ultrasonic Inspection Systems. Sonotron ISonic AUT 16 Available at URL: http://ndtsupply.com/sonotron-isonic-aut-16/32
[11] Krautkrämer J and Krautkrämer H *Ultrasonic Testing of Materials. Handbook* (M: Moscow) p 752
[12] Ultrasonic Flaw Detector Tomograph UD4-76 Available at URL: https://www.ndtprompribor.ru/product/ultrasonic-testing/ud4-76-ultrasonic-flaw-detector-tomograph
[13] Ultrasonic Equipment Available at URL: http://www.kip-guide.ru/info/5744-76
[14] KROPUS Research and Production Enterprise Ultrasonic Equipment Available at URL: http://www.kropus.ru/products/usound/ud2np.php
[15] Helling GMBH Instruments Available at URL: http://www.geo-ndt.ru/pribor-203-yltrazvykovoi-defektoskop-usn-60.htm
[16] Sonatest C-scan inspection instrument RapidScan2 Available at URL: http://pdf.directindustry.com/pdf/sonatest-ltd/rapidscan-2/21671-328147.html
[17] Nondestructive Testing Systems Research and Development Enterprise Instruments Available at URL: http://pvp-snk.ru/ultrazvukovoy-defektoskop-dryscan-410
[18] PANATEST, LLC Equipment Available at URL:http://www.panatest.ru/items?id=100030
[19] Schribre D S 1965 Ultrasonic Defectoscopy (M: Mettalurgy) p 416
[20] Dube N 2007 Introduction to Phased Array Ultrasonic Technology Applications (Olympus) p 376
[21] Acoustic Control Systems – ACS Group Instruments. A1040 MIRA Portable Handheld Ultrasonic Tomograph Available at URL: http://acs-international.com/product/a1040-mira/
[22] Acoustic Control Systems – ACS Group Instruments. A1040 MIRA Low Frequency Ultrasonic Tomograph Available at URL: http://www.acsys.ru/production/detail/a1050-planescan/
[23] Acoustic Control Systems – ACS Group Instruments. A1550 IntroVisor High Frequency Ultrasonic Flaw Detector-Tomograph Available at URL: http://acs-international.com/product/a1550-introvisor/
[24] Tecnatom Inspection Services and Equipment Available at URL: https://www.tecnatom.es/en/

Use of nanostructured materials for the sorption of heavy metals ions

M T Gabdullin[1,2,a], K K Khamitova[1,b], D V Ismailov[1,c], M N Sultangazina[1,d], D S Kerimbekov[1,e], S S Yegemova[3,f], A Chernoshtan[3,g] and D V Schur[4,h]

[1]National nanotechnology laboratory of open type of Al-Farabi Kazakh National University, Almaty, Kazakhstan
[2]Kazakh – British Technical University, Almaty, Kazakhstan
[3]Research Centre of Ecology and Environment of Central Asia, Almaty, Kazakhstan
[4]IPM Frantsevich, Kiev, Ukraine

[a]gabdullin@physics.kz, [b]khamitovakorlan@gmail.com, [c]ismailov_daniyar_v@bk.ru, [d]marjan_0309@mail.ru, [e]daurenks@bk.ru, [f]s.yegemova@gmail.com, [g]nas.lucky.96@mail.ru, [h]dmitry.schur@gmail.com

Abstract. Recently, carbon materials containing fullerenes have been used as sorbents. The purpose of the study was to explore and compare the sorption properties with respect to heavy metal ions of fullerene-containing materials. This will allow to obtain in the most effective way the fullerene-containing sorbents for use in processes of cleaning of environmental objects. Several types of sorbents obtained from shungite and graphite were selected for our research. For a month, the soil had been contaminated with the lead and zinc salts in a concentration exceeding the maximum permissible value of 30 times. Sorbents had been then added in a ratio of 1: 5. A week later, the soil extracts were analyzed for salt and pH content. The results indicated that fullerene-containing sorbents bind a significant part of pollutants and reduce the toxicity of the soil. Nevertheless, complete purification from metals was not observed. The obtained results are the basis for the further studying of sorption properties of carbon nanomaterials and are of practical importance in the field of nanotechnology, ecology, ecotoxicology, chemical technology of inorganic substances and others.

Introduction

Over the last 100 years, the speed of development of technology and industry has increased in several times, which has led to a tremendous growth in the number of enterprises around the world. It is obvious that with increasing volumes of production of all kinds of goods, the technogenic impact that humanity exerts on the environment grows, which leads to a number of environmental changes, some of which have an impact not only on the local level but also in the world.

One of the results of an active industrial impact in a region is the accumulation of heavy metals into the environment, which content exceeds the maximum of permissible level in many times. Most often, this problem is associated with inadequate technologies and methods of cleaning of industrial waste, which, is released into the environment, and mainly into water and soil, and then, as a result brings to the contamination. The regions are mostly exposed to the heavy metal contamination including cities and large industrial areas due to the activity of thermal power stations and road transport. Particularly damaging effects are caused by ash dumps formed as a result of the operation of the CHP plant, since the ash can be carried by the wind over many kilometers around, having a detrimental effect on water structure, rivers and soil

Heavy metals are one of the most harmful types of pollution and, according to the state standards, they are ranked as second in terms of the degree of danger, staying in one line with pesticides. These include elements such as vanadium, germanium, manganese, fluorine, nickel, zinc, molybdenum, strontium, mercury, lead, etc. – all in all more than 40 elements of periodic table.

At present, the negative impact of heavy metals is not fully taken into account in comparison with radioactive substances. Therefore, in the future they can become more dangerous due to excessive accumulation in the environment and can destabilize the ecosystems of different territories and then the planet itself.

Soil is mainly encountered medium in which heavy metals are absorbed, as well as atmosphere and hydrosphere. It serves as a source of secondary pollution entering from the surface air and water into the ocean. From the soil, heavy metals are assimilated into the plants, then they are found in food and in the light of an increase in the anthropogenic effects on soils, the question of quick and effective methods for cleaning soils from various pollutants is topical and justified. Therefore, great attention is paid to the search for universal means for cleaning soils from polluting substances based on both natural and technogenic origin.

The volume of wastewater formation at enterprises and the number of harmful components contained there define the requirements for clearing methods. These methods should be located in the first place on technological and hardware registration, providing increase of degree of sewage treatment. At the same time, the initial raw materials used in the course of extraction of harmful components should be effective enough, extended and have a low cost. The methods of sewage treatment with natural sorbents find their distribution at the preliminary and final stages of the technological purification chain. They are also used as separate methods.

The purpose of scientific work is research the effectivity in sorption of ions of heavy nonferrous metals from water solutions of the new carbon – mineral sorbents of a natural origin. The final stage of the research is to find the sorbents with the mostly effective sorption properties. They should interact weakly with water molecules and well with organic substances, being relatively large-porous. During the short time being in contact with water, they must have a high sorption capacity, high selectivity and low retention capacity while regeneration. In this case, the cost of regeneration should be low. Sorbents must be strong and not abrade, quickly wetted with water, and they have a certain granulometric composition.

In addition, they must have low catalytic activity with respect to oxidation reactions and condensation. Finally, their production should be low of cost as well as their sorption capacity after regeneration should vary insignificantly and provide a large number of work cycles.

At the moment porous materials with a high specific surface area are used as sorbents for a wide range of substances. Among them the most frequent are carbon materials, sometimes referred to as "active coal", as well as silica gel, alumogel and a number of others. The porosity of carbon materials can be increased by activation, for example by steam treatment, which increases their specific surface area and appropriately increases their sorption capacity. At the same time, there is an upper limit of the specific surface achieved in such activation processes, which creates limitations on the sorption properties of these carbon materials. Recently, carbon materials containing fullerenes-spherical molecules of several dozen carbon atoms have been used as sorbents. Such fullerene-containing carbon materials exhibit increased sorption properties with respect to a number of sorbed substances. At the same time, there are many ways of obtaining fullerene raw materials - by burning graphite, enriching shungite rocks, etc.

Meanwhile, despite the wide interest in fullerenes in various fields, the study of sorption properties with respect to metal ions is carried out a little.

The purpose of the study was to explore and compare the sorption properties with respect to heavy metal ions of fullerene-containing materials. For this purpose, the following tasks were identified:

IOP Conf. Series: Materials Science and Engineering **511** (2019) 012044 doi:10.1088/1757-899X/511/1/012044

1) to make the contamination of soil samples with zinc and lead salts of known concentration in the laboratory;

2) to make a comparative analysis of the sorption efficiency of different types of fullerene-containing carbon sorbents.

Materials and Methods of research

The following types of sorbents were selected for the investigation: initial shungite (Shr) (VKO, Bakyrchik deposit), shungite concentrate after flotation process (SHF), readyshungite sorbent (Shs), fullerene soot (F), birch activated carbon (BAC) and pharmacy activated charbon (AC).

Fullerene soot was formed by burning a graphite rod in a fullerene synthesis reactor operated in the National Nanotechnology Laboratory of an open type.

The concentration of metal ions before and after sorption was determined by electrochemical analysis method on the INESA DDSJ-308a TDS meter, the hydrogen indicator was determined on Mettler Toledo G20, the toxicity of the soil extract before and after sorption measured by the biotesting method.

For the model of pollution, zinc and lead were taken as the most common pollutants. A week later, the soil extracts were analyzed and found out the total dissolved solids and pH level.

Results and Discussion

According to the results, the most effective sorbent in an aqueous solution of zinc is AC, then group of materials in series of descending: AC –Shf– BAC – Shs– Full – Shr. The best sorption in solutions containing lead fractions was shown by flotation-concentrated shungite, then others in descending order: Shf-BAC-AC-Full-Shs-Shi. Generally, the concentration of salts was decreased in all solutions. So, the crushed raw shungite proved to be the worst sorbent. Shungite sorbent is lagging behind the indicators due to the smaller surface area as compared with the fleet-concentrated shungite.

Flotation-concentrated shungite lowered the acidity and removed the color of the extract, which indicates the absorption of organic matter from the soil.

The acidity of soil solutions decreased to the optimum values (pH 6.5-7.5) by all sorbents.

The soil was tested for phytotoxicity. Radish (RaphanusSativus) "18 days" and "ultra-early" was used as an object test to determine the phytotoxicity of the soil. Radish seeds were soaked in soil extracts. Radish was germinated. Repetition of germination was threefold. After the service life soil solutions with sorbents were compared with the reference value (figures 1-5).

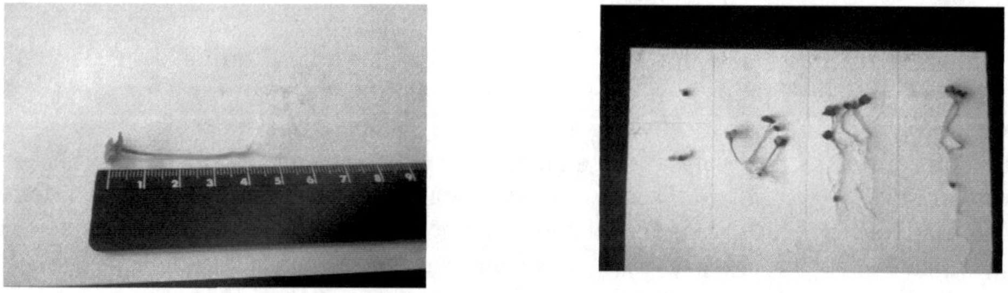

a *b*

Figure 1. Measuring the length of the shoots: (№1 0-0.49 cm, №2 - 0,5-0,9 cm, №3 - 1-4,9 cm, №4 - 5-9cm, №5 - 10-20 cm).

IOP Conf. Series: Materials Science and Engineering **511** (2019) 012044 doi:10.1088/1757-899X/511/1/012044

Figure 2. Bioindication of soil extracts by Raphanussativus.

Proceeding from the diagrams, one can clearly trace the decrease in germination, strength and speed of germination, as well as the effect of the presence of sorbents in contaminated soils on these parameters.

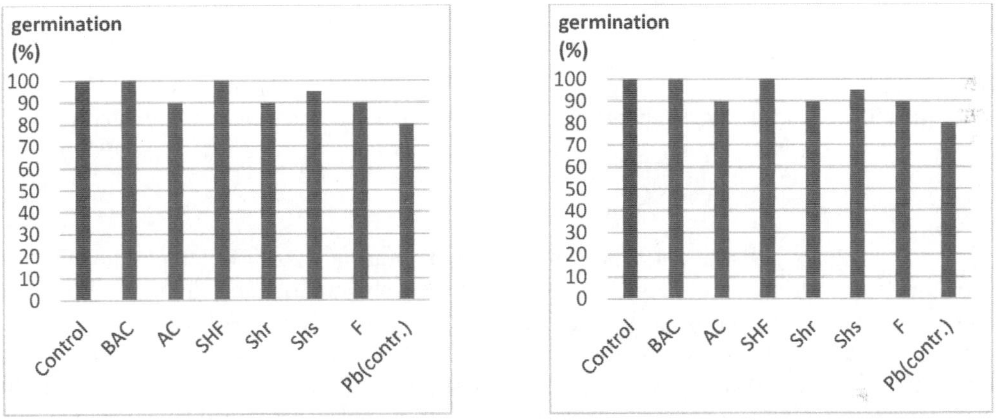

a *b*

Figure 3. General germination of plants in mixed with sorbents of contaminated soils Pb (a) and Zn (b).

On average, compared to control contaminated soil, all samples with sorbents gave inflated indices. In both cases, the best results gave the flotation-concentrated shungite.

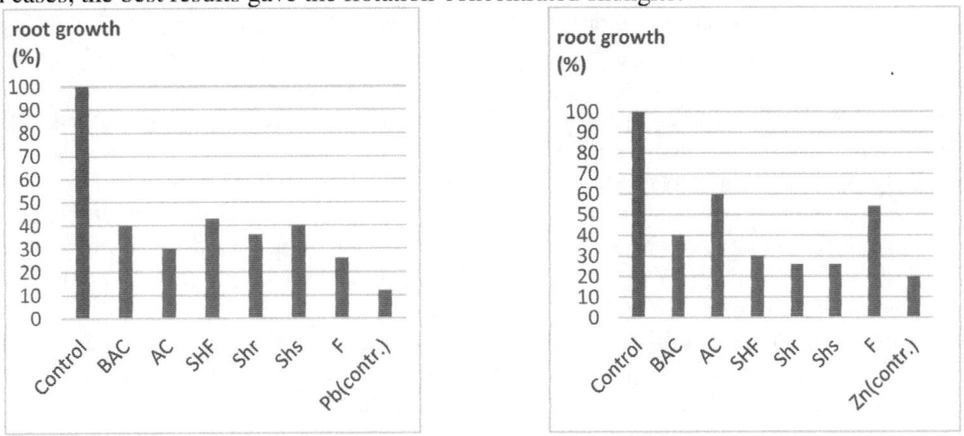

a *b*

Figure 4. Correlation of root growth in mixed with sorbents of contaminated soils Pb (a) and Zn (b).

When comparing the length of the roots of contaminated samples with a net control blank value, the best performance there were demonstrated by samples with activated carbon and fullerene soot (in exemplary zinc contamination). In soils contaminated with lead, there is a general decrease in root growth. The best performance was distinguished by shungite floto-concentrated and shungite sorbent.

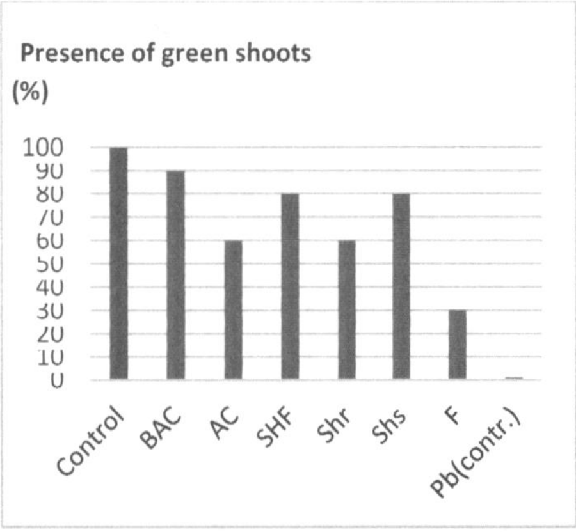

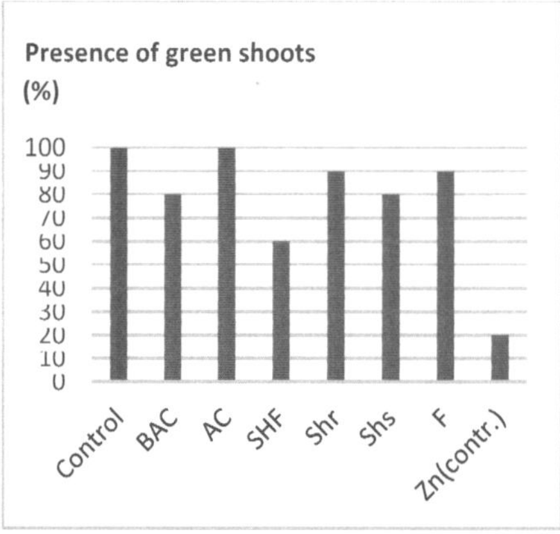

a *b*

Figure 5. Presence of green shoots grown in mixed with sorbents of contaminated soils with lead (a) and zinc (b).

Conclusion

Currently, a large number of studies on the methods of obtaining and use of fullerenes are being carried out. The initial raw material for obtaining fullerenes can be both graphites and shungites. There are two opposing opinions about the properties of fullerene materials. The majority of researchers believe and prove by scientific experiments that solutions, insisted on fullerene-containing materials, become not just pure drinking water, but also a molecular-colloidal solution of fullerenes. This water is attributed to a new generation of medical and preventive agents with a multifaceted effect on the body. Other researchers argue that water infused simply with a piece of carbon-containing mineral can be harmful, because it brings to a chemical reaction and has a possibility to form a low-concentration acid solution.

Based on figure 3, it is evident that fullerene-containing sorbents bind a significant part of pollutants and reduce the toxicity of the soil. Nevertheless, despite the presence of nanoparticles, complete purification of soils from metals is not observed and the soil remains toxic to plants.

Among the shungite materials, flotation-concentrated shungite was the most effective cleaning agent in the aggregate of indices. Next comes the shungite sorbent and crushed raw shungite. However, when compared with activated carbons, shungite demonstrates lower efficiency in soil cleaning.

The results of our studies showed that aqueous solutions with fullerenes obtained by infusion of water do not have highly effective cleaning properties.

Most likely, it is necessary to change the procedure for using fullerene-containing materials in sorption processes. Perhaps one can get positive results when passing water or soil extracts contaminated with heavy metals through a fullerene sorbent under dynamic conditions. To do this, the contaminated working solution passes through a sorption column filled with fullerene-containing materials. However, this method is no longer suitable for cleaning soil samples.

In the future, it is necessary to determine the content of metals in the soil and in plants entirely to obtain a complete overview. It is also necessary to investigate the cleaning rate of each sorbent containing fullerenes for the development of further recommendations.

References

[1] Hirsch A 2005 *Fullerenes: Chemistry and Reactions* (Weinheim: WILEY-VCH) p 423

IOP Conf. Series: Materials Science and Engineering **511** (2019) 012044 doi:10.1088/1757-899X/511/1/012044

[2] Mosin O, Ignatov I 2013 Composition and structural properties of the natural fullerene-bearing shungite mineral *Biotekhnospera* ISSN: 2073-4824
[3] Piotrovsky L B, Kiselev O I 2004 Fullerenes and viruses *Fullerenes, nanotubes, and carbon nanostructures* vol 12 pp 397–403
[4] Wharton T 2001 New non-ionic, highly water-soluble derivatives of C60 designed for biological compatibility *Tetrahedron Lett.* vol 42 pp 5159–5162
[5] Khamitova K K, Ismailov D V, Sultangazina M N, Kanat R 2018 Sorption Extraction of Heavy Metal Ions from Industrial Sewage Key Engineering Materials vol 769 pp 187–192.
[6] Khamitova K K, Kerimbekov D S, Ismailov D V, Gabdullin M T, Schur D V 2018 Analysis of the toxic effects of fullerene soot on radish seedlings *Modern problems of condensed matter physics* pp 159-163